Lecture Notes in Mathematics 1527

M. I. Freidlin J. F. Le Gall

Ecole d'Eté de Probabilités de Saint-Flour XX – 1990

Editor: P. L. Hennequin

Springer-Verlag
Berlin Heidelberg New York
London Paris Tokyo
Hong Kong Barcelona
Budapest

Authors

Mark I. Freidlin
Department of Mathematics
University of Maryland
College Park, MD 20742, USA

Jean-François Le Gall
Université Pierre et Marie Curie
Laboratoire de Probabilités
4, Place Jussieu, Tour 56
F-75230 Paris Cedex 05, France

Editor

Paul Louis Hennequin
Université Blaise Pascal
Clermont-Ferrand
Mathématiques Appliquées
F-63177 Aubière Cedex, France

Mathematics Subject Classification (1991): 60-02, 35A25, 35B40, 35C20, 35K55, 60G17, 60J65, 60J80

ISBN 3-540-56250-8 Springer-Verlag Berlin Heidelberg New York
ISBN 0-387-56250-8 Springer-Verlag New York Berlin Heidelberg

Printed in Germany

Typesetting: Camera ready by author/editor
Printing and binding: Druckhaus Beltz, Hemsbach/Bergstr.
46/3140-543210 - Printed on acid-free paper

INTRODUCTION

Ce volume contient deux des cours donnés à l'Ecole d'Eté de Calcul des Probabilités de Saint-Flour du 1er au 18 Juillet 1990.

Nous avons choisi de les publier sans attendre le troisième cours, "Function Estimation and the White Noise Model" de Monsieur DONOHO, dont la rédaction n'est pas encore complètement achevée et figurera dans le volume suivant.

Nous remercions les auteurs qui ont effectué un gros travail de rédaction définitive qui fait de leurs cours un texte de référence.

L'Ecole a rassemblé soixante six participants dont 32 ont présenté, dans un exposé, leur travail de recherche.

On trouvera ci-dessous la liste des participants et de ces exposés dont un résumé pourra être obtenu sur demande.

Afin de faciliter les recherches concernant les écoles antérieures, nous redonnons ici le numéro du volume des "Lecture Notes" qui leur est consacré :

Lecture Notes in Mathematics

1971 : n° 307 - 1973 : n° 390 - 1974 : n° 480 - 1975 : n° 539 -
1976 : n° 598 - 1977 : n° 678 - 1978 : n° 774 - 1979 : n° 876 -
1980 : n° 929 - 1981 : n° 976 - 1982 : n° 1097 - 1983 : n° 1117 -
1984 : n° 1180 - 1985 - 1986 et 1987 : n° 1362 - 1988 : n° 1427 -
1989 : n° 1464

Lecture Notes in Statistics

1986 : n° 50

TABLE DES MATIERES

Mark I. FREIDLIN : "SEMI-LINEAR PDE'S AND LIMIT THEOREMS FOR LARGE DEVIATIONS"

Jean-François LE GALL : "SOME PROPERTIES OF PLANAR BROWNIAN MOTION"

SEMI-LINEAR PDE'S AND LIMIT THEOREMS FOR LARGE DEVIATIONS

Mark I. FREIDLIN

SEMI-LINEAR PDE'S AND LIMIT THEOREMS FOR LARGE DEVIATIONS

Mark I. FREIDLIN
Department of Mathematics
University of Maryland
College Park

MARYLAND 20742

Introduction.

We consider two classes of asymptotic problems concerning semi-linear parabolic equations. The common element in both these classes is not only the connections with semi-linear PDE's, but the utilization of different kinds of limit theorems for random processes and fields. The limit theorems for large deviations are especially useful in the problems under consideration.

It is well known that a Markov process X_t with continuous trajectories can be connected with any second order elliptic, maybe degenerate, operator $L = \frac{1}{2} \sum_{i,j=1}^{r} a^{ij}(x) \frac{\partial^2}{\partial x^i \partial x^j} + \sum_{i=1}^{r} b^i(x) \frac{\partial}{\partial x^i}$. The most convenient, but not unique, way to construct this process is given by stochastic differential equations. The solutions of the natural boundary problems for L or of the initial-boundary problems for the operator $\frac{\partial}{\partial t} - L$ can be written as expectations of the proper functionals of the process X_t. These expectations are often called functional integrals. They, together with the stochastic equations, give more or less in an explicit way the dependence of the solutions on the coefficients of operator or on initial-boundary conditions. This makes the probabilistic representations very convenient instruments for studying the PDE's. The probabilistic approach turns out to be especially useful in many asymptotic problems for PDE's. Limit theorems, which is a traditional area of probability theory, help to solve the asymptotic problems for PDE's.

The probabilistic approach turns out to be useful for nonlinear second order parabolic equations, too.

The first class of problems which we consider here concerns some asymptotic problems for semi-linear parabolic equations and systems of such equations. The main attention is paid to wave front propagation in reaction-diffusion equations (RDE's) and systems (see, for example, [20]).

By an RDE we mean one equation or a system of equations of the following form:

$$\frac{\partial u_k}{\partial t} = L_k u_k(t,x) + f_k(x, u_1, \dots, u_n), \quad x \in D \subseteq R^r, \quad t > 0, \tag{0.1}$$

$$u_k(0,x) = g_k, \quad k = 1, \dots, n.$$

Here L_k, $k = 1, \dots, n$, are second order elliptic, maybe degenerate, linear operators. Some boundary conditions should be supplemented to the problem if $D \neq R^r$.

The simplest example of an RDE is the Kolmogorov-Petrovskii-Piskunov (KPP) equation:

$$\frac{\partial u}{\partial t} = D\frac{\partial^2 u}{\partial x^2} + u(1-u), \quad x \in R^1, \quad t > 0, \quad u(0,x) = g(x).$$

It was proved in [17] that for certain initial functions the solution of the KPP equation for large t is close to a running wave solution $v(x-\alpha t)$. The shape $v(z)$ of the wave and its speed α are defined by the equation.

We consider various generalizations of this result ([6]-[8], [10], [11]) in the first part of these lectures. These generalizations lead to some new effects in the behavior of the solutions, such as jumps of the wave fronts and breaking of the Huygens principle in slowly changing non-homogeneous media, or an increase of the speed of the fronts in the weakly coupled RDE's. In simple situations the motion of the wave front can be described by Huygens principle, in the proper Riemannian or Finsler metric. KPP equation and some generalizations of this equation are considered in [1]-[3], [14], [18].

The RDE system defines a semi-flow $U_t = (u_1(t,\cdot),\dots,u_n(t,\cdot))$ in the space of continuous functions of x. This semi-flow in general has a rich ω-limit set, which consists of the stationary points of the semi-flow, the periodic-in-time solutions, and more complicated subsets of the phase space. Suppose now that the semi-flow is subjected to small random perturbations. Then the solution of the perturbed RDE system $u^\varepsilon(t,x)$, (ε characterizes the "strength" of the perturbations) will be a random field. We can look on $u^\varepsilon(t,x)$ as on a random process in the functional space, which is a perturbation of the initial semi-flow.

The second class of problems which we consider here concerns the deviations of $u^\varepsilon(t,\cdot)$ from U_t. In the case of PDE's there are more ways to introduce perturbations than in the case of finite-dimensional dynamical systems. For example, an interesting problem is the consideration of perturbations of the boundary conditions. We study several classes of perturbations of the semi-flow and establish results of law-of-large-numbers type, of central-limit-theorem type, and limit theorems for large deviations (see [5], [9], [13], [16], [19], [21], [24]).

It is my pleasure to thank Richard Sowers for his assistance in the preparation of the manuscript and many useful remarks.

§1. Markov Processes and Differential Equations.

Let

$$L = \frac{1}{2}\sum_{i,j=1}^{r} a^{ij}(x)\frac{\partial^2}{\partial x^i \partial x^j} + \sum_{i=1}^{r} b^i(x)\frac{\partial}{\partial x^i}$$

be an elliptic, maybe degenerate, operator. This means that $\sum_{i,j=1}^{r} a^{ij}(x)\lambda_i\lambda_j \ge 0$ for any real $\lambda_1,\dots,\lambda_r$ and any $x \in R^r$.

We assume that the coefficients are bounded and at least Lipschitz continuous. If the matrix $(a^{ij}(x))$ degenerates we assume that the entries $a^{ij}(x)$ have bounded second order derivatives. This last assumption provides the existence of a matrix $\sigma(x) = (\sigma^i_j(x))^r_1$ with Lipschitz-continuous elements such that $\sigma(x)\sigma^*(x) = (a^{ij}(x))$ (see [8] Ch. 1). In the non-degenerate case, of course, the existence of such a matrix $\sigma(x)$ is provided by the Lipschitz continuity of the entries $a^{ij}(x)$.

Let W_t, $t \ge 0$, is the r-dimensional Wiener process. Consider the stochastic differential equation (SDE)

$$dX_t = b(X_t)dt + \sigma(X_t)dW_t, \quad X_0 = x.$$

Here $b(x) = (b^1(x),\dots,b^r(x))$; $\sigma(x)$ is introduced above. Since the coefficients of the equation are bounded and Lipschitz continuous, there exists a unique solution X^x_t (the index x points out the starting point $x \in R^r$). The set of random processes $\{X^x_t,\ x \in R^r\}$ form the Markov family X^x_t corresponding to the operator L in the phase space R^r. The family of probability measures P_x in the space $C_{0\infty}(R^r)$ of continuous functions on $[0,\infty)$ with values in R^r induced by the processes X^x_t in $C_{0,\infty}(R^r)$ is called the Markov process corresponding to the operator L.

For any smooth enough function $U(t,x)$, one can write the Ito formula

$$(1.1)\quad U(t,X^x_t) - U(0,x) = \int_0^t \Big[\frac{\partial U}{\partial t}(s,X^x_s) + LU(s,X^x_s)\Big]ds + \int_0^t (\nabla_x U(s,X^x_s),\ \sigma(X^x_s)dW_s).$$

Let B be the Banach space of functions $f(x)$, $x \in R^r$, which are bounded and measurable with respect to the Borel σ-field. We denote $\|f\| = \sup_{x\in R^r} |f(x)|$. Consider the semi-group T_t corresponding to the family X^x_t (to the Markov process $\{P_x\}$)

$$(T_t f)(x) = Ef(X_t^x) = E_x f(X_t).$$

The subscript x in the sign of expectation points out that we consider integral with respect to the measure P_x. We will use both notations: notations connected with the Markov family $(Ef(X_t^x))$ as well as the notation connected with the process $\{P_x\}$ $(E_x f(X_t))$. The family T_t is a positive contracting semi-group; $T_t f(x) \geq 0$ if $f(x) \geq 0$, and $\|T_t f(x)\| \leq \|f(x)\|$. If $f(x)$ is continuous, then $T_t f(x)$ is also continuous (Feller property). From here, taking into account that the trajectories X_t^x are continuous in t with probability 1 for any x, we conclude (see [8]) that the process $\{P_x\}$ (family X_t^x) has the strong Markov property.

Using (1.1) one can check that the generator $\mathcal{A}$: $\mathcal{A}f = \lim_{t\downarrow 0} \frac{T_t f - f}{t}$ of the semi-group is defined at least for the functions $f = f(x)$ having bounded uniformly continuous second derivatives, and $\mathcal{A}f = Lf$ for such f.

Consider the Cauchy problem

$$(1.2)\qquad \begin{aligned} &\frac{\partial u(t,x)}{\partial t} = Lu(t,x) + c(t,x)u(t,x), \quad t > 0, \quad x \in R^r, \\ &u(0,x) = g(x). \end{aligned}$$

Here $c(t,x)$ is a continuous bounded function. The famous Feynman-Kac formula gives the representation of the solution $u(t,x)$ of the problem (1.2) in the form of a functional integral

$$(1.3)\qquad u(t,x) = E_x g(X_t)\exp\left\{\int_0^t c(t-s,X_s)ds\right\}.$$

To prove (1.3) assume for a moment that $g(x)$ has bounded continuous second order derivatives, and that the operator L is uniformly elliptic. Then the problem (1.2) has a unique solution $u(t,x)$ which is a bounded function having a first derivative in t and a second derivative in x which is continuous and bounded for $0 \leq t \leq T$, $x \in R^r$. Let $Y_t^x = \int_0^t c(t-s,X_s^x)ds$ and consider the function $u(t-s,X_s^x)\exp\{Y_s^x\}$. By Ito's formula we have

$$(1.4)\qquad u(0,X_t^x)e^{\int_0^t c(t-s,X_s^x)ds} - u(t,x) = \int_0^t (\nabla_x u(t-s,X_s^x), \sigma(X_s^x)dW_s)$$

$$+\int_0^t [Lu(t-s,X_s) - \frac{\partial u}{\partial t}(t-s,X_s)$$

$$+ c(t-s,X_s^x)u(t-s,X_s^x)]\exp\left\{\int_0^t c(t-s,X_s^x)ds\right\}ds.$$

The first integral in the right side of the last equality has zero expectation. The last integral is equal to zero since $u(t,x)$ is the solution of problem (1.2). Taking into account that $u(0,x) = g(x)$, we derive (1.3) from (1.4).

If $g(x)$ is a continuous bounded function it can be uniformly approximated by a sequence $g_n(x)$ of smooth functions. For solutions $u_n(t,x)$ of problem (1.2) with the initial functions $g_n(x)$ we have the Feynman-Kac formula. From the maximum principle we conclude that $u_n(t,x) \to u(t,x)$ when $g_n \to g$. On the other hand, $E_x g_n(X_t)\exp\left\{\int_0^t c(t-s,X_s)ds\right\} \to$ $E_x g(X_t)\exp\left\{\int_0^t c(t-s,X_s)ds\right\}$ when $n \to \infty$. This leads to (1.3) for continuous bounded $g(x)$.

In general, if the operator L degenerates for some $x \in R^r$, problem (1.2) may not have a classical solution. In this case equality (1.3) defines the generalized solution. It is easy to check that this generalized solution coincides with the generalized solution in the small viscosity sense (see [7] Ch. 3).

Let τ be a Markov time with respect to the family of σ-fields $\mathcal{F}_t = \sigma(X_s, s \le t)$, $\tau_t = \tau \wedge t$. Then taking into account strong Markov property we can conclude from (1.3) that

$$u(t,x) = E_x u(t-\tau_t, X_t)\exp\left\{\int_0^{\tau_t} c(t-s,X_s)ds\right\}. \tag{1.5}$$

Let us consider now an initial-boundary problem

$$\frac{\partial u(t,x)}{\partial t} = Lu + c(t,x)u, \quad t > 0, \quad x \in D \subset R^r$$
$$u(0,x) = g(x), \quad u(t,x)|_{x \in \partial D} = \psi(x), \tag{1.6}$$

where D is a bounded domain in R^r with smooth boundary ∂D, and $\psi(x)$ is a continuous function on ∂D. Assume for brevity that the operator L is non-degenerate. Then the problem (1.6) has a unique solution. Denote $\tau = \inf\{t : X_t \notin D\}$, the first exit time from D. The following representation holds for the solution of the problem (1.6)

(1.7)

$$u(t,x) = E_x g(X_t)\chi_{\{\tau>t\}}\exp\left\{\int_0^t c(t-s,X_s)ds\right\} + E_x\psi(X_\tau)\chi_{\{\tau\le t\}}\exp\left\{\int_0^\tau c(t-s,X_s)ds\right\}.$$

This representation can be proved in the same way as (1.3) (see [7], Ch. 2).

There are representations in the form of functional integrals for the solutions of stationary problems for the operator L. For example, the solution of the Dirichlet problem $Lu(x) = 0$, $x \in D \subset R^r$, $u(x)|_{\partial D} = \psi(x)$ can be written in the form

$$u(x) = E_x\psi(X_\tau),$$

where $\tau = \inf\{t : X_t \notin D\}$. The domain D is assumed, for brevity, bounded with smooth boundary; the operator L is elliptic.

One can consider boundary and initial-boundary problems with the Neyman condition or with some more general boundary conditions. In this case the representations in the form of functional integrals are also available, but we should consider random processes in the domain with corresponding boundary conditions (see [7], Ch. 2).

Consider now the Cauchy problem for a quasi-linear parabolic equation of the form

$$(1.8) \quad \begin{cases} \dfrac{\partial u}{\partial t} = \dfrac{1}{2}\displaystyle\sum_{i,j=1}^{r} a^{ij}(x,u)\frac{\partial^2 u}{\partial x^i \partial x^j} + \sum_{i=1}^{r} b^i(x,u)\frac{\partial u}{\partial x^i} + c(x,u) \\ u(0,x) = g(x). \end{cases}$$

We assume that $\sum_{i,j=1}^{r} a^{ij}(x)\lambda_i\lambda_j \ge 0$, that the coefficients of the equation are bounded and Lipschitz continuous, and that the initial function $g(x)$ is continuous and bounded.

Suppose that there exists a solution $u(t,x)$ of the problem (1.8). Then we can consider the Markov family

$$(1.9)\qquad X_s^{t,x} - x = \int_0^s \sigma(X_{s_1}^{t,x}, u(t-s_1, X_{s_1}^{t,x}))dW_{s_1} + \int_0^s \sigma(X_{s_1}^{t,x}, u(t-s_1, X_{s_1}^{t,x}))ds_1.$$

Here $\sigma(x,u)\sigma^*(x,u) = (a^{ij}(x,u))$, $u(s,x) \equiv g(x)$ for $s \le 0$. Using the Feynman-Kac formula we can write down for $u(t,x)$ the following equality

$$(1.10)\qquad u(t,x) = Eg(X_t^{t,x})\exp\left\{\int_0^t c(X_s^{t,x}, u(t-s, X_s^{t,x}))ds\right\}.$$

So if $u(t,x)$ is the unique solution of (1.8), then $u(t,x)$ together with $X_s^{t,x}$ satisfies the system (1.9) - (1.10). Therefore we can introduce the generalized solution of (1.8) as a function $u(t,x)$ which, together with $X_s^{t,x}$, satisfies the system (1.9) - (1.10). Of course, if the matrix $(a^{ij}(x,u))$ is non-degenerate, the existence and the uniqueness of the classical (and thus generalized) solution follows from the a priori bounds for the solution of linear parabolic equations. In the degenerate case the solution $u(t,x)$ (in some weak sense) of problem (1.8) will be continuous for $t > 0$ small enough. The solution may have in general discontinuities for t greater than some $t_0 > 0$. We can give some sufficient conditions for existence and uniqueness and continuity of the generalized solution for all $t > 0$. One can also check that under some additional assumptions this solution will be classical.

I shall mention two classes of conditions which provide the existence of the continuous solution for all $t > 0$ (see [7] Ch. 5).

Let the diffusion matrix in (1.8) be independent of u, and assume for brevity that $c(x,u) = 0$. Consider the Markov family $\tilde{X}_t^x$ which is defined by the following equation

$$(1.11)\qquad d\tilde{X}_t^x = \sigma(\tilde{X}_t^x)dW_t,\quad \tilde{X}_0^x = x,$$

where $\sigma(x)\sigma^*(x) = (a^{ij}(x))$. Assume that there exists a bounded Lipschitz continuous vector field $\varphi(x,u) = (\varphi_1(x,u),\dots,\varphi_r(x,u))$ such that

$$(1.12)\qquad \sigma(x)\varphi(x,u) = b(x,u),$$

where $b(x,u) = (b^1(x,u),\dots,b^r(x,u))$. Then according to the Cameron-Martin-Girsanov formula, measures P_x and $\tilde{P}_x$ in C_{0T}, corresponding to the families X_t^x and $\tilde{X}_t^x$, are absolutely continuous each with respect to the other and

$$\frac{dP_x}{d\tilde{P}_x}(\tilde{X}) = \exp\left\{\int_0^T (\varphi(\tilde{X}_s^x, u(t-s,\tilde{X}_s^x)), dW_s) - \frac{1}{2}\int_0^T |\varphi(\tilde{X}_s^x, u(t-s,\tilde{X}_s^x))|^2 ds\right\}.$$

Taking into account the last formula we can write that

$$(1.13)\quad u(t,x) = Eg(X_t^x) = Eg(\tilde{X}_t^x)\exp\left\{\int_0^T (\varphi(\tilde{X}_s^x, u), dW_s) - \frac{1}{2}\int_0^T |\varphi(\tilde{X}_s, u)|^2 ds\right\}.$$

The system (1.11) - (1.13) is equivalent to (1.9) - (1.10).

After the change of variables $(X_t^x, u) \rightarrow (\tilde{X}_t, u)$, the system of equations became triangular. The first equation (1.11) can be solved independently of the second one. We can consider the family $\tilde{X}_t^x$ as a given one, and then equation (1.13) will be the equation for generalized solution $u(t,x)$ of the problem.

The equation (1.13), under condition that $\varphi(x,u)$ is Lipschitz continuous in u, can be solved by the successive approximations (see [7], Ch. 5): The approximations

$$u_0(t,x) \equiv g(x),\quad u_{n+1}(t,x) = Eg(\tilde{X}_t^x)\exp\left\{\int_0^t (\varphi(\tilde{X}_s^x, u_n(t-s,\tilde{X}_s)), dW_s)\right.$$
$$\left. - \frac{1}{2}\int_0^t |\varphi(\tilde{X}_s^x, u_n(t-s,\tilde{X}_s))|^2 ds\right\}$$

converge uniformly to the unique solution of the equation (1.13). Under some additional assumptions one can prove smoothness of the generalized solution.

Another type of assumption which also provides the existence and uniqueness and continuity of the generalized solution for all $t > 0$ is as follows. Let, for brevity, the last term in equation (1.8) be linear: $c(x,u) = cu$, $c = \text{const}$. It is not difficult to prove that there exists $t_0 > 0$ such that the generalized solution exists for $t \in [0,t_0]$ provided the coefficients and the initial function are smooth enough. The greater the constant $-c$, the bigger t_0 is. It turns out that a constant $c_{\text{critical}} < 0$ exists such that if $c < c_{\text{critical}}$ the continuous generalized solution exists for all $t \geq 0$. The exact statement and the proof one can find in §5.2 of [7].

A similar approach is also useful in the initial-boundary problems for quasi-linear equations.

Our goal here is to study RDE's. By an RDE we mean one equation or a system of equations of the form (0.1). As in the case of one equation, we shall use probabilistic representations of the solutions of the linear problems for studying nonlinear equationss. We shall consider now the probabilistic representations for the solutions of linear RDE systems.

Consider the following linear Cauchy problem

$$\frac{\partial u_k(t,x)}{\partial t} = L_k u_k(t,x) + \sum_{j=1}^{n} c_{kj}(x) u_j(t,x), \quad t > 0, \quad x \in R^r$$

(1.14)

$$u_k(0,x) = g_k(x), \quad k = 1,\dots,n.$$

Here $L_k = \frac{1}{2}\sum_{i,j=1}^{r} a_k^{ij}(x)\frac{\partial^2}{\partial x^i \partial x^j} + \sum_{i=1}^{r} b_k^i(x)\frac{\partial}{\partial x^i}$, $k = 1,\dots,n$, are elliptic, maybe degenerate, operators. We assume that the coefficients a_k^{ij}, b_k^i satisfy the same conditions as the coefficients of the operator L in the beginning of this section. The functions $c_{kj}(x)$ are supposed to be bounded and continuous as well as initial functions $g_k(x)$.

Assume that $c_{kj}(x) \ge 0$ for $k \ne j$. We shall explain later how one can get rid of this assumption.

Consider the Markov family $(X_t^{x,k}, \nu_t^{x,k})$ in the phase space $R^r \times \{1,\dots,n\}$ defined as follows

$$dX_t^{x,k} = \sigma_{\nu_t^{x,k}}(X_t^{x,k})dW_t + b_{\nu_t^{x,k}}(X_t^{x,k})dt, \quad X_0^{x,k} = x$$

(1.15)

$$P\{\nu_{t+\Delta}^{x,k} = j \mid \nu_t^{x,k} = i\} = c_{ij}(X_t^{x,k})\Delta + o(\Delta), \quad \Delta \downarrow 0, \quad i \ne j, \quad \nu_0^{x,k} = k.$$

Here $\sigma_k(x)\sigma_k^*(x) = (a_k^{ij}(x))$, $b_k(x) = (b_k^1(x),\dots,b_k^r(x))$, W_t is the Wiener process in R^r. it is easy to prove that under our assumptions on the coefficients the processes $(X_t^{x,k}, \nu_t^{x,k})$ exist for any initial point (x,k). The component $X_t^{x,k}$ is continuous. The family $(X_t^{x,k}, \nu_t^{x,k})$ has the strong Markov property.

In the space B of bounded measurable functions $f(x,k)$ on $R^r \times \{1,\dots,n\}$ provided by uniform norm, consider the family of operators

$$T_t f(x,k) = Ef(X_t^{x,k}, \nu_t^{x,k}) e^{\int_0^t c_{\nu_s^{x,k}}(X_s^{x,k})ds}, \quad t \ge 0, \quad x \in R^r, \quad k \in \{1,\dots,n\},$$

where $c_k(x) = \sum_j c_{kj}(x)$. Taking into account the Markov property of the family $(X_t^{x,k}, \nu_t^{x,k})$ it is easy to check that the family T_t is a semi-group: $T_{s+t} = T_s T_t$. Using Ito's formula on (1.15) it is easy to show that any function $g(x,k)$ which is twice uniformly continuously differentiable in x belongs to domain D_A of the generator $\mathcal{A}$ of the semi-group T_t, and that $\mathcal{A}g(x,k) = L_k g(x,k) + \sum_{j=1}^{n} c_{kj}(x) g(x,k)$. As is known from the general theory of semi-groups, if $g \in D_A$ then $u(t,x,k) = T_t g(x,k)$ is a solution of the abstract Cauchy problem

$$(1.16) \qquad \frac{\partial u}{\partial t} = \mathcal{A}u, \quad u(0,x,k) = g(x,k).$$

The solution of problem (1.16) is unique in the class of functions $u(t,x,k)$ such that $\sup_{x,k}|u(t,x,k)| < \text{const.} \exp\{ct\}$, for some $c > 0$. On the other hand, if $\{u_k(t,x)\}$ is a smooth enough solution of the problem (1.14) then the function $u(t,x,k) = u_k(t,x)$ satisfies (1.16) since in this case $u \in D_A$ and $\mathcal{A}u(t,x,k) = L_k u(t,x,k) + \sum c_j(x) u(t,x,k)$. From the uniqueness of the solution of the abstract Cauchy problem we conclude that the solution of (1.14) coincides with $T_t g(x,k)$

$$(1.17) \qquad u_k(t,x) = E g_{\nu_t^{x,k}}(X_t^{x,k}) \exp\left\{\int_0^t c_{\nu_s}(X_s^{x,k})\, ds\right\}.$$

Thus the solution of system (1.14), provided that solution exists, has the probabilistic representation (1.17). In general, formulas (1.17) define the generalized solution of (1.14). It is not difficult to prove that this generalized solution coincides with the generalized solution in the small viscosity sense.

If instead of (1.14) we have a non-homogeneous-in-time problem

$$(1.18) \qquad \frac{\partial u_k(t,x)}{\partial t} = \frac{1}{2} \sum_{i,j=1}^{r} a_k^{ij}(t,x) \frac{\partial^2 u_k}{\partial x^i \partial x^j} + \sum_{i=1}^{r} b_k^i(t,x) \frac{\partial u_k}{\partial x_i} + \sum_{i=1}^{r} c_{kj}(t,x) u_j,$$

$$u_k(0,x) = g_k(x), \quad x \in R^r, \quad k = 1,\dots,n,$$

we should consider the Markov family $X_s^{t,x,k}, \nu_s^{t,x,k}$

$$(1.19) \qquad dX_s^{t,x,k} = \sigma_{\nu_s^{x},k}(t-s, X_s^{t,x,k}) dW_s + b_{\nu_s^{x},k,t}(t-s, X_s^{t,x,k}) ds, \quad X_0^{t,x,k} = x$$

$P\{\nu_{s+\Delta}^{t,x,k} = j|\nu_s^{t,x,k} = i\} = c_{ij}(t-s,X_s^{t,x,k})\Delta + o(\Delta), \quad \Delta\downarrow 0, \quad i \neq j, \quad \nu_0^{t,x,k} = k.$

Then the solution of (1.18) can be written as follows:

$$(1.20) \qquad u_k(t,x) = Eg_{\nu_t^{t,x,k}}(X_t^{t,x,k})\exp\left\{\int_0^t c_{\nu_s^{t,x,k}}(t-s,X_s^{t,x,k})ds\right\}.$$

Here $c_k(t,x) = \sum_{j=1}^{n} c_{kj}(t,x)$, $\sigma_k(t,x)\sigma_k^*(t,x) = (a_k^{ij}(t,x))$, $b_k(t,x) = (b_k^1(t,x),\dots,b_k^r(t,x))$.

The solution of an initial boundary problem for system (1.18) with the Dirichlet conditions on the boundary can be written as expectation of the proper functional of the family $(X_s^{t,x,k},\nu_s^{t,x,k})$ also. Consider an auxiliary family $(\tilde{X}_s^{t,x,k},\tilde{\nu}_s^{t,x,k})$:

$$(1.21) \qquad \begin{aligned} d\tilde{X}_s^{t,x,k} &= \sigma_{\tilde{\nu}_s}(t-s,\tilde{X}_s^{t,x,k})dW_s + b_{\tilde{\nu}_s}(t-s,\tilde{X}_s^{t,x,k})ds, \quad \tilde{X}_0^{t,x,k} = x, \\ P\{\tilde{\nu}_{s+\Delta}^{t,x,k} &= j|\tilde{\nu}_s^{t,x,k} = i\} = \Delta + o(\Delta), \quad \Delta\downarrow 0, \quad \tilde{\nu}_0^{t,x,k} = k. \end{aligned}$$

So the difference between the processes (1.19) and (1.21) consists of different intensities of transitions for the second component. For the family (1.21) all these intensities are equal to 1. It is easy to prove that the measures μ and $\tilde{\mu}$ in the space of trajectories for this process (with the same initial conditions and the same Wiener process W_t) are absolutely continuous, each with respect to the other. One can write an explicit formula for the density $\frac{d\mu}{d\tilde{\mu}}(\tilde{X}_\cdot,\tilde{\nu}_\cdot)$. This density is, of course, a functional depending on $c_{ij}(t,x)$. Then we can rewrite (1.20) as the expectation with respect to the measure corresponding to the process $(\tilde{X}_s^{t,x,k},\tilde{\nu}_s^{t,x,k})$:

$$(1.22) \qquad u_k(t,x) = Eg_{\tilde{\nu}_t}(\tilde{X}_t^{t,x,k})\exp\left\{\int_0^t c_{\tilde{\nu}_s^{t,x,k}}(t-s,\tilde{X}_s^{t,x,k})\right\}\frac{\partial\mu}{\partial\tilde{\mu}}(\tilde{X}_\cdot^{t,x,k},\tilde{\nu}_\cdot^{t,x,k}).$$

It turns out that this representation of $u_k(t,x)$ as the expectation of the functional of the process $(\tilde{X}_s^{t,x,k},\tilde{\nu}_s^{t,x,k})$ is true in the case of $\{c_{ij}(t,x)\}$ of any sign, not necessarily positive. We will not write an exact formula for the density $\frac{d\mu}{d\tilde{\mu}}$ here, since we will not use it. One can find it in [6].

§2. Generalized KPP-equations and Large Deviations.

The first class of problems which will be considered here consists of generalizations of the KPP-equation [17]

$$\frac{\partial u(t,x)}{\partial t} = \frac{D}{2}\frac{\partial^2 u(t,x)}{\partial x^2} + f(u(t,x)), \quad t > 0, \quad x \in R^1,$$

(2.1)

$$u(0,x) = \chi_-(x).$$

Here $f(u)$ is a C^1-function such that $f(0) = f(1) = 0$, $f(u) > 0$ for $u \in (0,1)$, $f(u) < 0$ for $u \notin [0,1]$, and $f'(0) = \max_{0\le u\le 1} f'(u)$. The class of such functions we call $\mathcal{F}_1$. As the initial function in (2.1) we take the indicator of the set $\{x \in R^1 : x < 0\}$.

It was proved in [17] that the solution $u(t,x)$ of (2.1) for large t will be close to a running wave type solution $v(x-\alpha t)$. The speed of the wave is $\alpha = \sqrt{2Df'(0)}$, and the shape $v(z)$ is a solution of the problem

$$\frac{D}{2}v''(z) + \alpha v(z) + f(v(z)) = 0, \quad -\infty < z < \infty$$

(2.2)

$$v(-\infty) = 1, \quad v(+\infty) = 0.$$

Problem (2.2) has a solution which is unique up to a shift of the argument. So the limiting behavior of the solution of problem (2.1) is defined by the speed α and by the shape $v(z)$ of the running wave.

Denote $c(u) = u^{-1}f(u)$. Since $f \in \mathcal{F}_1$ we have that $c = c(0) = \max_{u\ge 0} c(u)$. Using the Feynman-Kac formula we can write

$$u(t,x) = E_x\chi_-(X_t)e^{\int_0^t c(u(t-s,X_s))ds} \le e^{ct}P_x\{X_t < 0\}, \tag{2.3}$$

where X_t is the Markov process in R^1 corresponding to $\frac{D}{2}\frac{d^2}{dx^2}$. It is easy to check that

$$P_x\{X_t < 0\} = \frac{1}{\sqrt{2\pi Dt}}\int_x^\infty e^{-\frac{z^2}{2Dt}}dz = \frac{1}{\sqrt{2\pi}}\int_{x/\sqrt{Dt}}^\infty e^{-\frac{z^2}{2}}dz \tag{2.4}$$

$$\sim \frac{1}{\sqrt{2\pi}}e^{-\frac{x^2}{2tD}} \quad \text{when} \quad \frac{x}{\sqrt{t}} \longrightarrow \infty.$$

We conclude from (2.3) and (2.4) that

$$u(t,x) \le e^{ct} P_x\{X_t < 0\} \sim \text{const.}\ e^{ct-\frac{x^2}{2Dt}}.$$

From here we see that $u(t,x) \to 0$ when $t \to \infty$ and $x > t(\sqrt{2cD}+\delta)$, $\delta > 0$. A little more delicate bounds show that $u(t,x) \to 1$ when $t \to \infty$ and $x < t(\sqrt{2cD}-\delta)$, $\delta > 0$. The main *input* in the expectation (2.3) is given by trajectories X_t such that their position at time t differs from the starting point by the distance close to $t\sqrt{2cD}$. The normal deviations of X_t from $X_0 = x$ have the order $\sqrt{t}$ for large t. Thus the main *input* in (2.3) is given by large deviations. I shall recall later the main results on large deviations which I use here, but now I'll formulate one of the generations of KPP-equation, which we will study in the next sections.

First of all, I would like to separate the problems of calculating the speed and the shape of the limiting wave.

Consider the function $u^\varepsilon(t,x) = u(\frac{t}{\varepsilon},\frac{x}{\varepsilon})$ where $u(t,x)$ is the solution of (2.1) Here ε is a small positive parameter. The function $u^\varepsilon(t,x)$ is the solution of the problem

$$(2.5) \qquad \frac{\partial u^\varepsilon(t,x)}{\partial t} = \frac{\varepsilon D}{2}\frac{\partial^2 u^\varepsilon}{\partial x^2} + \frac{1}{\varepsilon} f(u^\varepsilon), \quad t > 0,\ x \in R^1, \quad u^\varepsilon(0,x) = \chi_-(x).$$

Since $u(t,x) \approx v(x-\alpha t)$ for large t we can expect that

$$u^\varepsilon(t,x) = u(\frac{t}{\varepsilon},\frac{x}{\varepsilon}) \sim v(\frac{x-\alpha t}{\varepsilon}) \quad \text{for} \quad \varepsilon \downarrow 0.$$

Thus we have $u^\varepsilon(t,x) \to 0$ if $x > \alpha t$, and $u^\varepsilon(t,x) \to 1$ when $x < \alpha t$. Thus the solution of problem (2.5) tends to the step function $\chi_-(x-\alpha t)$, when $\varepsilon \downarrow 0$. The zero approximation of $u^\varepsilon(t,x)$, $\varepsilon \downarrow 0$, is characterized by the speed only. The shape will appear in the next approximations.

Now we can formulate the general problem. Let $L = \frac{1}{2}\sum_{i,j=1}^{r} a^{ij}(x)\frac{\partial^2}{\partial x^i \partial x^j}$ be an elliptic operator with Lipschitz continuous coefficients (we omitted the first derivatives for brevity). Assume that $f(x,0) \in \mathcal{F}_1$ for any $x \in R^1$ and let $f(x,u) = c(x,u)u$ such that $c(x,u)$ is bounded, Lipschitz continuous, $c(x,u) > \alpha(u) > 0$ for $u \in [0,1)$, $x \in R^r$, and $c(x,u) < \alpha(u) < 0$ for $u > 1$, $x \in R^r$, where $\alpha(u)$, $u \ge 0$, is a continuous function. Consider the Cauchy problem

$$\frac{\partial u^{\varepsilon}(t,x)}{\partial t} = \varepsilon L u^{\varepsilon} + \frac{1}{\varepsilon} f(x,u^{\varepsilon}), \quad t > 0, \quad x \in R^r$$

(2.6)

$$u^{\varepsilon}(t,x) = g(x).$$

Here the initial function $g(x)$ is supposed to be bounded, non-negative and having support $G_0 = \{x \in R^1 : g(x) > 0\}$ such that $[G_0] = [(G_0)]$, where (A) is the interior of the set A and $[A]$ is the closure of $A \subset R^r$. Our goal is to study the behavior of $u^{\varepsilon}(t,x)$ when $\varepsilon \downarrow 0$. If X_t^{ε} is the process corresponding to the operator εL, using the Feynman-Kac formula we can write an integral equation in the space C for $u^{\varepsilon}(t,x)$

$$u^{\varepsilon}(t,x) = E_x g(X_t^{\varepsilon}) e^{\frac{1}{\varepsilon}\int_0^t c(X_s^{\varepsilon}, u^{\varepsilon}(t-s, X_s^{\varepsilon}))ds}. \tag{2.7}$$

The asymptotics of the expectation in the right side of this last equation when $\varepsilon \downarrow 0$ is defined by large deviations for the process X_t^{ε} when $\varepsilon \downarrow 0$. I follow here the book [12], where the proofs can be found.

Remark. We come to problem (2.6) when we consider the Cauchy problem for the equation with small parameter

$$\frac{\partial \tilde{u}^{\varepsilon}(t,x)}{\partial t} = \varepsilon^2 L \tilde{u}^{\varepsilon}(t,x) + f(x, \tilde{u}^{\varepsilon}(t,x)), \quad \tilde{u}^{\varepsilon}(0,x) = g(x).$$

After rescaling the time we get equation (2.6) for the function $u^{\varepsilon}(t,x) = \tilde{u}^{\varepsilon}(t/\varepsilon, x)$.

We also get problem (2.6) when we consider equation with slowly changing coefficients

$$\frac{\partial \bar{u}^{\varepsilon}(t,x)}{\partial t} = \frac{1}{2}\sum a^{ij}(\varepsilon x) \frac{\partial^2 \bar{u}^{\varepsilon}}{\partial x^i \partial x^j} + f(\varepsilon x, \bar{u}^{\varepsilon}).$$

Rescaling of the space and the time gives equation (2.6) for the function $u^{\varepsilon}(t,x) = \bar{u}(\frac{t}{\varepsilon}, \frac{x}{\varepsilon})$.

In the rest of this section, I shall recall the main results on large deviations for random processes.

Suppose $\{X, \rho\}$ is a metric space, and let μ^{ε} be a family of probability measures on the Borel σ-field of the space X, depending on a parameter $\varepsilon > 0$. Let $\lambda(\varepsilon)$ be a positive real-valued function such that $\lim_{\varepsilon \downarrow 0} \lambda(\varepsilon) = \infty$, and let $S(x)$ be a function on X with values in $[0, \infty]$. We shall say that $\lambda(\varepsilon)S(x)$ is an action function (rate function) for μ^{ε} as $\varepsilon \downarrow 0$ if the following assertions hold:

0) the set $\Phi(s) = \{x \in X : S(x) \le s\}$ is compact for any $s \ge 0$;

1) for any $\delta > 0$ and $\gamma > 0$ and any $x \in X$, there exists an $\varepsilon_0 > 0$ such that

$$\mu^h\{y \in X : \rho(x,y) < \delta\} \ge \exp\{-\lambda(\varepsilon)[S(x)+\gamma]\}$$

for $\varepsilon \le \varepsilon_0$;

2) for any $\delta > 0$, any $\gamma > 0$ and any $\zeta > 0$, there exists an $\varepsilon_0 > 0$ such that

$$\mu\{y \in X : \rho(y,\phi(\zeta) \ge \delta\} \le \exp\{-\lambda(\varepsilon)(\zeta-\gamma)\}$$

for $\varepsilon \le \varepsilon_0$.

If X is a function space we shall use the term action functional. Separately, the functions $S(x)$ and $\lambda(\varepsilon)$ will be called the normalized action function and normalizing coefficient. Of course, the decomposition of an action functional into two factors $\lambda(\varepsilon)$ and $S(x)$ is not unique. Nevertheless, one can prove that for a given normalizing coefficient, the normalized action function is uniquely defined. There are some other definitions of the action function which are essentially equivalent to this given here (see §3.3 in [12]).

If we have a family of random processes $X_t^\varepsilon(\omega)$, $0 \le t \le T$, (maybe even defined on different Ω-spaces) with trajectories X_t^ε, $0 \le t \le T$, belonging to a Banach psace $\mathcal{B}$, the action function corresponding to the family of measures $\mu^\varepsilon : \mu^\varepsilon(B) = P\{X^\varepsilon \in B\}$, is called action functional for $X^\varepsilon_\cdot$ in the space $\mathcal{B}$ when $\varepsilon\downarrow 0$. The usefulness of the notion of the action functional can be explained as follows. Suppose $B \in \mathcal{B}$ is a Borel set such that $\inf\{S(\varphi) : \varphi \in (B)\} = \inf\{S(\varphi) : \varphi \in [B]\}$, where (B) is the interior and $[B]$ is the closure of the set B. Then

$$\lim_{\varepsilon\downarrow 0} \lambda^{-1}(\varepsilon)\ell n P\{X_\cdot^\varepsilon \in B\} = -\inf\{S(x) : x \in B\}. \tag{2.8}$$

Another important relation: Let $F(x)$, $x \in \mathcal{B}$, is a bounded continuous functional. Then

$$\lim_{\varepsilon\downarrow 0} \lambda^{-1}(\varepsilon)\ell n E e^{\lambda(\varepsilon)F(X^\varepsilon)} = \sup_{y\in\mathcal{B}}[F(\zeta) - S(\zeta)]. \tag{2.9}$$

There is a slight generalization of (2.9) which we shall use. Let $A \subset \mathcal{B}$ be a Borel set and let χ_A be its indicator. Assume that the set A is such that

$$\sup_{\zeta\in(A)} [F(\zeta) - S(\zeta)] = \sup_{\zeta\in[A]} [F(\zeta) - S(\zeta)].$$

Then

$$\lim_{\varepsilon\downarrow 0} \lambda^{-1}(\varepsilon)\ln E\chi_A e^{\lambda(\varepsilon)F(X^\varepsilon)} = \sup_{\zeta\in A}[F(\zeta) - S(y)]. \tag{2.10}$$

Thus the action function allows us to calculate the logarithmic asymptotic of probabilities and expectations connected with the family X^ε.

Consider several examples. First of all, let $X_t^\varepsilon = \varepsilon W_t$, where W_t is the Wiener process in R^r, $W_0 = 0$. As the Banach space X we can take the space C_{0T} of continuous functions on $[0,T]$ with values in R^r. Introduce the functional

$$S_{0T}(\varphi) = \begin{cases} \frac{1}{2}\int_0^T |\dot\varphi_s|^2 ds, & \text{if } \varphi_s,\ 0 \le s \le T, \text{ is absolutely continuous,} \\ +\infty, & \text{for the rest of } C_{0T}. \end{cases} \tag{2.11}$$

Here $\dot\varphi_s = \frac{d\varphi_s}{ds}$. One can prove that the product $\varepsilon^{-2}S_{0T}(\varphi)$ is the action functional for the family εW_t in the space C_{0T} when $\varepsilon\downarrow 0$. This result was proved by M. Shilder in 1966.

As we have seen the logarithmic asymptotics of the probabilities and expectations connected with the processes εW_t is defined by infimum of $S_{0T}(\varphi)$ on corresonding sets. It follows from condition (0) that $S_{0T}(\varphi)$ attains its minimum on every non-empty closed set.

The following simple result (Theorem 3.1 in Chapter 3 of [12]) allows us to consider a large number of examples:

<u>Proposition 2.1</u>. Let $\lambda(\varepsilon)S^\mu(x)$ be the action function for a family of measures μ^ε on a metric space (X,ρ_X). Let φ be a continuous mapping of X into space Y with a metric ρ_y and let a measure ν^ε on Y be given by the formula $\nu^\varepsilon(A) = \mu^\varepsilon(\varphi^{-1}(A))$. The asymptotics of the family ν^ε as $\varepsilon\downarrow 0$ is given by the action function $\lambda(\varepsilon)S^\nu(y)$, where $S^\nu(y) = \min\{S^\mu(x) : x \in \varphi^{-1}(y)\}$ (the minimum over the empty set is said to be equal to $+\infty$).

Let us apply the last result to calculate the action functional for the family of random processes X_t^ε defined as follows:

$$dX_t^{\varepsilon} = b(X_t^{\varepsilon})dt + \varepsilon dW_t, \quad X_0^{\varepsilon} = x \in R^r.$$

Here $b(x)$ is a Lipschitz-continuous and bounded function. It is easy to check that the mapping $B_x : \varphi_{\cdot} \to X_{\cdot}$ defined by the equation

$$X_t - x = \int_0^t b(X_s)ds + \varphi_t,$$

is continuous in the space C_{OT}. Taking into account that $(B_x^{-1}(X_{\cdot}))_t = X_t - x - \int_0^t b(X_s)ds = \varphi_t$, we have from (2.11) and Proposition 2.1 that the action functional for the family X_t^{ε} in $\mathcal{C}_{OT}$ has the form $\varepsilon^{-2}S_{OT}^X(\varphi)$, where

$$S_{OT}^X(\varphi) = S_{OT}(B^{-1}(\varphi)) = \begin{cases} \frac{1}{2}\int_0^T |\dot{\varphi}_s - b(\varphi_s)|^2 ds, & \text{if } \varphi_s \text{ is absolutely continuous and } \varphi_0 = x, \\ +\infty, & \text{for the rest of } C_{OT}. \end{cases}$$

The general stochastic differential equation of the form

$$dX_t^{\varepsilon} = b(X_t^{\varepsilon})dt + \varepsilon\sigma(X_t^{\varepsilon})dW_t, \quad X_0 = x \in R^r, \tag{2.12}$$

with bounded Lipschitz continuous coefficients defines the mapping $W_{\cdot} \to X_{\cdot}$ which, in general, will not be continuous as a mapping from C_{OT} to C_{OT}. But, nevertheless, the result will be true: the action functional for the family X_t^{ε}, defined by (2.12), in the space C_{OT} as $\varepsilon \downarrow 0$ has the form $\varepsilon^{-2}S_{OT}^X(\varphi)$, where

$$S_{OT}^X(\varphi) = \begin{cases} \frac{1}{2}\int_0^T \sum_{i,j=1}^{r} a_{ij}(\varphi_s)(\dot{\varphi}_s^i - b^i(\varphi_s))(\dot{\varphi}_s^j - b^j(\varphi_s))ds & \text{for } \varphi_s \text{ absolutely continuous and } \varphi_0 = x, \\ +\infty & \text{for the rest of } C_{OT}. \end{cases} \tag{2.13}$$

Here $(a_{ij}(x)) = (a^{ij}(x))^{-1} = (\sigma(x)\sigma^*(x))^{-1}$ and we assume that the diffusion matrix $(a_{ij}(x))$ is non-degenerate. The formula (2.13) was proved by SRS. Varadhan in the case $b^i(x) \equiv 0$ and by M. Freidlin and A. Wentzell in the general case (see, for example, [22]).

Another example concerns the family of random processes defined by an ordinary differential equation with a fast oscillating noise

(2.14) $$\dot{X}_t^\varepsilon = b(X_t^\varepsilon, \xi_{t/\varepsilon}), \quad X_0^\varepsilon = x \in R^r.$$

Assume that ξ_t is a non-degenerate diffusion process on a compact manifold S (or in a bounded domain with a regular reflection on the boundary). Let L be the differential operator, corresponding to ξ_t (with boundary conditions if S is a compact domain). Consider the eigenvalue problem

(2.15) $$L\varphi(z) + (\beta, b(x,z))\varphi(z) = \lambda\varphi(z), \quad z \in S.$$

Here $\beta, x \in R^r$ are parameters.

We assume that the operator L has smooth enough coefficients and that the function $b(x,z)$ is also smooth. Under these conditions, problem (2.15) has a positive eigenfunction $\varphi(z) = \varphi_{x,\beta}(z)$. The corresponding eigenvalue $\lambda = \lambda(x,\beta)$ is simple, real and smooth as a function of the parameters x and β. One can prove that $\lambda(x,\beta)$ is convex in $\beta \in R^r$. Denote $L(x,\alpha)$ the Legendre transformation of $\lambda(x,\beta)$ in β : $L(x,\alpha) = \sup_{\beta \in R^r} [(\alpha,\beta) - \lambda(x,\beta)]$.

One can prove (see [12] Ch. 7) that the action functional for the family X_t^ε defined by (2.14) in the space C_{0T} as $\varepsilon \downarrow 0$ is given by the formula $\varepsilon^{-1} S_{0T}^X(\varphi)$, where

(2.16) $$S_{0T}^X(\varphi) = \begin{cases} \int_0^T L(\varphi_s, \dot{\varphi}_s)ds, & \text{if } \varphi_s \text{ is absolutely continuous,} \\ +\infty & \text{for the rest of } C_{0T}. \end{cases}$$

If the process ξ_t in (2.14) is a Markov process with finite state space $\{1,\dots,m\}$, and $P\{\xi_{t+\Delta} = j | \xi_t = i\} = c_{ij}\Delta + o(\Delta)$, $\Delta \downarrow 0$, $c_{ij} > 0$ for $i \neq j$, then the action functional for the family X_t^ε in C_{0T} has the same form, but the function $L(x,\alpha)$ in (2.16) is defined as the Legendre transformation of the function $\lambda(x,\beta)$ which is the first eigenvalue of the matrix $Q^{x,\beta} = (q_{ij}^{x,\beta})$, where $q_{ij}^{x,\beta} = c_{ij} + \delta_{ij}(b(x,i),\beta)$, δ_{ij} is the Kronecker symbol, and $c_{ii} = -\sum_{j: j\neq i} c_{ij}$.

We shall deal later with the action functionals for some other families of random processes and fields.

§3. Generalized KPP-equation Under Condition (N).

Here we start to study equation (2.6). Since we assume that $f(x,\cdot) \in \mathcal{F}_1$ for any $x \in R^r$, the function $f(x,u)$ can be represented as follows: $f(x,u) = c(x,u)u$, where $c(x,1) = 0$, $c(x,u) > 0$ for $u \in [0,1]$, $c(x,0) = \max_{u\ge 0} c(x,u)$. Denote $c(x) = c(x,0)$. The problem (2.6) can be written in the form

$$(3.1)\qquad \frac{\partial u^{\varepsilon}(t,x)}{\partial t} = \varepsilon L u^{\varepsilon} + \frac{1}{\varepsilon} c(x,u^{\varepsilon})u^{\varepsilon}, \quad x \in R^r, \quad t > 0,$$
$$u^{\varepsilon}(0,x) = g(x).$$

The assumptions on the coefficients and the initial function were listed in §2.

If $X_t^{\varepsilon,x}$ is the Markov family corresponding to the operator εL, we can write the following equation for unknown function $u^{\varepsilon}(t,x)$:

$$(3.2)\qquad u^{\varepsilon}(t,x) = Eg(X_t^{\varepsilon,x})\exp\left\{\frac{1}{\varepsilon}\int_0^t c(X_s^{\varepsilon,x}, u^{\varepsilon}(t-s, X_s^{\varepsilon,x}))ds\right\}.$$

The family $X_t^{\varepsilon,x}$, $x \in R^r$, can be constructed as the solutions of the stochastic differential equation

$$(3.3)\qquad dX_t^{\varepsilon,x} = \sqrt{\varepsilon}\sigma(X_t^{\varepsilon,x})dW_t, \quad X_0^{\varepsilon,x} = x.$$

Here $\sigma(x)$ is such a matrix that $\sigma(x)\sigma^*(x) = (a^{ij}(x))$ is the diffusion matrix in the operator L. We assume for simplicity that the drift is equal to zero. Of course, we can write down (3.2) in the notations connected with Markov process, corresponding to εL

$$(3.4)\qquad u^{\varepsilon}(t,x) = E_x g(X_t^{\varepsilon})\exp\left\{\frac{1}{\varepsilon}\int_0^t c(X_s^{\varepsilon}, u^{\varepsilon}(t-s, X_s^{\varepsilon}))ds\right\}.$$

Equation (3.4) is an integral equation in the space C_{0T}. Since we assumed that $g(x) \ge 0$ we have from (3.4) that $u^{\varepsilon}(t,x) \ge 0$ for all $t \ge 0$. Taking into account that $\max_{u\ge 0} c(x,u) = c(x,0) = c(x)$, we derive from (3.4)

$$(3.5)\qquad u^{\varepsilon}(t,x) \le E_x g(X_t^{\varepsilon})\exp\left\{\frac{1}{\varepsilon}\int_0^t c(X_s^{\varepsilon})ds\right\}.$$

The functional $F[\varphi_s,\ 0 \le s \le t] = \int_0^t c(\varphi_s)ds$ is continuous in C_{0T} since the function $c(x)$ is continuous. We assumed that the set $G_0 = \{x \in R^r : g(x) > 0\}$ is such that $[G_0] = [(G_0)]$, therefore

$$(3.6)\qquad \sup_{\varphi \in C_{0t}:\varphi_0 = x,\, \varphi_y \in [G_0]} \left\{ \int_0^t \left[c(\varphi_s) - \frac{1}{2} \sum_{i,j=1}^{r} a_{ij}(\varphi_s)\dot{\varphi}_s^i \dot{\varphi}_s^j \right] ds \right\}$$

$$= \sup_{\varphi \in C_{0t}:\varphi_0 = x,\, \varphi_y \in (G_0)} \left\{ \int_0^t \left[c(\varphi_s) - \frac{1}{2} \sum_{i,j=1}^{r} a_{ij}(\varphi_s)\dot{\varphi}_s^i \dot{\varphi}_s^j \right] ds \right\},$$

where $(a_{ij}(x)) = (a^{ij}(x))^{-1}$. The normalized action functional for the family X_s^ε, $0 \le s \le t$, in the space C_{0T} is given by the formula (2.13). Because of (3.6) we can apply formula (2.10) to calculate the logarithmic asymptotics of the right side of (3.5) as $\varepsilon \downarrow 0$. Denote

$$R_{0t}(\varphi) = \int_0^t [c(\varphi_s) - \tfrac{1}{2}\textstyle\sum a_{ij}(\varphi_s)\dot{\varphi}_s^i \dot{\varphi}_s^j]ds$$

$$V(t,x) = \sup\{R_{0t}(\varphi) : \varphi \in C_{0T},\ \ \varphi_0 = x,\ \ \varphi_t \in G_0\}.$$

It is easy to check that $V(t,x)$ is a continuous function. We have from (2.9)

$$(3.7)\qquad \lim_{\varepsilon \downarrow 0} \varepsilon \ln E_x g(X_t^\varepsilon) \exp\left\{ \frac{1}{\varepsilon} \int_0^t c(X_s^\varepsilon) ds \right\} = V(t,x).$$

Equality (3.7), together with (3.5), implies that

$$(3.8)\qquad \overline{\lim_{\varepsilon \downarrow 0}}\, \varepsilon \ln u^\varepsilon(t,x) \le V(t,x).$$

We conclude from (3.8) that

$$(3.9)\qquad \lim_{\varepsilon \downarrow 0} u^\varepsilon(t,x) = 0 \quad \text{if} \quad V(t,x) < 0.$$

If we had been able to prove that

$$(3.10)\qquad \lim_{\varepsilon \downarrow 0} u^\varepsilon(t,x) = 1 \quad \text{for} \quad (t,x) \quad \text{such that} \quad V(t,x) > 0,$$

then the equation $V(t,x) = 0$ would define the position of the wave front at time t: the surface defined by this equation would separate the set where $u^{\varepsilon}(t,x) \to 0$ and the set where $u^{\varepsilon}(t,x) \to 1$.

It turns out that without additional assumptions (3.10) is not true. The equation defining the position of the front in the general case is more complicated and we shall introduce it later. Now I want to formulate a condition which provides (3.10).

We say that condition (N) is fulfilled if

$$(3.11) \qquad V(t,x) = \sup\{R_{0t}(\varphi) : \varphi \in C_{0T}, \ \varphi_0 = x, \ \varphi_t \in G_0, \ V(t-s,\varphi_s) < 0 \text{ for } 0 < s < t\}$$

for (t,x) such that $V(t,x) = 0$.

In the right side of (3.11) we have supremum over a set which is smaller than the set involved in the definition of the function $V(t,x)$. Nevertheless, we assume that supremum over this set coincides with supremum involved in the definition of $V(t,x)$. As we shall see later, condition (N) is not always true, but it is true in many important cases.

Lemma 3.1. Assume that condition (N) is true. Then for any $T, \delta > 0$ there exists an $\varepsilon_0 > 0$ such that for all $(t,x) \in (0,T]\times R^r$ with $V(t,x) = 0$

$$(3.12) \qquad u^{\varepsilon}(t,x) \ge \exp\{-\frac{\delta}{\varepsilon}\} \quad \text{for} \quad 0 < \varepsilon < \varepsilon_0.$$

Proof. By the definition of $V(t,x)$ and condition (N) there is a φ_t in C_{0t} with $\varphi_0 = x$, $\varphi_t \in G_0$,

$$R_{0t}(\varphi) = \int_0^t [c(\varphi_s) - \frac{1}{2} \sum_{i,j=1}^{r} a_{ij}(\varphi_s)\dot{\varphi}_s^i\dot{\varphi}_s^j]ds > -\delta,$$

and such that $V(t-s,\varphi_s) < 0$ for $0 < s < t$. Now, for small $\hat{t} > 0$, we can alter φ_s near $s = t$ to find a function $\tilde{\varphi} \in C_{0t}$ with $\tilde{\varphi}_0 = x$, $\tilde{\varphi}_t \in (G_0)$, $\rho_{0,t}(\varphi,\tilde{\varphi}) < \delta$, and $R_{0t}(\tilde{\varphi}) > -2\delta$ such that $V(t-s,\tilde{\varphi}_s) < 0$ for $\hat{t} < s < t-\hat{t}$ (see Figure 1). Define

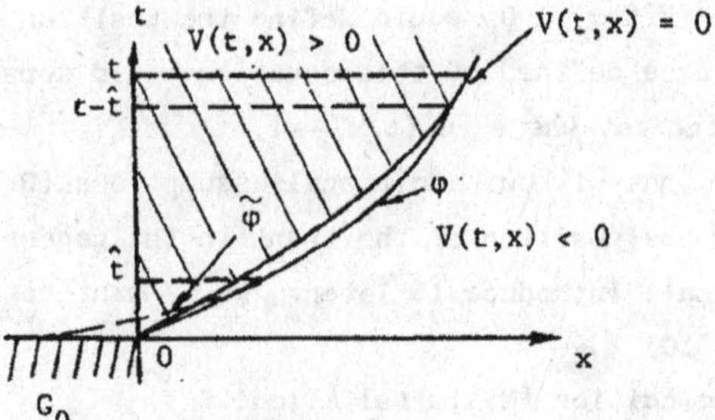

Figure 1.

$$\kappa_{\hat{t}} = \text{distance}\left[\{(t-s,\tilde{\varphi}_s),\ \ \hat{t} \le s \le t-\hat{t}\},\ \{(s,y) \in [0,T]\times R^r : V(s,y) = 0\}\right]$$

(distance in the space $[0,\infty]\times R^r$). By our construction of $\tilde{\varphi}$, the distance $\kappa_{\hat{t}}$ will be positive. Because of (3.9), $u^{\varepsilon}(t-s,y)$ tends to zero uniformly for all (s,y) such that $|y-\tilde{\varphi}_s| < \kappa_{\hat{t}}/2$ and $\hat{t} < s < t-\hat{t}$. Thus there is an $\varepsilon_0 > 0$ such that

$$|c(\hat{\varphi}_s) - c(\hat{\varphi}_s, u^{\varepsilon}(t-s,\hat{\varphi}_s))| < \frac{\delta}{2T} \tag{3.13}$$

if $\hat{t} < s < t-\hat{t}$ and $|\hat{\varphi}_s - \tilde{\varphi}_s| < \frac{1}{2}\kappa_{\hat{t}}$ and $0 < \varepsilon < \varepsilon_0$.

Define $\kappa = \hat{\kappa}_{\hat{t}} = \min(\kappa_{\hat{t}}, \rho(\tilde{\varphi}_t, R^r\setminus G_0))$. Taking into account (3.4), (3.13), we can calculate

$$u^{\varepsilon}(t,x) = E_x g(X_t^{\varepsilon})\exp\left\{\frac{1}{\varepsilon}\int_0^t c(X_s^{\varepsilon}, u^{\varepsilon}(t-s,X_s^{\varepsilon}))ds\right\} \tag{3.14}$$

$$\ge E_x g(X_t^{\varepsilon})\chi_{\{\rho_{0t}(X^{\varepsilon},\tilde{\varphi})<\hat{\kappa}\}}\exp\left\{\frac{1}{\varepsilon}\int_{\hat{t}}^{t-\hat{t}} c(X_s^{\varepsilon})ds - \frac{\delta T}{\varepsilon}\right\}$$

$$\ge \underline{g}E_x\chi_{\{\rho_{0t}(X^{\varepsilon},\tilde{\varphi})<\hat{\kappa}\}}\exp\left\{\frac{1}{\varepsilon}\int_0^t c(\tilde{\varphi}_s)ds - \frac{\delta T}{\varepsilon} - \frac{2\bar{c}\hat{t}}{\varepsilon}\right\}$$

$$\ge \underline{g}\ \exp\left\{\frac{R_{0t}(\tilde{\varphi}_s)-\delta\tau-2\bar{c}\hat{t}}{\varepsilon}\right\},$$

where $\underline{g} = \min\{g(y) : |y-\tilde{\varphi}_t| \le \hat{\kappa}_{\hat{t}}$, $\bar{c} = \sup\{c(y) : |y| \le \|\varphi\| + \delta\}$. We used in (3.14) the lower bound for $E_x\chi_{\{\rho_{0t}(X^{\varepsilon},\tilde{\varphi})<\hat{\kappa}\}} = P_x\{\|X^{\varepsilon} - \chi\| < \hat{\kappa}\}$ given by property (1) in the definition of the action functional: for any $\bar{\delta}, \hat{\kappa} > 0$ there exists $\varepsilon_0 > 0$ such that

$$P_x\{\|X^\varepsilon - \varphi\| < \hat{\kappa}\} \ge \exp\{-(S^x_{0T}(\varphi)+\bar{\delta})\varepsilon^{-1}\} \quad \text{for} \quad \varepsilon \le \varepsilon_0.$$

Since $\hat{t}$ and $\delta > 0$ can be chosen arbitrarily small, we get (3.12) from (3.14). □

Now we can formulate the main result of this section (see [6], [7]).

<u>Theorem 3.1</u>. a) If $V(t,x) < 0$ then $\lim_{\varepsilon\downarrow 0} u^\varepsilon(t,x) = 0$. The convergence is uniform for (t,x) from any compact set $K \subset \{(s,y) : V(s,y) < 0\}$.

b) Assume that condition (N) is fulfilled. If $V(t,x) > 0$ then $\lim_{\varepsilon\downarrow 0} u^\varepsilon(t,x) = 1$. The convergence is uniform for (t,x) from any compact set $K \subset \{(s,y) : V(s,y) > 0\}$.

<u>Proof</u>. The first statement follows from (3.9). To prove statement b), we need the following result:

$$\overline{\lim_{\varepsilon\downarrow 0}}\, u^\varepsilon(t,x) \le 1 \quad \text{for} \quad t > 0. \tag{3.15}$$

To prove (3.15) assume that $\overline{\lim}_{\varepsilon\downarrow 0}\, u^\varepsilon(t_0,x_0) = 1 + h > 1$ for some $t_0 > 0$, $x_0 \in R^r$. Denote $D^\varepsilon = \{(t,x) : u^\varepsilon(t,x) > 1 + \frac{h}{2}\}$, $\tau^\varepsilon = \inf\{t : X^\varepsilon_t \notin D^\varepsilon\}$. The set D^ε contains the point (t_0,x_0) for arbitrary small $\varepsilon > 0$. Let χ_1 be the indicator of the set of trajectories for which $\tau^\varepsilon < t_0$. From (3.4) using strong Markov property we have

$$\begin{aligned} u^\varepsilon(t_0,x_0) &= E_x u^\varepsilon(t_0-\tau^\varepsilon \wedge t_0, X^\varepsilon_{\tau^\varepsilon\wedge t_0})\exp\left\{\frac{1}{\varepsilon}\int_0^{\tau^\varepsilon\wedge t_0} c(X^\varepsilon_s, u^\varepsilon(t_0 0s, X^\varepsilon_s))ds\right\} \\ &= E_x\chi_1 u^\varepsilon(t_0-\tau^\varepsilon, X^\varepsilon_{\tau^\varepsilon})\exp\left\{\frac{1}{\varepsilon}\int_0^{\tau^\varepsilon} c(X^\varepsilon_s, u^\varepsilon(t_0-s, X^\varepsilon_s))ds\right\} \\ &\quad + E_x(1-\chi_1)g(X^\varepsilon_{t_0})\exp\left\{\frac{1}{\varepsilon}\int_0^{t_0} c(X^\varepsilon_s, u(t_0-s, X^\varepsilon_s))ds\right\} \\ &\le \left(1+\frac{h}{2}\right)P_x\{\tau^\varepsilon < t_0\} + P_x\{\tau^\varepsilon \ge t_0\} \max g(x)\cdot e^{-\alpha t_0/\varepsilon}, \end{aligned} \tag{3.16}$$

where $\alpha = -\inf_{\substack{u\ge 1+\frac{h}{2}\\ x\in R^r}} c(x,u) > 0$. For ε small enough, $\max g(x)\cdot\exp\{-\frac{\alpha t_0}{\varepsilon}\} <$ $1+\frac{h}{2}$, and we get from (3.16)

$$u^\varepsilon(t_0,x_0) < \left[1+\frac{h}{2}\right][P_x\{\tau^\varepsilon < t_0\} + P_x\{\tau^\varepsilon \ge t_0\}] = 1+\frac{h}{2}.$$

This last inequality is a contradiction to our assumption that $u^\varepsilon(t_0,x_0) > 1+h$ for a sequence of $\varepsilon > 0$ tending to zero. The contradiction proves (3.15).

Let us show now that $\underline{\lim}_{\varepsilon\downarrow 0}\, u^\varepsilon(t,x) \ge 1$ if $V(t,x) > 0$. From here and (3.15) the statement b) follows.

Suppose there is a point (t_0,x_0) with $V(t_0,x_0) > 0$, $h > 0$ and a sequence $\{\varepsilon_n\}$, $\varepsilon_n \downarrow 0$ as $n\to\infty$, such that $u^{\varepsilon_n}(t_0,x_0) < 1-h$ for all n. Define a set

$$D_\varepsilon = \{(t,x) : u^\varepsilon(t,x) < 1-\frac{h}{2},\ \ V(t,x) > 0\},\quad \varepsilon > 0,$$

and corresponding exit times (Figure 2)

$$\tau^\varepsilon = \min\{s : (t_0-s, X_s^\varepsilon) \notin D_\varepsilon\}.$$

Denote

$$c_0 = \inf\{c(x,u),\ 0 \le u \le 1-\frac{h}{2},\ \ x\in R^r\}.$$

It follows from our assumptions about $c(x,u)$ that $c_0 > 0$. Then for $\varepsilon = \varepsilon_n$ small enough

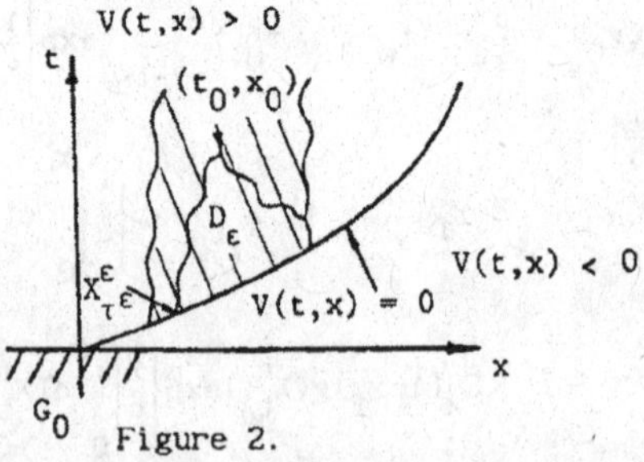

Figure 2.

$$(3.17)\qquad u^\varepsilon(t_0,x_0) = E_x u^\varepsilon(t_0-\tau^\varepsilon, X^\varepsilon_{\tau^\varepsilon})\exp\left\{\frac{1}{\varepsilon}\int_0^{\tau^\varepsilon} c(X_s^\varepsilon, u^\varepsilon(t-s,X_s^\varepsilon))ds\right\}$$

$$\geq E_x u^\varepsilon(t_0-\tau^\varepsilon, X^\varepsilon_{\tau^\varepsilon})\exp\left\{\frac{c_0\tau^\varepsilon}{\varepsilon}\right\}$$

$$\geq E_x \chi_{\{V(t_0-\tau^\varepsilon, X^\varepsilon_{\tau^\varepsilon})>0\}} u^\varepsilon(t_0-\tau^\varepsilon, X^\varepsilon_{\tau^\varepsilon})$$

$$+ E_x \chi_{\{V(t_0-\tau^\varepsilon, X^\varepsilon_{\tau^\varepsilon})=0\}} u^\varepsilon(t_0-\tau^\varepsilon, X^\varepsilon_{\tau^\varepsilon}) e^{c_0\tau^\varepsilon/\varepsilon}.$$

Let $2d = \text{distance}\{(t_0,x_0), \{(t,x) : V(t,x) = 0\}\}$, $d > 0$, since $V(t_0,x_0) > 0$. The first term in the right hand side of (3.17) is equal to

$$\left[1-\frac{h}{2}\right]P_x\{V(t_0-\tau^\varepsilon, X^\varepsilon_{\tau^\varepsilon}) > 0\}.$$

The second term can be bounded from below by

$$(3.18) \qquad E_x \chi_{\{V(t_0-\tau^\varepsilon, X^\varepsilon_{\tau^\varepsilon})=0\}} u^\varepsilon(t_0-\tau^\varepsilon, X^\varepsilon_{\tau^\varepsilon})\chi_{\{\tau^\varepsilon>d\}} e^{c_0 d/\varepsilon}$$

$$+ E_x \chi_{\{V(t_0-\tau^\varepsilon, X^\varepsilon_{\tau^\varepsilon})=0\}} u^\varepsilon(t_0-\tau^\varepsilon, X^\varepsilon_{\tau^\varepsilon})\chi_{\{\tau^\varepsilon\leq d\}}.$$

According to Lemma 3.1 there is an $\varepsilon_0 > 0$ such that $u^\varepsilon(t_0-\tau^\varepsilon, X^\varepsilon_{\tau^\varepsilon}) > \exp\{-\frac{dc_0}{2\varepsilon}\}$, if $V(t_0-\tau_\varepsilon, X_{\tau_\varepsilon}) = 0$ and $\varepsilon < \varepsilon_0$. Therefore the first term in (3.18) bounded from below by

$$e^{c_0 d/2\varepsilon}\, P_x\{V(t_0-\tau^\varepsilon, X^\varepsilon_{\tau^\varepsilon}) = 0, \quad \tau^\varepsilon > d\}$$

for ε small enough.

Taking into account (3.15), one can check that (3.18) is bigger than

$$\left[1-\frac{h}{2}\right]P_x\{V(t_0-\tau^\varepsilon, X^\varepsilon_{\tau^\varepsilon}) = 0\} - 2P_x\{V(t_0-\tau^\varepsilon, X^\varepsilon_{\tau^\varepsilon}) = 0, \quad \tau^\varepsilon \leq d\}.$$

Combining all these estimates we have from (3.17)

$$(3.19) \qquad u^\varepsilon(t_0,x_0) \geq 1-\frac{h}{2} - 2P_x\{V(t_0-\tau^\varepsilon, X^\varepsilon_{\tau^\varepsilon}) = 0, \quad \tau^\varepsilon \leq d\}$$

for ε small enough. Since $P_x\{V(t_0-\tau^\varepsilon, X^\varepsilon_{\tau^\varepsilon}) = 0, \ \tau^\varepsilon \leq d\} \to 0$ as $\varepsilon\downarrow 0$, we conclude from (3.19) that $u^{\varepsilon}(t_0,x_0) \geq 1-\frac{2h}{3}$ for ε small enough. The last inequality is a contradiction to the assumption that $u^{\varepsilon}(t_0,x_0) < 1-h$ for $\varepsilon = \varepsilon_n \to 0$ as $n\to\infty$. Thus $\lim_{\varepsilon\downarrow 0} u^{\varepsilon}(t,x) = 1$ if $V(t,x) > 0$. The uniformity of the convergence follows from uniformity of the bounds used above. □

Thus under condition (N) the equation $V(t,x) = 0$ defines the border between the set of (t,x) where $u^{\varepsilon}(t,x)$ tends to 1 and to 0. This border we call the wave front.

It follows from the definition of the function $V(t,x)$ that $V(t,x)$ is a monotone increasing function of the variable t for any $x \in R^r$. Then for any $x \in R^r$ we can in the unique way define $t^* = t^*(x)$ such that $V(t^*,x) = 0$. If we interpret the domain where $u^{\varepsilon}(t,x)$ is close to one as an excited domain, then $t^*(x)$ is time when the excitation comes to the point x from the domain G_0.

Denote $V(t,x,y) = \inf\{R_{0t}(\varphi) : \varphi \in C_{0t},\ \varphi_0 = x,\ \varphi_t = y\}$. Then $V(t,x) = \inf\{V(t,x,y) : y \in G_0\}$. One can write the Hamilton-Jacobi equation for the function $V(t,x,y)$:

$$\frac{\partial V}{\partial t} = c(y) - \frac{1}{2}\sum_{i,j=1}^{r} a^{ij}(y)\,\frac{\partial V}{\partial y^i}\,\frac{\partial V}{\partial y^j}. \tag{3.20}$$

But we prefer to deal with variational problems rather than with generalized solutions of (3.20).

In the next section we shall consider a number of examples which show that in some more or less simple situation the motion of the wave front can be described by a Huygens principle with a proper speed field. In more complicated media the motion of the front will be more complicated and it is impossible to describe it by Huygens principle.

§4. Examples.

1. Let $c(x) = c(x,0) = c = \text{const.}$ independent of $x \in R^r$. In this case

$$V(t,x) = \sup\{R_{0t}(\varphi) : \varphi \in C_{0t},\ \varphi_0 = x,\ \varphi_t \in G_0\} \tag{4.1}$$

$$= ct - \inf\left\{\frac{1}{2}\int_0^t \sum_{i,j=1}^r a_{ij}(\varphi_s)\dot\varphi_s^i\dot\varphi_s^j ds \ :\ \varphi_0 = x,\quad \varphi_y \in G_0\right\}.$$

Denote by $\rho(\cdot,\cdot)$ the Riemannian distance in R^r corresponding to the metric form $ds^2 = \sum_{i,j=1}^r a_{ij}(x)dx^i dx^j$.

Lemma 4.1. The following equality holds:

$$\inf\left\{\int_0^t \sum_{i,j=1}^r a_{ij}(\varphi_s)\dot\varphi_s^i\dot\varphi_s^j ds \ :\varphi \in C_{0t},\quad \varphi_0 = x,\quad \varphi_y \in G_0\right\} = \frac{\rho^2(x,G_0)}{t}. \tag{4.2}$$

This infimum is attained on the set of minimal geodesics which connect the point x and the set G_0, provided the parameter along these geodesics is proportional to the length.

Proof. It is sufficient to prove (4.2) for $G_0 = \{y\}$. Let us denote by $\ell_{0t}(\varphi)$ the length of the curve φ between the point φ_0 and φ_t:

$$\ell_{0t}(\varphi) = \int_0^t \sqrt{\sum a_{ij}(\varphi_s)\dot\varphi_s^i\dot\varphi_s^j}\, ds.$$

Suppose that γ_s, $0 \le s \le t$, is the minimal geodesic connecting the points $x, y \in R^r$. Let the parameter s along the curve be proportional to the arc length. Then

$$\sum_{i,j=1}^r a_{ij}(\gamma_s)\dot\gamma_s^i\dot\gamma_s^j ds = \frac{d^2(x,y)}{t^2} = \frac{1}{t^2}\ell_{0t}^2(\gamma), \quad s \in [0,t], \tag{4.3}$$

$$\int_0^t \sum a_{ij}(\gamma_s)\dot\gamma_s^i\dot\gamma_s^j ds = \frac{1}{t}d^2(x,y).$$

Thus the value $\frac{1}{t}d^2(x,y)$ can be attained if we take $\varphi_s = \gamma_s$. Now let us prove that the functional cannot be less than $\frac{1}{t}d^2(x,y)$: Using the Schwarz inequality we have for arbitrary φ_s, $0 \le s \le t$, connecting the points x and y:

$$t\cdot\int_0^t \sum_{i,j=1}^r a_{ij}(\gamma_s)\dot\gamma_s^i\dot\gamma_s^j ds \ge \left[\int_0^t \sqrt{\sum_{i,j=1}^r a_{ij}(\varphi_s)\dot\varphi_s^i\dot\varphi_s^j}\,ds\right]^2 = \ell_{0t}^2(\varphi) \ge d^2(x,y). \tag{4.4}$$

From (4.3) and (4.4), equality (4.2) follows. The first inequality in (4.4) turns into an equality only provided the parameter along the curve φ is proportional to the arc length (then $\sum a_{ij}(\varphi)\dot\varphi_s^i\dot\varphi_s^j \equiv \text{const.}$). The last inequality becomes an equality if φ is the minimal geodesic connecting the points x and y. □

From (4.1) and Lemma 4.1 we have

$$V(t,x) = ct - \frac{\rho^2(x, G_0)}{2t}.$$

Thus

$$\lim_{\varepsilon\downarrow 0} u^\varepsilon(t,x) = 0, \quad \text{if} \quad \rho^2(x,G_0) > t\sqrt{2c}.$$

If $V(t,x) = 0$ then $\rho(x,G_0) = t\sqrt{2c}$. For any $\delta \in (0,t)$ define $\varphi^\delta(s) = x$ for $s \in [0,\delta]$, and $\varphi^\delta_{(s)} = \gamma_{s-\delta}$ for $s \in [\delta,t]$, where γ_s, $0 \le s \le t-\delta$, is the minimal geodesic connecting points x and G_0. It is easy to see that $V(t-s,\varphi^\delta_{(s)}) < 0$ for $0 < s < t$, and that

$$\frac{1}{2}\int_0^t \sum_{i,j=1}^r a_{ij}(\varphi_s^\delta)\dot\varphi_s^{\delta,i}\dot\varphi_s^{\delta,j}\,ds = \frac{\rho^2(x,G_0)}{2(t-\delta)} \to \frac{\rho^2(x,G_0)}{2t} \quad \text{as} \quad \delta\downarrow 0.$$

This means that condition (N) is fulfilled in this example. Thus, according to Theorem 3.1, we have

$$\lim_{\varepsilon\downarrow 0} u^\varepsilon(t,x) = 1 \ \text{if} \ V(t,x) > 0.$$

The manifold $\{x \in R^r : \rho(x,G_0) = t\sqrt{2c}\}$ separates the set where $u^\varepsilon(t,x)$ is close to 1 from the set where $u^\varepsilon(t,x)$ is close to to 0 for $\varepsilon \ll 1$. This manifold is a wave front at time t. The geodesics of the metric $\rho(\cdot,\cdot)$ play the role of rays, and the front travels in accordance with the Huygens principle. The statement of Theorem 3.1 means that if $c(x) \equiv c = \text{const.}$, then the velocity $v(x,e)$ of the spreading of the excitation at a point x in the direction of the unit vector e in the Euclidean metric in R^r is equal to

$$v(x,e) = \sqrt{2c \sum_{i,j=1}^r a_{ij}(x)e^i e^j}.$$

In the Riemannian metric $\rho(\cdot,\cdot)$ the velocity field is homogeneous isotropic and equal to $\sqrt{2c}$. In particular, if equation (2.61 has the form

$$\frac{\partial u^\varepsilon}{\partial t} = \frac{\varepsilon}{2}\Delta u^\varepsilon + \frac{1}{\varepsilon}c(x,u)u,$$

and $c(x,0) \equiv c =$ const., then $v(x,e) = \sqrt{2Dc}$ as we saw in the KPP-equation.

Even if G_0 is simple, for example. is a ball, the wave front at time $t > 0$ can have a complicated structure. But in the case $c(x) \equiv$ const. the excited area Increases continuously: the set $\{x \in R^r : V(t+\Delta,x) > 0\}$ contains only points close to $\{x \in R^r : V(t,x) > 0\}$, if A is small. As we shall see in Example 3, for variable $c(x)$ the situation is more complicated.

2. Let $x \in R^1$, $a^{11}(x) \equiv 1$, $f(x,u) = c(x)u(1-u)$. Suppose that $c(x) = 1$ for $x < 0$ and $c(x) = 1+x$ for $x > 0$. Let the support of the initial function be $G_0 = G_0^a = \{x \quad R^1, \ x < a\}$. In this case

$$V(t,x) = V_a(t,x) = \sup\left\{\int_0^t \left[1+\varphi_s - \frac{1\cdot 2}{s}\dot\varphi_s\right]ds : \ \varphi_0 = x, \quad \varphi_t = a\right\}, \ x \ge a.$$

The Euler equation for this variational problem has the form $\ddot\varphi = 1$. This means that $\varphi(s)$ is a second order polynomial, and taking into account the boundary conditions, we can find the extremal $\hat\varphi_s$ on which the supremum is attained:

$$\hat\varphi_s = -\frac{s^2}{2} + \left(\frac{t}{2} + \frac{(a-x)}{t}\right)s + x, \quad 0 \le s \le t.$$

Then for $V_a(t,x)$ we have the expression

$$V_a(t,x) = \int_0^t [1+\hat\varphi_s - \frac{1\cdot 2}{s}\hat\varphi_s^2]ds = \frac{t^3}{24} + t\left(1+\frac{a-x}{2}\right) - \frac{(a-x)^2}{2t}.$$

By equating $V_a(t,x)$ to zero, we find the expression for the front position $X_a^*(t)$ at time t:

$$X_a^*(t) = \frac{t^2}{2} + a + \sqrt{\frac{t^4}{3} + 2t^2(1+a)}.$$

Note that since $X_a^*(t)$ is convex and the extremals are concave, condition (N) is fulfilled. Therefore In this case Theorem 3.1 can be applied.

For every $a \ge 0$ the function $X_a^*(t)$ is strictly increasing, and at each time t its derivative in t can be represented in the form of a function of $X_a^*(t)$. It may seem that in this case the front propagation also admits description with the help of the Huygens principle with an appropriate velocity field. But the velocity field here turns out be depend on the initial condition. It is not universal as in the case $c(x) \equiv c \equiv \text{const}$. To see this it is sufficient to evaluate $\tilde a = X_0^*(1)$ and $X_0^*(2)$:

$$X_0^*(1) \equiv \tilde a \equiv \frac{1}{2} + \sqrt{7/3}, \quad X_0^*(2) = 2 + \sqrt{40/3} \approx 5.6,$$

$$X_{\tilde a}^*(1) = 1 + \sqrt{7/3} + \sqrt{10/3 + \sqrt{28/3}} \approx 5.1.$$

If the velocity field did not depend on the initial condition, then the equality $X_0^*(2) \equiv X_{\tilde a}^*(1)$ would be valid. In our case $X_0^*(2) > X_{\tilde a}^*(1)$, and thus the velocity field is not of such a universal nature.

The motion of the wave front in this example does not satisfy the Markov property (semi-group property): knowledge of the position of the front $X_0^*(1)$ at time 1 is not enough to predict $X_0^*(2)$. To have the Markovian property we should extend the phase space. We should remember at time t not only the position of the wave front, but the exponentially small tails ahead of the front in the area where $u^\varepsilon(t,x) \to 0$ as $\varepsilon \downarrow 0$.

One more particularity of the wave front propagation which one can see in this example is non-symmetry. We calculated how much time it takes for the front to reach the point 1, if $G_0 = \{x \in R^1, \ x < 0\}$. This time t_1^* is equal to the solution of the equation $X_0^*(t_1^*) = 1$. It turns out that if $G_0 = \{x \in R^1 : x > 1\}$, then it will take more time for the front to come from 1 to 0 than t_1^*. This follows from Example 4. It is one more feature showing that in the case $c(x) \neq \text{const.}$, the propagation of the fronts is more complicated.

3. In this example we show that in the case of variable $c(x) \equiv c(x,0)$ the wave front may have jumps. We will first choose a discontinuous function for $c(x,u)$. Note that this is done solely for the sake of simplification of the computations. In the final part of the example we will see that "new sources" also arise in the case where $c(x,u)$ is continuous, provided $c(x,0)$ increases sufficiently quickly in some finite interval.

Let $x \in R^2$, $a^{11}(x) \equiv 1$, $f(x,u) = c(x)u(1-u)$, $c(x) \equiv c_1 > 0$ for $x < h$, and $c(x) = c_2 > 2c_1$ for $x \geq h > 0$. Assume that the support of the initial function is $G_0 \equiv \{x < 0\}$. Inside each of the domains $\{x < h\}$ and $\{x > h\}$ the Euler equation for corresponding functional $R_{0,t}(\varphi) \equiv \int [c(\varphi_2) - \frac{1}{2}\dot\varphi_s^2]ds$ takes the form $\ddot\varphi \equiv 0$. This means that the extremals of the functional $R_{0t}(\varphi)$ will be either segments of lines or broken lines with vertices on the line $x = h$ (see Figure 3). Let us compute

$$V(t,x) = \sup\{R_{0t}(\varphi) : \varphi \in \mathrm{COT},\ \varphi_0 = x,\ \varphi_t \le 0\}$$

for the points of the line $x = h$. On the broken line connecting points (t_0,h), (t_1,h) and $(0,0)$ the functional $R_{0t_0}(\varphi)$ takes the value

$$R^{t_1} = c_2(t_0-t_1) + c_1t_1 - \frac{h^2}{2t_1} = c_2t_0 - (c_2-c_1)t_1 + \frac{h^2}{2t_1}.$$

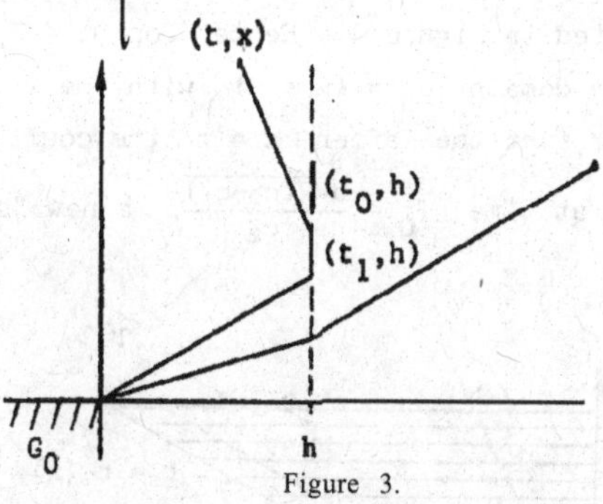

Figure 3.

Let us find t_1, for which this quantity is maximal. It is easily seen that

$$\max_{0 \le t_1 \le t_0} R^{t_1} = c_2t_0 - h\sqrt{2(c_2-c_1)}.$$

The maximum is attained for $t_1 = \hat{t}_1 = \frac{h}{\sqrt{2(c_2-c_1)}}$. Thus if $t_0 > \hat{t}_1$, then the absolute maximum is attained on the broken line (rather than on the segment connecting the points (t_0,h) and (0,011, and

$$V(t_0,h) = R^{\hat{t}_1} = c_2t_0 - h\sqrt{2(c_2-c_1)}.$$

The condition $V(t,x) = 0$ yields that the wave front reaches the point $x = h$ at the time $T_0 = \frac{h\sqrt{2(c_2-c_1)}}{c_2}$. We observe that $T_0 > \hat{t}_1$ since $c_2 > 2c_1$.

So $\lim_{\varepsilon\downarrow 0} u^\varepsilon(t,x) = 1$ for $t > T_0$. It is not difficult to check that for $x < \frac{1}{2}[h + T_0\sqrt{2c_1}] = \bar{x}$ the upper bound involved in the definition of $V(t,x)$ is attained on the linear segments connecting the points (t,x) and $(0,0)$. Therefore, $V(t,x) = c_1t - \frac{x^2}{2t}$ for $x < \bar{x}$, and the wave front in this domain travels according to the law $x = t\sqrt{2c_1}$. For $x > \bar{x}$ the upper bound is attained on the broken lines having vertices on the line $x = h$. In particular, if $x \in (\bar{x}, h)$ then the extremal is not monotonic: first it reaches the point $x = h$, spends a certain time at this point, and then in the remaining time it reaches zero. One can interpret this as that the points $x \in (\bar{x}, h)$ will be excited by the new source which has appeared at time T_0 at the point $x = h$. The shape of the curve $t^*(x)$, determined by the equation $V(t^*, x) = 0$ is represented in Figure 4. Hence, for $t < T_0$, the wave propagates to the right of the domain $G_0 = \{x < 0\}$ with the velocity $\sqrt{2c_1}$, "taking no notice" of the fact that after $x = h$ the coefficient $c(x)$ takes a larger value c_2. But at time $T_0 = \frac{h\sqrt{2(c_2-c_1)}}{c_2}$, a new "source" arises at the point $x = h$, away

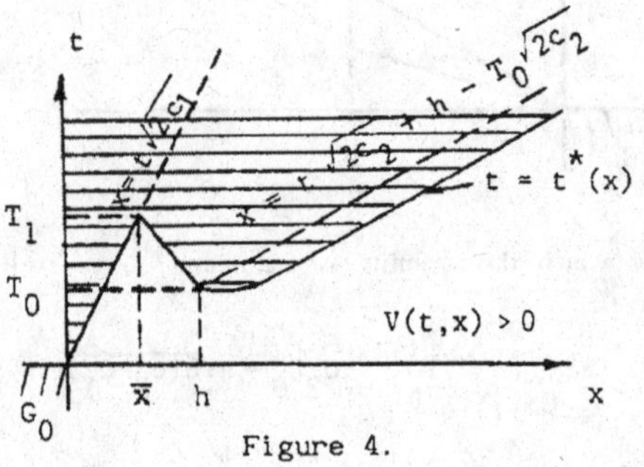

Figure 4.

from which the front starts propagating in both directions: to the left with velocity $\sqrt{2c_1}$, and to the right with the velocity close to $\sqrt{2c_2}$.

It is not hard to verify that condition (N)$'$ is fulfilled here. Below the curve $t = t^*(x)$ the function $u^\varepsilon(t,x)$ tends to zero as $\varepsilon \downarrow 0$. Above this curve $\lim u^\varepsilon(t,x) = 1$.

This example shows that the domain where $\lim\limits_{\varepsilon\downarrow 0} u^\varepsilon(t,x) = 1$ can expand in a non-continuous way. At the time T_0 a jump of the wave front occurs and the "excited region" has two connected components for $t \in (T_0, T_1)$.

Now let $c(x)$ be a smooth monotone function which is equal to c_1 for $x < \bar{x} + \frac{1}{2}(h + T_0\sqrt{2c_1}) + \delta$, where δ is small enough, and is equal to c_2 for $x > h$. Using comparison theorems, one can show that for such a function $c(x)$ the front also has a jump. □

In the first example the wave front propagates according to the Huygens principle with velocity field homogeneous and isotropic, if we calculate in the corresponding Riemannian metric. One can check that in the second and third examples, the wave front moves faster than it would move if it was governed by the local law. More precisely, let us introduce the velocity field

$$v(x,e) = \sqrt{2c(x)}\left[\sum_{i,j=1}^{r} a_{ij}(x)e^i e^j\right]^{-1/2}, \quad x \in R^r, \quad |e| = 1.$$

We put

$$\tau_{G_0}(x) = \inf\left\{\int_0^t \frac{|\dot{\varphi}_s|ds}{v(\varphi_s, \dot{\varphi}_s|\dot{\varphi}_s|^{-1})} : \varphi \in C_{0T}, \quad \varphi_0 = x, \quad \varphi_t \in G_0\right\},$$

$$G_t = \{y \in R^r; \quad \tau_{G_0}(y) < t\}.$$

So G_t is exactly the domain which would be occupied by excitation if the excitation propagated according to Huygens principle with velocity field $v(x,e)$ in R^r.

Using the maximum principle for parabolic equations and the result of Example 1, one can prove the following proposition.

<u>Proposition 4.1</u>. Let $u^\varepsilon(t,x)$ be the solution of problem (3.1), $c(x,u)$ is bounded from above and continuously differentiable. Then $\lim_{\varepsilon\downarrow 0} u^\varepsilon(t,x) = 1$ for $x \in G_t$.

The proof of this statement one can find in [7] (Lemma 6.2.4).

We underline that this bound from below is true without condition (N). The upper bound is given by statement a) of Theorem 3.1. Under condition (N) the bound given by statement a) of Theorem 3.1 is precise. The next examples show that without (N) we can have $\lim_{\varepsilon\downarrow 0} u^\varepsilon(t,x) = 0$ in a bigger area than $\{x : V(t,x) < 0\}$.

4. Let $x \in R^1$, $a^{11}(x) \equiv 1$, $G_0 = \{x \in R^1, \ x < 0\}$, and $c(x)$ be monotonically decreasing for $x > 0$. Consider the function $\psi(s)$, $s > 0$, defined by the differential equation

$$\dot{\psi}(s) = \sqrt{2c(\psi(s))}, \quad \psi_0 = 0.$$

This function increases monotonically; its derivative $\dot{\psi}(s)$ decreases monotonically, remaining positive (see Figure 5). One can deduce from Proposition

4.1 that $\lim_{\varepsilon\downarrow 0} u^{\varepsilon}(t,x) = 1$ in the shaded domain above the curve ψ. Let us show that $\lim_{\varepsilon\downarrow 0} u^{\varepsilon}(t,x) = 0$ if $x > \psi(t)$. This means that $x = \psi(t)$ is exactly the position of the front at time t. Denote $\tau_t^{\psi} = \tau_t = \inf\{s : X_s \geq \psi(t-s)\} \wedge t$. It is a Markov time and we can write

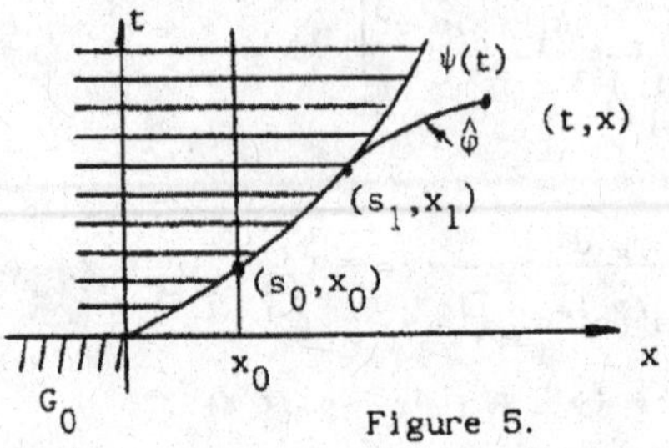

Figure 5.

(4.5) $$u^{\varepsilon}(t,x) = E_x u^{\varepsilon}(t-\tau_t, X^{\varepsilon}_{\tau_t}) \exp\left\{\frac{1}{\varepsilon}\int_0^{\tau_t} c(X_s^{\varepsilon}, u^{\varepsilon}(t-s, X_s^{\varepsilon}))ds\right\}$$

$$\leq (1 \vee \sup_{x\in R^1} g(x)) E_x \chi_{\{\tau_t<t\}} \exp\left\{\frac{1}{\varepsilon}\int_0^{\tau_t} c(X_s^{\varepsilon})ds\right\},$$

where $g(x)$ is the initial function which is assumed bounded, and $\chi_{\{\tau_t<t\}}$ is the indicator of the set $\{\tau_t < t\}$. Using large deviation estimates one can check (see the proof of Theorem 5.1 in the next section) that the right hand side of (4.5) is logarithmically equivalent as $\varepsilon\downarrow 0$ to

$$\exp\{\varepsilon^{-1}\tilde{V}(t,x)\},$$

where

$$\tilde{V}(t,x) = \sup\left\{\int_0^s [c(\varphi_\sigma) - \frac{1}{2}\dot{\varphi}_\sigma^2]d\sigma : \varphi \in C_{0s},\ \varphi_0 = x,\ \varphi_\sigma > \psi(t-\sigma) \text{ for } 0 \leq \sigma < s,\ \varphi_s = \psi(t-s),\ s \leq t\right\}.$$

Let us show now that $\tilde{V}(t,x) < 0$ for $x > \psi(t)$. Then we can derive from (4.5) that $\lim_{\varepsilon\downarrow 0} u^{\varepsilon}(t,x) = 0$ for $x > \psi(t)$. Suppose the functional involved in the definition of $\tilde{V}(t,x)$ attains its upper bound on an extremal $\hat{\varphi}_s$, $s \in [0,t-s_1]$

$$(4.6) \qquad \tilde V(t,x) = \int_0^{s_1} [c(\hat\varphi_\sigma) - \frac{1}{2}\dot{\hat\varphi}_\sigma^2]ds.$$

Along with $\hat\varphi$ we shall consider the function $\hat{\hat\varphi}_s$ which is just a segment of ψ : $\hat{\hat\varphi}_s = \psi(t-s)$, $0 \le s \le t - s_1$. Since $c(x)$ monotonically decreases and $\hat{\hat\varphi}_s \le \hat\varphi_s$, we conclude that $c(\hat{\hat\varphi}_s) > c(\hat\varphi_s)$ for $0 \le s \le t - s_1$. Next we observe that

$$(4.7) \qquad \int_0^{t-s_1} \frac{1}{2}\dot{\hat{\hat\varphi}}_\sigma^2\, ds \ge \frac{(x-x_1)^2}{2(t-s_1)} > \int_0^{t-s_1} \frac{1}{2}\dot{\hat\varphi}_\sigma^2\, ds.$$

In the last inequality we have made use of the fact that $\dot{\hat{\hat\varphi}} < \frac{x-x_1}{t-s_1}$ for $0 \le s \le t - s_1$. From (4.6) and (4.7) we have

$$\tilde V(t,x) = \int_0^{t-s_1} [c(\hat\varphi_s) - \frac{1}{2}\dot{\hat\varphi}_\sigma^2]ds < \int_0^{t-s_1} c(\hat{\hat\varphi}_s)ds - \int_0^{t-s_1} \frac{1}{2}\dot{\hat{\hat\varphi}}_\sigma^2\, ds = 0$$

for $x > \psi(t)$. □

§5. General Result.

In this section we describe the behavior of the solution of Problem 3.1 without assumption (N).

Let us conside the "heat" process (t_s, X_s^ε). The first component is the deterministic motion with the speed -1, $t_s = t_0 - s$. The second component X_s^ε is the Markov process corresponding to the operator εL. The phase space of the heat process is $(-\infty,\infty)\times R^r = \mathcal{H}$. Let F be a closed subset of $\mathcal{H}$. Define the functional $\tau = \tau_F[t,\varphi]$ on $(-\infty,\infty)\times C_{0\infty}$ with the values in $[0,\infty]$ by the formula

$$\tau = \tau_F(t,\varphi) = \inf\{s : (t-s,\varphi_s) \in F\}.$$

It is clear that $\tau_F(t,X^\varepsilon)$ is the first time when the heat process touches F; τ_F is a Markov time with respect to the family of σ-fields $\{\mathcal{F}_s, s \ge 0\}$; $\mathcal{F}_s$ is the minimal σ-field in the probability space, such that $X_{s_1}^\varepsilon$ is $\mathcal{F}_s$-measurable for any $s_1 \le s$. The functionals τ_F we call Markov functionals.

Denote by θ the set of all Markov functionals.

Let us introduce the function $V^*(t,x)$, $t>0$, $x\in R^r$.

$$V^*(t,x) = \inf_{\tau\in\theta} \sup\left\{\int_0^{\tau\wedge t} \left[c(\varphi_s) - \frac{1}{2}\sum_{i,j=1}^{r} a_{ij}(\varphi_s)\dot\varphi_s^i\dot\varphi_s^j\right]ds \ :\ \varphi\in C_{0t},\ \varphi_0 = x,\ \varphi_t \in G_0\right\}.$$

It is clear that $V^*(t,x) \le (0\wedge V(t,x))$, where $V(t,x)$ was introduced in §3.

Since $\tau = \tau(t,X^\varepsilon)$ and $\tau\wedge t$ are Markov times, using the strong Markov property of the process X_t^ε we derive from (3.2) that the following equation is fulfilled for the function $u^\varepsilon(t,x)$:

$$(5.1)\qquad u^\varepsilon(t,x) = E_x u^\varepsilon(t-(\tau\wedge t), X^\varepsilon_{\tau\wedge t}) \exp\left\{\frac{1}{\varepsilon}\int_0^{\tau\wedge t} c(X_s^\varepsilon, u^\varepsilon(t-s,X_s^\varepsilon))ds\right\}.$$

Of course equation (5.1) is true for any Markov time τ with respect to the family of σ-fields $\mathcal{F}_s$ not only for the above-defined functionals of the heat process. Consideration of the equation (5.1) instead of (3.2) is actually the main modification which allows us to describe the motion of the wave fronts in the general situation without condition (N).

Lemma 5.1. If $V^*(t,x) < 0$ then $\lim_{\varepsilon\downarrow 0} \varepsilon \ln u^\varepsilon(t,x) < 0$, and $\lim_{\varepsilon\downarrow 0} u^\varepsilon(t,x) = 0$ uniformly in (t,x) from any compact set $F \subset \{(s,y),\ s>0,\ V^*(s,y)<0\}$.

Proof. Since $V^*(t,x) < 0$ there exist $\tau^* \in \theta$ such that

$$\sup\left\{\int_0^{\tau^*\wedge t} \left[c(\varphi_s) - \frac{1}{2}\sum_{i,j=1}^{r} a_{ij}(\varphi_s)\dot\varphi_s^i\dot\varphi_s^j\right]ds \ :\ \varphi\in C_{0t},\ \ \varphi_0 = x,\ \ \varphi_t \in G_0\right\} = -\beta < 0.$$

Taking into account that $c(x,u) \le c(x) = c(x,0)$, for every integer $n>0$ we have from (5.1)

$$(5.2)\qquad u^\varepsilon(t,x) \le \sum_{k=1}^{n} E_x\chi_k u^\varepsilon(t-\tau^*\wedge t, X^\varepsilon_{t\wedge\tau^*}) \exp\left\{\frac{1}{\varepsilon}\int_0^{\tau^*\wedge t} c(X_s^\varepsilon)ds\right\}$$

$$+ E_x\chi_{\{\tau\ge t\}} g(X_t^\varepsilon) \exp\left\{\frac{1}{\varepsilon}\int_0^t c(X_s^\varepsilon)ds\right\},$$

where χ_k is the indicator of the set $\left\{\frac{t(k-1)}{n} \le \tau^* < \frac{kt}{n}\right\} = 1,\cdots,n$ and $\chi_{\{\tau\ge t\}}$

is the indicator of the set $\{\tau^* \geq t\}$. Using the Laplace asymptotic formula for functional integrals, we get:

$$(5.3)\qquad E_x\,\chi_k\, u^{\varepsilon}(t-(\tau^*\wedge t), X^{\varepsilon}_{t\wedge\tau^*})\,\exp\left\{\frac{1}{\varepsilon}\int_0^{\tau^*\wedge t} c(X^{\varepsilon}_s)ds\right\}$$

$$\leq [1+\sup_x g(x)]E_x\chi_k \exp\left\{\frac{1}{\varepsilon}\int_0^{tk/n} c(X^{\varepsilon}_s)ds\right\}$$

$$\approx \exp\left\{\frac{1}{\varepsilon}\cdot\sup\left\{\int_0^{tk/n} c(\varphi^{\varepsilon}_s)ds - S_{0,tk/n}(\varphi)\ :\ \varphi_0 = x,\ \frac{k-1}{n}t \leq \tau^*(\varphi) \leq \frac{kt}{n}\right\}\right\},\ \varepsilon\downarrow 0.$$

Here the sign "≈" means logarithmic equivalence for $\varepsilon\downarrow 0$. Note that

$$(5.4)\qquad \sup\left\{\int_0^{tk/n} c(\varphi^{\varepsilon}_s)ds - S_{0,tk/n}(\varphi)\ :\ \varphi_0 = x,\ \frac{t(k-1)}{n} \leq \tau^*(t,\varphi) \leq \frac{kt}{n}\right\}$$

$$\leq \sup\left\{\int_0^{\tau^*\wedge t}\left[c(\varphi_s) - \frac{1}{2}\sum_{i,j=1}^{r} a_{ij}(\varphi_s)\dot\varphi^i_s\dot\varphi^j_s\right]ds\ :\ \frac{(k-1)t}{n} \leq \tau^*(t,\varphi) \leq \frac{kt}{n}\right\}$$

$$+\frac{t}{n}\sup_{x\in R^r} c(x) \leq -\beta + \frac{t}{n}\sup_{x\in R^r} c(x)$$

for $k = 1,\cdots,n$. Now choosing $n > \frac{1}{\beta}2t\cdot\sup_{x\in R^r} c(x)$, we derive from (5.3) and (5.4) that

$$(5.5)\qquad \overline{\lim_{\varepsilon\downarrow 0}}\ \varepsilon \ln E_x\chi_k\ u^{\varepsilon}(t-(\tau^*\wedge t), X^{\varepsilon}_{t\wedge\tau^*})\ \exp\left\{\frac{1}{\varepsilon}\int_0^{\tau^*\wedge t} c(X^{\varepsilon}_s)ds\right\} \leq -\frac{\beta}{2}.$$

Similar bound holds for the last term in (5.2)

$$(5.6)\qquad \overline{\lim_{\varepsilon\downarrow 0}}\ \varepsilon \ln E_x\chi_{\tau^*\geq t}\, g(X^{\varepsilon}_t)\ \exp\left\{\frac{1}{\varepsilon}\int_0^{t} c(X^{\varepsilon}_s)ds\right\} \leq -\frac{\beta}{2}.$$

From (5.2), (5.5) and (5.6) we have the first statement of Lemma 5.1. The second statement follows from the first one and uniformity of the convergence in (5.5) and (5.6). □

Lemma 5.2. Suppose that $\lim_{\varepsilon\downarrow 0} \varepsilon \ln u^{\varepsilon}(t_0,x_0) = 0$, $t_0 > 0$. Then a constant A exists such that $\lim_{\varepsilon\downarrow 0} u^{\varepsilon}(t,x) = 1$ uniformly in (t,x) from any compact subset of the cone $K^{A}_{t_0,x_0} = \{(s,y) : s > t_0,\ |x-x_0| < A(s-t_0)\}$.

Proof. Using the a priori bound for the Hölder norm of a bounded solution of a uniformly parabolic equation with bounded coefficients, we can derive from the conditions of the Lemma 5.2, that for any $\delta > 0$ there exist $\varepsilon_0, \delta_1 > 0$ such that

$$u^{\varepsilon}(t_0,x) > e^{-\delta/\varepsilon} \quad \text{for} \quad |x-x_0| \le e^{-\delta_1/\varepsilon}, \quad 0 < \varepsilon < \varepsilon_0.$$

Now using properties 1 and 2 of problem (3.1), we conclude that $u^{\varepsilon}(t,x) \ge \tilde{u}^{\varepsilon,\delta}(t-t_0,x)$, where $\tilde{u}^{\varepsilon,\delta}(t,x)$ is the solution of the problem (3.1) with the initial function

$$g = g^{\varepsilon,\delta}(x) = \begin{cases} e^{-\delta/\varepsilon}, & \text{for} \quad |x-x_0| \le e^{-\delta_1/\varepsilon} \\ 0, & \text{for} \quad |x-x_0| > e^{-\delta_2/\varepsilon} \end{cases} \tag{5.7}$$

and $c(x,u)$ is replaced by $\tilde{c}(u) = \inf_{x\in R^r} c(x,u)$. □

In the case $c = \tilde{c}(u)$ independent of x we can use Theorem 3.1. We should just take into account that now our initial function depends on ε. It does not influence the proof of the upper bound. So we can conclude that $\lim_{\varepsilon\downarrow 0} \tilde{u}^{\varepsilon,\delta}(t,x) = 0$ for $t > 0$ and $\rho(x,x_0) > t\sqrt{2\tilde{c}(0)}$, where ρ is the Riemannian metric corresponding to the form $ds^2 = = \sum_{i,j=1}^{r} a_{ij}(x)dx^i dx^j$, $(a_{ij}(x)) = (a^{ij}(x))^{-1}$. The convergence of $\tilde{u}^{\varepsilon,\delta}(t,x)$ to zero is uniform in any compact subset of the set $\mathcal{E} = \{(s,y) : s > 0,\ \rho(x,y) > s\sqrt{2\tilde{c}(0)}\}$.

To prove that $\lim_{\varepsilon\downarrow 0} \tilde{u}^{\varepsilon}(t,x) = 1$ for (t,x) such that $t > 0$, $\rho(x,x_0) < t\sqrt{2\tilde{c}(0)}$ we use the following bound for the transition density $p_{\varepsilon}(t,x,y)$ of the process X^{ε}_t: for any $\delta_1, t > 0$ there exist $\varepsilon_0, \delta_0 > 0$ such that

$$p_{\varepsilon}(t,x,y) > e^{-\delta_1/\varepsilon} \quad \text{for} \quad |x-y| < \delta_0, \quad 0 < \varepsilon < \varepsilon_0. \tag{5.8}$$

From (5.7) and (5.8), we derive that for any $\delta_2 > 0$ one can find $s_1 \in (0,\delta_2)$ and $\delta, \varepsilon_0, \delta_3 > 0$ that the following bound holds:

$$(5.9)\qquad \tilde{u}^{\varepsilon,\delta}(s,y) > e^{-\delta_2/\varepsilon} \quad \text{for} \quad |y-x_0| < \delta_3, \quad 0 < \varepsilon < \varepsilon_0.$$

Now we can prove that $\lim_{\varepsilon\downarrow 0} \varepsilon \ln \tilde{u}^{\varepsilon,\delta}(t,x) = 0$ for points (t,x) with $t > 0$, $\rho(x,x_0) < t\sqrt{2\tilde{c}(0)}$, as was done in Theorem 3.1, and then we can check that $\lim_{\varepsilon\downarrow 0} \tilde{u}^{\varepsilon,\delta}(t,x) = 1$ uniformly on any compact subset of the set $\{(t,x),\ \rho(x,x_0) < t\sqrt{2\tilde{c}(0)}\}$. Since $\{(t,x) : \rho(x,x_0) < t\sqrt{2\tilde{c}(0)}\} \supset \{(t,x) : |x-x_0| < At\}$ for some $A > 0$, and $u^{\varepsilon}(t,x) \geq \tilde{u}^{\varepsilon,\delta}(t-t_0,x)$, we derive from here the statement of the Lemma 5.2. □

Lemma 5.3. Assume that $\lim_{\varepsilon'\downarrow 0} u^{\varepsilon'}(t_0,x_0) = 0$ for some sequence $\varepsilon'\downarrow 0$. Then there exists $A > 0$ such that $\overline{\lim}_{\varepsilon'\downarrow 0} \varepsilon' \ln u^{\varepsilon'}(t,x) < 0$ for any point

$$(t,x) \in D^{A}_{t_0,x_0} = \{(s,y) : 0 < s < t_0,\ \ |x_0-y| < A(t_0-s)\}.$$

Let $\mathcal{E}^{(\varepsilon')} = \{(t,x) : \lim_{\varepsilon'\downarrow 0} u^{\varepsilon'}(t,x) = 0,\ \ t > 0\}$. For every compact F belonging to the interior $(\mathcal{E}^{(\varepsilon')})$ of $\mathcal{E}^{(\varepsilon')}$, $\lim_{\varepsilon'\downarrow 0} u^{\varepsilon'}(t,x) = 0$ uniformly in $(t,x) \in F$.

Proof. The first statement follows immediately from Lemma 5.2. To prove the second statement note that compact F can be covered by a finite number of cones $D^{A/2}_{t_k,x_k}$ with vertices $(t_k,x_k) \in (\mathcal{E}^{(\varepsilon')})\backslash F$. The uniformity follows from the uniformity of the bound in Lemma 5.2.

Remark. It follows from Lemma 5.3 that the set $\mathcal{E}^{(\varepsilon')}$ belongs to the closure of its interior $(\mathcal{E}^{(\varepsilon')})$. If $(t,x) \in \mathcal{E}^{(\varepsilon')}$, then $(t-h,x) \in (\mathcal{E}^{(\varepsilon')})$ for small $h > 0$.

Lemma 5.4. Let F be a compact subset of the interior (M) of the set $M = \{(t,x) : t > 0,\ \ x \in R^r,\ \ V^*(t,x) = 0\}$. Then $\lim_{\varepsilon\downarrow 0} \varepsilon \ln u^{\varepsilon}(t,x) = 0$ uniformly in $(t,x) \in F$.

Proof. Suppose that for a point $(t,x) \in (M)$ there exists a sequence $\varepsilon'\downarrow 0$ such that $\lim_{\varepsilon'\downarrow 0} \varepsilon' \ln u^{\varepsilon'}(t,x) = -\beta < 0$. Then $\lim_{\varepsilon'\downarrow 0} u^{\varepsilon'}(t,x) = 0$ and $(t,x) \in \mathcal{E}^{(\varepsilon')}$, where $\mathcal{E}^{(\varepsilon')}$ was introduced in Lemma 5.3. Without loss of generality we can assume that $(t,x) \in (\mathcal{E}^{(\varepsilon')})$. If not, one can take a point

$(t-h, x)$ with small enough $h > 0$. This new point belongs to $(\mathcal{E}^{(\varepsilon')})$ according to the remark above, and belongs to (M) since (M) is open.

Define the Markov functional corresponding to the complement of the set $(\mathcal{E}^{(\varepsilon')})$

$$\tau = \tau(t, \varphi) = \min \{s : (t-s, \varphi_s) \notin (\mathcal{E}^{(\varepsilon')})\}.$$

Since $(t, x) \in M$

$$\sup\left\{\int_0^{\tau\wedge t} \left[c(\varphi_s) - \frac{1}{2}\sum_{i,j=1}^{r} a_{ij}(\varphi_s)\dot{\varphi}_s^i \dot{\varphi}_s^j\right] ds \; : \; \varphi_0 = x, \quad \varphi_t \in G_0\right\} \geq 0.$$

Therefore for any $\delta > 0$, there exists φ_s^δ, $s \in [0, t]$, with $\varphi_0^\delta = x$, $\varphi_t^\delta \in G_0$ such that

$$S_{0,\tau\wedge t}(\varphi^\delta) = \int_0^{\tau\wedge t} \left[c(\varphi_s^\delta) - \frac{1}{2}\sum_{i,j=1}^{r} a_{ij}(\varphi_s^\delta)\dot{\varphi}_s^{\delta,i} \dot{\varphi}_s^{\delta,j}\right] ds \geq -\frac{\delta}{4},$$

$(t-s, \varphi_s^\delta) \in \mathcal{E}^{(\varepsilon')}$ for $s \in [0, \tau(\varphi^\delta))$, and $(t-(t\wedge\tau), \varphi^\delta_{\tau(\varphi^\delta)\wedge t}) \in \partial\mathcal{E}^{(\varepsilon')}$.

Now we define a reconstruction of φ^δ. For any small $\lambda_1, \lambda_2 > 0$ we introduce the function $\varphi_s^{\delta,\lambda_1,\lambda_2}$:

$$\varphi_s^{\delta,\lambda_1,\lambda_2} = \begin{cases} \varphi_0^\delta, & \text{for } s \in [0, \lambda_1] \\ \varphi^{\delta,\lambda_1,\lambda_2}_{\frac{(s-\lambda_1)(T-\lambda_1)}{T-2\lambda_1}} & \text{for } s \in [\lambda_1, T-\lambda_1] \\ \varphi^\delta_{T-\lambda_1} + \frac{s-T+\lambda_1}{1-\lambda_2}, & s \in [T-\lambda_1, T-\lambda_1\lambda_2]\,. \end{cases}$$

Here $T = \tau(t, \varphi^\delta) \wedge t$; the function $\varphi_s^{\delta,\lambda_1,\lambda_2}$ is defined for $s \in [0, T-\lambda_1\lambda_2]$.

The second reconstruction is defined by the formula (h is again a small positive number, $h < \lambda_1\lambda_2$)

$$\bar{\varphi}_2 = \bar{\varphi}_s^{\delta_1,\lambda_1,\lambda_2,h} = \begin{cases} \varphi_0^{\delta,\lambda_1,\lambda_2}, & \text{for } s \in [0, h] \\ \varphi_{s-h}^{\delta,\lambda_1,\lambda_2}, & \text{for } s \in [h, T-\lambda_1\lambda_2+h]. \end{cases}$$

Denote $\bar{T} = T - \lambda_1\lambda_2 + h$, $z = \bar{\varphi}_{\bar{T}} = \varphi_T^\delta$. The positive numbers λ_1, λ_2, h one can choose so small that

$$(5.10)\quad \int_0^{\bar T}\left[c(\bar\varphi_s)-\frac12\sum_{i,j=1}^r a_{ij}(\bar\varphi_s)\dot{\bar\varphi}^i_s\,\dot{\bar\varphi}^j_s\right]ds \ge -\frac{\delta}{2},\quad \int_{\bar T-h}^{\bar T}\left[\sum_{i,j=1}^r a_{ij}(\bar\varphi_s)\dot{\bar\varphi}^i_s\,\dot{\bar\varphi}^j_s\right]ds \le \frac{\delta}{8}.$$

Note that the set $\{(t-s,\bar\varphi_s) : s\in[h,\bar T-h]\}$ is a compact subset of $(\mathcal{E}^{(\varepsilon')})$. Therefore, it follows from Lemma 5.3 that $u^{\varepsilon'}(t-s,\bar\varphi_s)\to 0$ when $\varepsilon'\downarrow 0$ uniformly in $s\in[h,\bar T-h]$. Since $(t-T,z)\notin\mathcal{E}^{(\varepsilon')}$ and $\bar T<T$, we have from Lemma 5.2:

$$\lim_{\varepsilon''\downarrow 0} u^{\varepsilon''}(t-\bar T,z)=1$$

at least for a subsequence $\{\varepsilon''\}$ of the sequence $\{\varepsilon'\}$. Moreover $\lim_{\varepsilon''\downarrow 0} u^{\varepsilon''}(s,y)=1$ uniformly in a neighborhood of the point $(t-\bar T,z)$.

Let α_1 be the Euclidian distance between the set $\{(s,y),\ s\in[h,t-h],\ y=\bar\varphi_s\}$ and the compliment of the set $(\mathcal{E}^{(\varepsilon')})$; α_2 is the size of the neighborhood of the point $(t-\bar T,z)$ where $\lim_{\varepsilon''\downarrow 0} u^{\varepsilon''}(s,y)=1$. Denote $t_y=\min\{s : |\bar\varphi_s-\bar\varphi_{\bar T}|<y\}$, $y>0$, and let $\alpha_3>0$ be so small that

$$\max_x c(x)(\bar T-t_{\alpha_3})+\int_{t_{\alpha_3}}^{\bar T}\sum_{i,j=1}^r a_{ij}(\bar\varphi_s)\dot{\bar\varphi}^i_s\,\dot{\bar\varphi}^j_s\,ds\le\frac{\delta}{8};$$

$\alpha_4>0$ is such that $|c(x)-c(y)|\le\frac{\delta}{8t}$ for $|x-y|<\alpha_4$.

Put

$$\Gamma=\{(s,y) : |s-\bar T|+|y-z|\le\tfrac12(\alpha_2\wedge\alpha_3)\},\quad \alpha=\alpha_1\wedge\alpha_2\wedge\alpha_3\wedge\alpha_4,$$

$$\zeta(t,\varphi)=\min\{(s : (t-s,\varphi_s)\in\Gamma\},\quad \zeta=\zeta(t,X^\varepsilon_\cdot).$$

Using (5.1) we have

$$(5.11)\quad u^\varepsilon(t,x)=E_x u(t-\zeta\wedge t,X^\varepsilon_{\zeta\wedge t})\exp\left\{\frac1\varepsilon\int_0^{\zeta\wedge t}c(X^\varepsilon_s,u^\varepsilon(t-s,X^\varepsilon_s))ds\right\}$$

$$\ge E_x u^\varepsilon(t-\zeta\wedge t,X^\varepsilon_{\zeta\wedge t})\chi_\alpha\exp\left\{\frac1\varepsilon\int_0^{\zeta\wedge t}c(X^\varepsilon_s,u^\varepsilon(t-s,X^\varepsilon_s))ds\right\}=I_1,$$

where χ_α is the indicator of the set $\left\{\sup_{0\le s\le\bar T}|X^\varepsilon_s-\bar\varphi_s|<\alpha\right\}=B_\alpha$.

For small enough $\varepsilon>0$, $u^\varepsilon(s,y)>\frac12$ in the α-neighborhood of the point $(\bar{T}, z)$, and

$$\int_0^{t\wedge\zeta} c(X_s^\varepsilon, u^\varepsilon(t-s, X_s^\varepsilon))ds \ge \int_0^{\bar{T}} c(X_s^\varepsilon)ds - \frac{\delta}{8} \quad \text{for} \quad X_\cdot^\varepsilon \in B_\alpha.$$

Therefore, using the lower bound for $P_x\{B_\alpha\}$ given by the large deviation principle, we get that

$$(5.12) \qquad I_1 \ge \frac{1}{2} E_x \chi_\alpha \exp\left\{\frac{1}{\varepsilon}\int_0^{\bar{T}} c(\bar{\varphi}_s)ds\right\}\cdot\exp(-\frac{\delta}{4\varepsilon})$$

$$\ge \exp\left\{\frac{1}{\varepsilon}\left[\int_0^{\bar{T}}\left[c(\bar{\varphi}_s) - \frac{1}{2}\sum_{i,j=1}^{r} a_{ij}(\bar{\varphi}_s)\dot{\bar{\varphi}}_s^i \dot{\bar{\varphi}}_s^j\right]ds - \frac{\delta}{2}\right]\right\},$$

for $\varepsilon > 0$ small enough. From (5.10), (5.11), and (5.12) we have that $\lim_{\varepsilon\downarrow 0} \varepsilon \ln u^\varepsilon(t,x) \ge -\delta$. Since δ is an arbitrary positive number, taking into account that $\overline{\lim_{\varepsilon\downarrow 0}}\, \varepsilon \ln u^\varepsilon(t,x) \le 0$, we conclude that $\lim_{\varepsilon\downarrow 0} \varepsilon \ln u^\varepsilon(t,x) = 0$.

Uniformity of the convergence $u^\varepsilon(t,x)$ to zero for points $(t,x) \in F \subset (M)$ follows from the fact that set F can be covered by finite number of the cones $D_{t_k,x_k}^{A/2}$, introduced in Lemma 5.3, with the vertices outside F. □

Theorem 5.1. Let $u^\varepsilon(t,x)$ be the solution of problem (3.1). Then $\lim_{\varepsilon\downarrow 0} u^\varepsilon(t,x) = 0$ uniformly for (t,x) belonging to any compact set $F_1 \subset \{(s,y) : V^*(s,y) < 0\}$. For any compact subset F_2 of the interior of the set $\{(s,y),\ s > 0,\ V^*(s,y) = 0\}$, $\lim_{\varepsilon\downarrow 0} u^\varepsilon(t,x) = 1$ uniformly in $(t,x) \in F_2$.

Proof. The first statement follows from Lemma 5.1, the second statement follows from Lemmas 5.3 and 5.4.

Remark. In general $V^*(t,x) \le (V(t,x)\wedge 0)$ and $\{(t,x) : V(t,x) < 0\} \subseteq \{(t,x) : V^*(t,x) < 0\}$. The inclusion may be strict. See the corresponding example at the end of the previous section. If the condition (N) is fulfilled, the inclusion becomes an equality. One can give a bound from below for the interior of the set $\{(t,x) : V^*(t,x) = 0\}$:

$$(5.13) \qquad (\{(t,x) : V^*(t,x) = 0\}) \supseteq \{(t,x) : \hat{\rho}(x, G_0) < t\},$$

where $\hat{\rho}$ is the Riemannian metric corresponding to the form $ds^2 =$

$\frac{1}{2c(x)} \sum_{i,j=1}^{r} a_{ij}(x)dx^i dx^j$. The proof of the inclusion (5.13) follows from Proposition 4.1. In the case $c(x) = c = \text{const.}$, (5.13) becomes an equality. Under condition (N) the interior of the set $\{(t,x) : t > 0,\ V^*(t,x) = 0\}$ is equal to the set $\{(t,x),\ V(t,x) > 0\}$.

An analytical proof of Theorem 5.1 was given by L.C. Evans and P.E. Souganidis in [4], where they generalized results of [6] and suggested a game-theoretical approach to the problem which, together with the small-viscosity solution method gives the description of the wave front. It turns out that one can give a more simple expression for the function $V^*(t,x)$ which defines the wave front in the general case

$$V^*(t,x) = \sup\left\{ \min_{0\le a\le t} \int_0^a \left[c(\varphi_s) - \frac{1}{2} \sum_{i,j=1}^{r} a_{ij}(\varphi_s)\dot{\varphi}_s^i \dot{\varphi}_s^j \right] ds : \varphi \in C_{0t},\ \varphi_0 = x,\ \varphi_s \in G_0 \right\}$$

The proof of this formula can be found in [11].

§6. Models of Evolution.

We consider in this section two problems which are interpreted as models of evolution (see [8]). Of course such equations describe some other problems in biophysics, chemical kinetics and other areas. Let x be the genotype of an individual. We assume that $x \in R^r$, and we interpret R^r as a space of all possible genotypes. The qualitative result in which we are interested actually does not depend heavily on the structure of the genotype space.

Let $u(t,x)$ be the density of the number of individuals with genotype $x \in R^r$ at time t. The evolution of $u(t,x)$ in time is a result of the interaction of two processes: multiplication and mutation. The multiplication is characterized by the fitness coefficient $c(x,u)$. This coefficient defines how $u(t,x)$ changes in the absence of mutations

$$\frac{1}{u(t,x)} \frac{du}{dt} = c(x,u),$$

for most of the genotypes $c(x,u) < 0$ for any u. This means that individuals with such genotypes cannot survive in a given environment. But there exist separate "islands," $K_1, K_2, \ldots, K_n$, in the genotype space R^r such that $c(x,u) > 0$ for $x \in \cup K_i$ and $u(t,x)$ not very large (see Figure 6). The genotypes can also change because of mutations. Let us assume that the mutation process is described as follows

$$X_t^\varepsilon = X + \varepsilon W_t,$$

where W_t is the Wiener process in R^r and ε is small parameter characterizing the intensity of mutations.

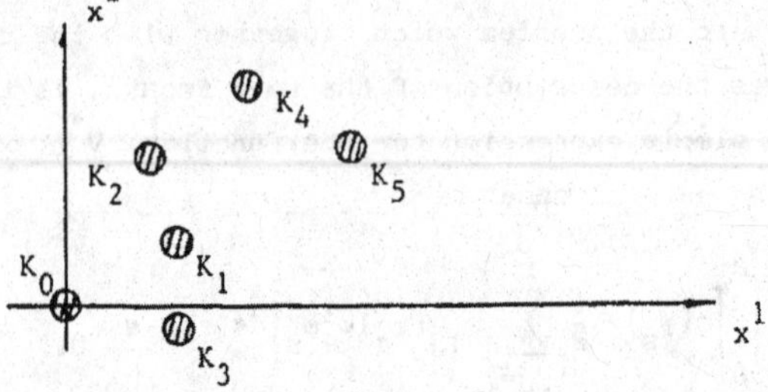

Figure 6

The interplay between multiplication and mutations leads to the following law of evolution of the function $u(t,x)$ in time

$$\frac{\partial u}{\partial t}(x) = \frac{\varepsilon^2}{2}\Delta u + c(x,u)u, \quad t > 0, \quad x \in R^r. \tag{6.1}$$

Suppose that at the initial time, $u(t,x)$ was positive only for $x \in K_0$; $u(0,x) = 0$ for $x \notin K_0$;

$$u(0,x) = g(x), \quad \text{supp}\, g(x) = K_0. \tag{6.2}$$

We shall consider the set of all individuals with genotype x belonging to an island K_i as one species. Assumption (6.2) means that at the initial time only one species K_0 existed. Our goal is to understand which (if any) species will appear next? When will it happen? Who will initiate the next species?

The function $c(x,u)$ has the shape (1) (see Figure 7) for $u \in \cup K_i$ and the shape (2) for $x \notin \cup K_i$. We assume that $c(x) = c(x,0) = \max c(x,u)$ for $x \in \cup K_i$, and $c(x,u) < 0$ for $u > a(x)$. These assumptions have some biological sense. Sometimes it is natural to assume that the fitness coefficient has shape (3). But we shall not consider this case here (see the next section and [7]).

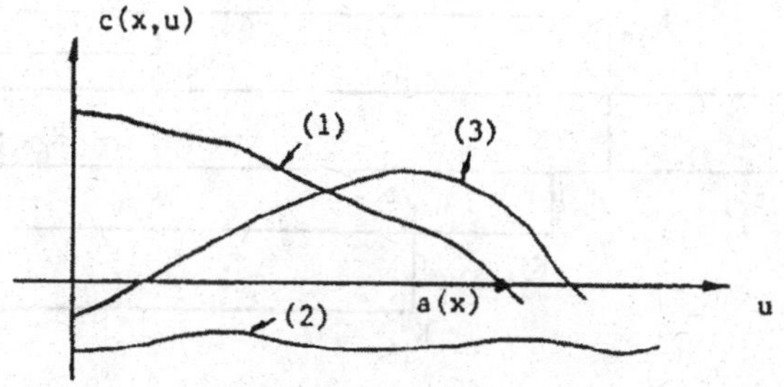

Figure 7.

It is easy to see that substantial changes of the domain in R^r where the solution $u(t,x)$ of Problems (6.1) - (6.2) is close to zero as $\varepsilon \ll 1$, happen in the time scale of order ε^{-1}. Therefore let us consider the function $u^\varepsilon(t/\varepsilon,x) = u^\varepsilon(t,x)$. The function $u^\varepsilon(t,x)$ is the solution of the problem

$$(6.3)\qquad \frac{\partial u^\varepsilon(t,x)}{\partial t} = \frac{\varepsilon}{2}\,\Delta u^\varepsilon + \frac{1}{\varepsilon}c(x,u^\varepsilon)u^\varepsilon, \quad u^\varepsilon(0,x) = g(x), \quad x \in R^r, \quad t > 0.$$

The only difference with the problem considered before is that $c(x,u)$ now can be negative for all $u \geq 0$. It is easy to see that limit behavior of $u^\varepsilon(t,x)$ as $\varepsilon \downarrow 0$ can be described in the same way as in Theorem 5.1.

Let us introduce the function $V^*(t,x)$

$$V^*(t,x) = \sup\left\{ \min_{0\le a\le t} \int_0^a [c(\varphi_s) - \tfrac{1}{2}|\dot\varphi_s|^2]ds \;:\; \varphi \in C_{0t},\; \varphi_0 = x,\; \varphi_t \in K_0 \right\}.$$

Denote $t_i = \inf\{t : \max_{x\in[K_i]} V^*(t,x) = 0\}$. It is easy to check that all t_i, $i = 1,2,\dots$, are finite. Let $t_1^* = \min\{t_i : i = 1,2,\dots\}$, $t_{k+1}^* = \min\{t_i : t_i > t_k^*\}$, and i^* be such that $t_i^* = t_{i^*}$.

One can see that $V^*(t,x)$ will be negative for all t if $x \notin \cup[K_i]$. This means that $\lim_{\varepsilon\downarrow 0} u^\varepsilon(t,x) = 0$ for such x and any $t > 0$. The species K_{1^*} lights up at time t_1^*, and for any t_i^* the new species K_{i^*} will light up at time t_i^*. And, what is more, we can say (in the case of general position) who switched on the species K_{i^*} with $i^* > 0$: if $\max_{x\in[K_i]} V^*(t_i^*,x)$ is attained for an extremal φ_s^*, and on its way from a point $x^* \in K_{i^*}$ the first island which φ^* touches is K_j, then the species K_{i^*} is generated by K_j.

Thus we can reconstruct the evolution tree

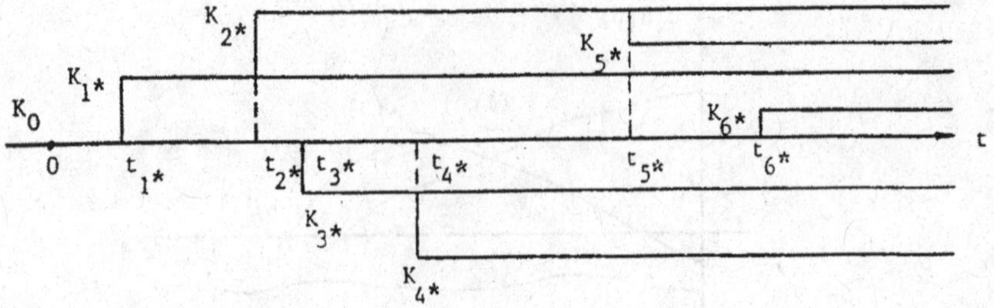

Figure 8.

Figure 8 shows that at time t_1^* the new species K_{1^*} was switched on by K_0. The species K_{2^*} was switched on at time t_2^* by K_{1^*}. The species K_{3^*} was switched on by K_0 and K_{5^*} was switched on by K_{2^*} at time t_5^*.

We see that, though the evolution is the result of random mutation, the sequence of new species and the times when they appear are not random at least in the main term.

This model does not take into account competition between species. We can include the competition by changing the fitness coefficient after appearance of each new species. We shall include competition in a different way in the next model [7]. As we shall see, the new assumptions will lead not to solutions with running wave fronts, but to running impulses. All the density will be concentrated near the "most advanced" genotype.

Assume that the fitness coefficient c depends on the genotype x and on the general number of specimens: $c = c(x, \int_{R^r} u(t,y)dy)$. Then the density $u^\varepsilon(t,x)$ (in rescaled time) satisfies the equation

$$\frac{\partial u^\varepsilon(t,x)}{\partial t} = \frac{\varepsilon}{2}\Delta u^\varepsilon(t,x) + \frac{1}{\varepsilon}\, c(x, v^\varepsilon(t))u^\varepsilon(t,x),$$

(6.4)

$$u^\varepsilon(0,x) = g(x), \quad v^\varepsilon(t) = \int_{R^r} u^\varepsilon(t,y)dy.$$

If X_t^ε is the Markov process corresponding to $\frac{\varepsilon}{2}\Delta$ we can write down the following equation for $u^\varepsilon(t,x)$:

$$u^\varepsilon(t,x) = E_x g(X_t^\varepsilon)\exp\left\{\frac{1}{\varepsilon}\int_0^t c(X_s^\varepsilon, v_{t-s}^\varepsilon)ds\right\} \qquad (6.5)$$

$$v_t^\varepsilon = \int_{R^r} u^\varepsilon(t,y)dy.$$

We assume that $c(x,v)$ is Lipschitz continuous. Then from equation (6.5) one can readily deduce the existence and uniqueness of the solution of problem (6.5).

We assume that the function $c(x,v)$ has the shape drawn on Figure 9: $c(x,v) > 0$ for $0 < v < a(x)$ and any $x \in R^r$, and $c(x,u) < 0$ for $x \in R^r$, $v > a(x)$. We assume that $\sup_{x\in R^r} a(x) < \infty$. It turns out that under these assumptions $u^\varepsilon(t,x)$ tends to $a(\hat{\varphi}_t)\delta(x-\hat{\varphi}_t)$ as $\varepsilon \downarrow 0$. Here $\delta(x)$ is Dirac δ-function, and $\hat{\varphi}_t$ is a piecewise continuous function such that $a(\hat{\varphi}_t)$ does

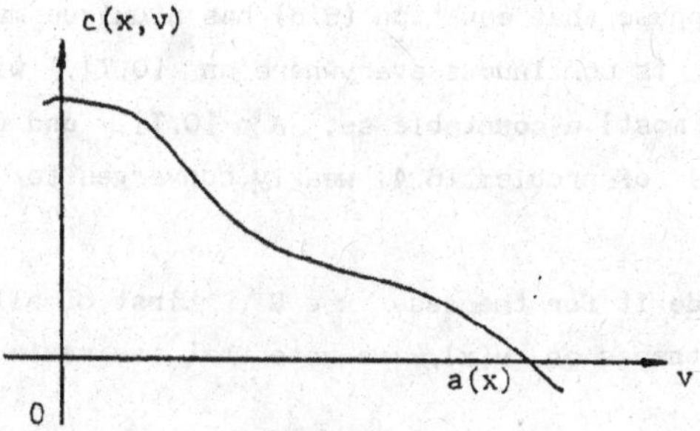

Figure 9

not decrease as t grows. We will evaluate $\hat{\varphi}$ later on. Therefore the impulse has a volume that is non-decreasing with time. It moves with finite velocity for all t with the possible exception of a countable number of times. At these times the impulse has jumps. For large t the impulse is concentrated near the point at which $\sup_{x\in R^r} a(x)$ is attained, provided such a point exists. Unless this supremum is achieved, the impulse goes to infinity.

For a strictly non-decraesing right continuous function $h(t) : [0,T] \to R^1$, $T > 0$, we will define the function $V_h(t,x)$, $t \in [0,T]$, $x \in R^r$

$$V_h(t,x) = \sup\left\{\int_0^t \left[c[\psi_s, h(s)] - \frac{|\dot{\psi}_s|^2}{2}\right]ds \; : \; \psi \in C_{0T}(R^r),\right.$$

$$\left.\psi_0 \in G_0, \quad \psi_t, = x, \quad \text{and} \quad \psi_s \text{ is absolutely continuous}\right\}.$$

The function $V_h(t,x)$ is continuous for $t \in [0,T]$ and $x \in R^r$. As $t\downarrow 0$, $V_h(t,x)$ tends to 0 provided $x \in G_0 \cup \partial G_0$, and $V_h(t,x) \to -\infty$ for

$x \notin G_0 \cup \partial G_0$.

By Ξ_T, $T > 0$, we denote the set of measurable functions $\varphi : [0,T] \to R^r$, such that $a(\varphi_t)$ does not decrease, $a(\varphi)_t$ is right continuous for $t > 0$, and $\lim_{t\downarrow 0} a(\varphi_t)$ exists.

A function $\hat{\varphi} \in \Xi_T$ is called a maximal solution of the equation

$$V_{a(\varphi)}(t,\varphi_t) = 0, \quad t \in [0,T], \tag{6.6}$$

whenever equation (6.6) is satisfied for $\varphi_t = \hat{\varphi}_t$, and

$$V_{a(\hat{\varphi})}(t,x) \le 0 \quad \text{for} \quad t \in [0,T], \quad x \in R^r.$$

<u>Theorem 6.1</u>. Suppose that equation (6.6) has a unique maximal solution $\hat{\varphi} \in \Xi_T$. Then $a(\hat{\varphi}_t)$ is continuous everywhere on $[0,T]$, with the possible exception of (at most) a countable set $\Lambda \subset [0,T]$, and for $t \in [0,T]\setminus\Lambda$, the solution $u^\varepsilon(t,x)$ of problem (6.4) weakly converges to $a(\hat{\varphi}_t)\delta(x-\hat{\varphi}_t)$ as $\varepsilon\downarrow 0$.

<u>Proof</u>. We provide it for the case $x \in R^1$. First of all, by taking into account the hypotheses on $g(x)$, we note that the maximum principle implies

$$0 < u^\varepsilon(t,x), \quad 0 < v_t^\varepsilon = \int_{R^r} u^\varepsilon(t,y)dy \le a_0 = \sup_{x\in R^r} a(x) < \infty. \tag{6.7}$$

For any $0 < t_1 < t_2$, one can find $\delta_\varepsilon = \delta_\varepsilon(t_1,t_2)$ such that

$$\lim_{\varepsilon\downarrow 0} \delta_\varepsilon = 0, \quad v_{t_1}^\varepsilon < v_{t_2}^\varepsilon + \delta_\varepsilon. \tag{6.8}$$

Indeed, if such is not the case, then there is a sequence of $\varepsilon\downarrow 0$ for which $\lim_{\varepsilon\downarrow 0} v_{t_1}^\varepsilon = v_{t_1}$, $\lim v_{t_2}^\varepsilon = v_{t_2}$, and $v_{t_1} > v_{t_2}$. Suppose that $b \in (v_{t_1}, v_{t_2})$. By the definition of v_{t_1},

$$\lim_{\varepsilon\downarrow 0}\left[\int_{\{a(x)<b\}} u^\varepsilon(t_1,y)dy + \int_{\{a(x)\ge b\}} u^\varepsilon(t_1,y)dy\right] = v_{t_1}.$$

If

$$\overline{\lim_{\varepsilon\downarrow 0}} \int_{\{a(x)\ge b\}} u^\varepsilon(t,y)dy = \delta_t > 0$$

for some $t \in (0,t_2)$, then

$$v_{t_2} = \overline{\lim_{\varepsilon\downarrow 0}} \int_{\{a(x)\ge b\}} u^{\varepsilon}(t_2,y)dy \ge b,$$

contradicting the choice of $b \in (v_{t_2}, v_{t_1})$. On the other hand, if $\delta_t = 0$ for $t \in (0,t_2)$, and

$$\int_{\{a(x)<b\}} g(x)dx < b,$$

then, for any $\lambda > 0$ and $t \in (0,t_2)$,

$$w_b^{\varepsilon}(t) = \int_{\{a(x)<b\}} u^{\varepsilon}(t,y)dy < b+\lambda,$$

provided ε is small enough. This bound can easily be deduced from equation (6.5) using the fact that $a(x)$ is Lipschitz continuous. Because $w_b^{\varepsilon}(t_1) < b+\lambda$ and $\delta_{t_1} = 0$, we obtain $v_{t_1} = \lim_{\varepsilon\downarrow 0} w_b^{\varepsilon}(t_1) + \delta_{t_1} < b+\lambda$, which contradicts the choice of $b \in (v_{t_2}, v_{t_1})$ for appropriately small λ. Therefore, equation (6.8) holds true. □

A sequence of functions $v^{\varepsilon}(t)$, $t \in [0,T]$, is said to converge c-weakly to $v(t)$ if $\lim_{\varepsilon\downarrow 0} v^{\varepsilon}(t) = v(t)$ at every continuity point of the limit function $v(t)$. Of course, for a limit in this sense to be unique, it is necessary that $v(t)$ have quite a lot of continuity points.

From equations (6.7) and (6.8) it follows that there is a subsequence $v_t^{\varepsilon'}$ of the family v_t^{ε} that c-weakly converges to a function v_t. (Henceforth, the prime in ε' will be dropped.) By equation (6.8) the function v_t does not decrease, and thus, at most, it has a countable number of discontinuity points. For the limit $\lim_{\varepsilon\downarrow 0} v^{\varepsilon}$ to be unique, we will consider v_t to be right continuous.

In the sequel, some bounds of $u^{\varepsilon}(t,x)$ in the uniform norm and some bounds of the continuity module of $u^{\varepsilon}(t,x)$ will be useful. Here, to a great degree, we use the fact that x is one-dimensional. For $r = 1$ from equations (6.4) and (6.7), we obtain

$$(6.9) \qquad 0 \le u^{\varepsilon}(t,x) = \frac{1}{\sqrt{2\pi\varepsilon t}} \int_{-\infty}^{\infty} g(y)\exp\left\{-\frac{(y-x)^2}{2\varepsilon t}\right\}dy$$

$$+\frac{1}{\varepsilon}\int_0^t\int_{-\infty}^{\infty}\frac{c(y,v_s^\varepsilon)u^\varepsilon(s,y)}{\sqrt{2\pi\varepsilon s}}\exp\left\{-\frac{(y-x)^2}{2\varepsilon t}\right\}dsdy$$

$$\le \max_{y\in R^1} g(y)+\frac{\sup_{y,v\in R^1}|c(y,v)|}{\sqrt{2\pi\varepsilon}^{3/2}}\int_0^t\frac{1}{\sqrt{s}}v_s^\varepsilon ds \le c_1+c_2\varepsilon^{-3/2},$$

where c_1 and c_2 are some constants independent of ε, with $t\in[0,T]$ and $x\in R^1$.

From equation (5.4), while relying on equation (6.9) we have

$$\|u_t^\varepsilon\|_{L^2_{[0,T]\times R^1}}+\|u_{xx}^\varepsilon\|_{L^2_{[0,T]\times R^1}}\le c_3\varepsilon^{-3/2}.$$

Because the norms of u_t' and u_{xx}' in $L^2_{[0,T]\times R^1}$ are bounded, we derive that $u^\varepsilon(t,x)$ is Hölder continuous of exponent (at least) $\frac{1}{4}$

(6.10) $$|u^\varepsilon(t+h,x+\delta)-u^\varepsilon(t,x)|\le\frac{c_4}{\varepsilon^{3/2}}(|h|^{1/4}+|\delta|^{1/4}).$$

We notice that the continuity module can be bounded more exactly; however, for us, the bound (6.10) is sufficient.

Therefore, suppose that as $\varepsilon\downarrow 0$, v_t^ε converges c-weakly to a bounded right-continuous and strictly non-decreasing function v_t, $t\in[0,T]$. We put $\mathcal{E}_t=\{x\in R^1 : a(x)=v_t\}$. We show that if t_0 is a continuity point of v_t and $x\in\mathcal{E}_{t_0}$, then

(6.11) $$\overline{\lim_{\varepsilon\downarrow 0}}\,\varepsilon\ln u^\varepsilon(t_0,x)<0.$$

Indeed, if $a(x)<v_{t_0}$, then one can find a small h, $\delta>0$, such that $a(y)<v_s$ for $s\in[t_0-h,t_0]$, $|x-y|<\delta$. Therefore, $c(y,v_s)\le -c_0<0$ for $(s,y)\in\Pi=\{(s,y) : t_0-h\le s\le t_0,\ |x-y|<\frac{\delta}{2}\}$. Consider the Markov time $\tau^\varepsilon=\tau^\varepsilon_{h,\delta}=h\wedge\min\{s : |X_s^\varepsilon-x|\ge\frac{\delta}{2}\}$. Using the strong Markov property of the process X_t^ε, we derive (from equation (6.5)) the bound

(6.12) $$0<u^\varepsilon(t_0,x)<P_x\{\tau^\varepsilon>h\}\exp\{-\frac{c_0h}{\varepsilon}\}+P_x\{\tau^\varepsilon\le h\}\max_{(t,y)\in\partial H}|u^\varepsilon(t,y)|,$$

where $\partial\Pi$ is the surface of the cylinder Π. By the properties of the Wiener process,

$$\lim_{\varepsilon\downarrow 0} \varepsilon \ln P_x\{\tau^\varepsilon < h\} = -\alpha < 0,$$

which together with equation (6.12) implies equation (6.11). With similar arguments and the bound from equation (6.10), one can prove equation (6.11) for the points x such that $a(x) > v_{t_0}$.

As is known, the action functional for the family of processes X_t^ε in the space C_{0T} has the form $\varepsilon^{-1}S_{0T}(\psi)$ as $\varepsilon\downarrow 0$ with

$$S_{0T}(\psi) = \begin{cases} \frac{1}{2}\int_0^T |\dot\psi_s|^2 ds, & \psi \in C_{0T}, \text{ where } \psi \text{ is absolutely continuous,} \\ +\infty, & \psi \in C_{0T}, \text{ where } \psi \text{ is not absolutely continuous.} \end{cases}$$

If v_t^ε c-weakly converges to v_t, then by relying on the results of §3.3 in [12] and by employing equation (6.5), we obtain the following expression for the logarithmic asymptotics of $u^\varepsilon(t,x)$:

$$\lim \varepsilon \ln u^\varepsilon(t,x) = V_v(t,x) = \sup\Big\{\int_0^t [c(\psi_s, v_s) - \tfrac{1}{2}|\dot\psi_s|^2]ds : \psi_0 \in G_0, \tag{6.13}$$
$$\psi_t = x, \text{ where } \psi \text{ is absolutely continuous}\Big\}.$$

From equation (6.11), it follows that $V_v(t,x) < 0$ outside the level set $\mathcal{E}_t = \{x \in R^1 : A(x) = v_t\}$. The function $V_v(t,x)$ cannot be negative everywhere in R^1; otherwise, it would contradict the condition

$$\lim_{\varepsilon\downarrow 0} \int_{R^1} u^\varepsilon(t,x)dx = v_t > 0.$$

Therefore, one can find a point $\hat\varphi_t$ in $\mathcal{E}_t$ such that $V_v(t,\hat\varphi_t) \ge 0$. However, $V_v(t,\hat\varphi_t)$ cannot be positive because this, together with equation (6.10) would lead to $\lim_{\varepsilon\downarrow 0} v_t^\varepsilon = \infty$. Hence, there is a point $\hat\varphi_t$ such that $V_v(t,\hat\varphi_t) = 0$ and $a(\hat\varphi_t) = v_t$. From equation (6.13) we derive the equation for $\hat\varphi_t$:

$$0 = V_{a(\hat\varphi)}(t,\hat\varphi_t) = \sup\Big\{\int_0^t c\Big[(\psi_s, a(\hat\varphi_s) - \frac{|\dot\psi_s|^2}{2}\Big]ds : \psi \in C_{0t}, \ \psi_0 \in G_0, \tag{6.14}$$
$$\psi_t = \hat\varphi_t, \text{ where } \psi \text{ is absolutely continuous}\Big\}.$$

This equality also holds at the discontinuity points of $a(\hat{\varphi}_t)$, provided that at these points $\hat{\varphi}_t$ is defined by

$$\hat{\varphi}_t = \overline{\lim_{\varepsilon\downarrow 0}}\ \hat{\varphi}_s.$$

The above reasoning yields $V_{a(\hat{\varphi})}(t,x) \le 0$, and thus $\hat{\varphi}$ is the maximal solution of the equation $V_{a(\hat{\varphi})}(t,\hat{\varphi}_t) = 0$.

By the condition of the theorem, such a solution is unique; this implies the existence of a unique limit point $v_t = a(\hat{\varphi}_t)$ for the family v_t^ε in the sense of c-weak convergence. From this we conclude that v_t^ε converges to v_t for all ways in which ε tends to 0. The uniqueness of the maximal solution also implies that there is only one point $\hat{\varphi}_t$ in $\mathcal{E}_t$ at which $V_{a(\hat{\varphi})}(t,\hat{\varphi}_t) = 0$. For $x \ne \hat{\varphi}_t$, we have $V_{a(\hat{\varphi})}(t,x) < 0$. From this

$$\lim_{\varepsilon\downarrow 0} u^\varepsilon(t,x) = 0 \quad \text{for} \quad x \ne \hat{\varphi}_t,$$

$$\lim_{\varepsilon\downarrow 0} \int_{\{|y-\hat{\varphi}_t|<\delta\}} u^\varepsilon(t,y)dy = v_t,$$

$$\lim_{\varepsilon\downarrow 0} \int_{\{|y-\hat{\varphi}_t|\ge\delta\}} u^\varepsilon(t,y)dy = 0$$

for any $\delta > 0$. This completes the proof of the theorem. □

Now, suppose that $c(x,v) = a(x) - v$ and let $a(x)$ satisfy the above-listed hypotheses. Then equation (6.6) takes the form

$$A(t,\hat{\varphi}_t) = \int_0^t a(\hat{\varphi}_s)ds, \quad t \in (0,T], \tag{6.15}$$

where the function $A(t,x)$, $t > 0$, $x \in R^r$, is defined by

$$A(t,x) = \sup\left\{\int_0^t \left[a(\psi_s) - \frac{|\dot{\psi}_s|^2}{2}\right]ds \;:\; \psi \in C_{0t}(R^r),\ \ \psi_0 \in G_0,\ \ \psi_t = x\right\}. \tag{6.16}$$

For $A(t,x)$, one can write the Hamilton-Jacobi equation which, along with (6.15) and the maximality condition, may be employed for calculating $\hat{\varphi}_t$.

Next, we consider in detail the case where $x \in R^1$, $c(x,v) = a(x) - v$, and $a(x)$ is a piecewise linear function. First, we assume that $a(x) = 1$

for $x < 0$ and $a(x) = x+1$ for $x > 0$. For the boundedness condition to be fulfilled, we will suppose that for appropriately large x, the function $a(x)$ is cut: $a(x) = a(N)$ for $x \ge N$, $N \gg 1$. Such a cutting does not at all affect the impulse movement until it reaches the point N, provided that the support G_0 of the initial function lies on the left of the point N. Let $G_0 = (-h,0)$, $h > 0$. For $a(x) = x+1$, the supremum in equation (6.16) can easily be calculated and we have

$$A(t,x) = \frac{t^3}{24} + t - \frac{tx}{2} - \frac{x^2}{2t}.$$

Equation (6.15) takes the form

$$\frac{t^4}{12} + t^2\hat{\varphi}_t - \hat{\varphi}_t^2 = 2t\int_0^t \hat{\varphi}_s ds.$$

We put $\phi_t = \int_0^t \hat{\varphi}_s ds$. For ϕ_t we obtain the equation

$$\frac{t^4}{12} + t^2\phi_t' - \phi_t'^2 = 2t\phi_t, \quad \phi_0 = 0. \tag{6.17}$$

From equation (6.17) we derive

$$\phi_t' = \frac{t^2}{2} + \sqrt{\frac{t^4}{3} - 2t\phi_t}, \quad \phi_0 = 0. \tag{6.18}$$

The minus sign before the root gives a solution that does not satisfy the maximality condition. By straightforward substitution, one can make sure that $\phi_t = t^3/6$ is a solution of problem (6.18). From equation (6.18) we derive that $\phi_t' \ge t^2/2$ and thus $\phi_t \ge t^3/6$. On the other hand, ϕ_t can exceed $t^3/6$ for no $t > 0$ because in this case, the expression under the root sign in equation (6.18) would be negative. Therefore, $\phi_t = t^3/6$ is a unique solution of equation (6.18), the hypotheses of Theorem 6.1 hold, and the impulse movement is controlled by the formula $\hat{\varphi}_t = \phi_t' = t^2/2$. By Theorem 6.1, $u^\varepsilon(t,x)$ weakly converges as $\varepsilon \downarrow 0$ to $(1+t^2/2)\delta(x-t^2/2)$ for every $t > 0$.

Now let $a(x)$ be the piecewise linear function shown in Figure 10. We will suppose that $0 < \alpha < 1$, $a(-h) = 1$, $a(A) < a(B)$, and $G_0 = (-h,0)$. Generally speaking, the impulse formed under the initial condition can move both to the right and to the left of $[-h,0]$. If $\alpha < 1$ and $a(A) < a(B)$, then one can deduce that the solution $\hat{\varphi}_t$ of equation (6.6) that takes negative values is not maximal. It is not hard to see that for small $t > 0$ for

$a(x)$ shown in Figure 10, the impulse will move to the right according to the same law as in the case of $a(x) = x+1$ for $x > 0$, that is, $\hat{\varphi}_t = t^2/2$.

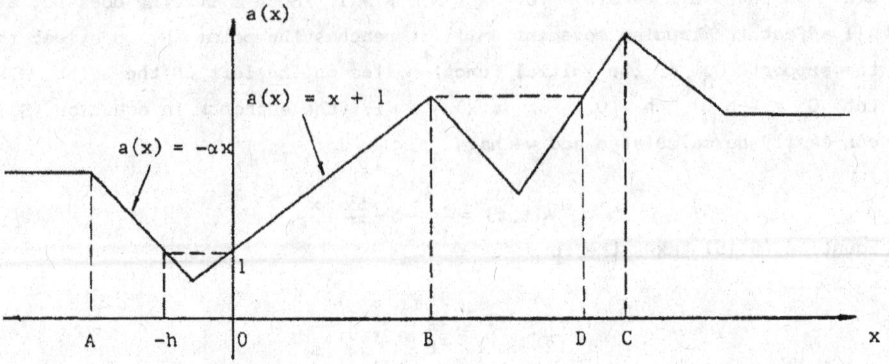

Figure 10.

If (for fixed B and C) $a(C) - a(B)$ is positive, but not too large, then at time $t = \sqrt{2B}$, the impulse will reach the point $x = B$ and will stay there for some time. Afterwards, the impulse will tunnel into the left neighborhood of the point C. If $a(C) - a(B)$ is large enough, then it will tunnel into a neighborhood of C without hitting B. To calculate the time of the jump and the impulse position after the jump, we denote $\lambda(t) = t^2/2$ for $t \le \sqrt{2B}$ and $\lambda(t) = B$ for $t > \sqrt{2B}$. Consider the function

$$A(t,x) = \sup\left\{\int_0^t \left[a(\psi_s) - \frac{|\dot{\psi}_s|^2}{w}\right]ds \; : \; \psi \in C_{-t}, \quad \psi_0 = 0, \quad \psi_t = x\right\},$$

where $a(x)$ is the piecewise linear function shown in Figure 10. If one writes down the equations of the lines whose segments make up the broken line $a(x)$, then $A(t,x)$ can be evaluated in an explicit way. We find t^* and x^* from the condition

$$t^* = \min\left\{t \; : \; A(t,x) = \int_0^t a[\lambda(s)]ds \quad \text{for some} \quad x > B\right\},$$

(6.19)

$$A(t^*,x^*) = \int_0^{t^*} a[\lambda(s)]ds.$$

For $a(C) > a(B)$, such a $t^* < \infty$ exists. In the case of general position, the point x^* is defined by equation (6.19) in a unique way. It is not difficult to check that $x^* \in (D,C]$. By Theorem 6.1, $\hat{\varphi}_t = t^2/2$ for $t \in (0,t^*)$. At time t^*, the tunneling occurs from the point $\lambda(t^*) \le B$ into x^*. If, as shown in Figure 10, $a(C) = \max a(x)$, then from the point x^*, the impulse arrives at the point C in a finite time and remains there forever. If $a(C) < \max a(x)$, then sooner or later there will be a tunneling into a neighborhood of the maximum of the function $a(x)$, that is, larger than $a(C)$.

We emphasize that the impulse may also have jumps in the case of strictly increasing $a(x)$ whenever $a(x)$ has regions of both slow and sufficiently quick growth.

§7. Some Remarks and Generalizations.

1. Let $X_t^{\varepsilon,x}$ be a Markov family with the phase space $\bar{D} \subset R^r$. Assume that $X_t^{\varepsilon,x} \to x$ as $\varepsilon \downarrow 0$ uniformly in $t \in [0,T]$ in probability, and let $\lambda(\varepsilon)S_{0T}(\varphi)$ be the corresponding action functional. Denote by $\mathcal{A}^\varepsilon$ the infinitesimal operator of the family $X_t^{\varepsilon,x}$. Consider the Cauchy problem

$$
(7.1)\qquad \begin{aligned} \frac{\partial u^\varepsilon(t,x)}{\partial t} &= \mathcal{A}^\varepsilon u^\varepsilon + \lambda(\varepsilon) f(x,u^\varepsilon), \quad t > 0, \quad x \in \bar{D}, \\ u^\varepsilon(0,x) &= g(x) \ge 0, \end{aligned}
$$

where $f(x,\cdot) \in \mathcal{F}_1$ for any $x \in \bar{D}$, $f(x,u) = c(x,u)u$. Then we can write the following equation for $u^\varepsilon(t,x)$:

$$
u^\varepsilon(t,x) = E_x g(X_t^\varepsilon)\exp\left\{\lambda(\varepsilon)\int_0^t X_s^\varepsilon, u^\varepsilon(t-s,X_s^\varepsilon))ds\right\}.
$$

Under some minor assumption one can repeat the arguments considered above to prove that

$$
\lim_{\varepsilon\downarrow 0} u^\varepsilon(t,x) = 0 \quad \text{if} \quad V(t,x) < 0,
$$

where

$$
(7.2)\qquad V(t,x) = \sup\left\{\int_0^t c(\varphi_s,0)ds - S_{0t}(\varphi) \; : \; \varphi_0 = x, \quad \varphi_t \in G_0 = \operatorname{supp} g\right\}.
$$

If the counterpart of condition (N) is fulfilled, one can prove that

$$\lim_{\varepsilon\downarrow 0} u^{\varepsilon}(t,x) = 1 \quad \text{when} \quad V(t,x) > 0.$$

Let us consider for example the family $X_t^{\varepsilon,x}$ corresponding to an elliptic second order operator εL in a domain $D \subset R^r$, with reflection on the boundary ∂D of the domain D in the direction of co-normal $n(x) = (n^1(x),\dots,n^r(x))$. We assume that the boundary ∂D and the coefficients of the operator are smooth enough.

One can construct the trajectories $X_t^{\varepsilon,x}$ using stochastic differential equations for the process with reflection (see §1.6 in [7])

$$dX_t^{\varepsilon,x} = \sqrt{\varepsilon}\sigma(X_t^{\varepsilon,x})dW_t + \chi_{\partial D}(X_t^{\varepsilon,x})n(X_t^{\varepsilon,x})d\xi_t^{\varepsilon,x}$$

$$X_0^{\varepsilon,x} = x, \quad \xi_0^{\varepsilon,x} = 0.$$

Here $\xi_t^{\varepsilon,x}$ is the local time on the boundary for $X_t^{\varepsilon,x}$, $\sigma(x)\sigma^*(x) = (a^{ij}(x))$ is the diffusion matrix in L, and $\chi_{\partial D}(x)$ is the indicator of the set $\partial D \subset R^r$. We consider for brevity the case of zero drift.

The action functional for the process $(X_t^{\varepsilon,x},\xi_t^{\varepsilon,x})$ as $\varepsilon\downarrow 0$ has the form $\varepsilon^{-1}S_{0,T}^{\varepsilon,\xi}(\varphi,\mu)$, where ([7] Ch. 6)

$$S_{0,T}^{\varepsilon,\xi}(\varphi,\mu) = \begin{cases} \frac{1}{2}\int_0^T \sum_{i,j=1}^{r} a_{ij}(\varphi_s)(\dot\varphi_s^i - \chi_{\partial D}(\varphi_s)n^i(\varphi_s)\dot\mu_s)(\dot\varphi_s^j - \chi_{\partial D}(\varphi_s)n^j(\varphi_s)\dot\mu_s)ds, \\ \text{if } \varphi \text{ is absolutely continuous with values in } D\cup\partial D,\ \varphi_0 = 0, \\ \text{and } \mu_s \text{ is non-negative, non-decreasing and continuous;} \\ +\infty \text{ for the rest of continuous functions.} \end{cases} \tag{7.3}$$

Problem (7.1) in this example has the form

$$\frac{\partial u^{\varepsilon}(t,x)}{\partial t} = \frac{\varepsilon}{2}\sum_{i,j=1}^{r} a^{ij}(x)\frac{\partial^2 u^{\varepsilon}}{\partial x^i \partial x^j} + \frac{1}{\varepsilon}c(x,u^{\varepsilon})u^{\varepsilon}, \quad t>0, \quad x\in D$$

$$\frac{\partial u^{\varepsilon}(t,x)}{\partial n(x)}\Big|_{x\in\partial D,\, t>0} = 0, \quad u^{\varepsilon}(0,x) = g(x). \tag{7.4}$$

To describe the wave front propagation for (7.4) we need the action functional $\varepsilon^{-1}S_0^x(\varphi)$ for the first component $X_t^{\varepsilon,x}$ only. It is easy to see that

$$S_{0T}^{x}(\varphi) = \sup_{\mu} S_{0T}^{x,\mu}(\varphi)(\varphi,\mu).$$

Then equality (7.2) defines $V(t,x)$ and the position of the wave front at time t if condition (N) is fulfilled. One can formulate for problem (7.4) the counterpart of Theorem 5.1, too.

If we replace the reflection conditions in (7.4) by the Dirichlet condition $u(t,x)|_{\partial D} = 0$, we can use formula (7.2) to describe the motion of the wave front. In this case $S_{0T}(\varphi)$ is defined as in §3, but the supremum in (7.3) should be considered only over continuous functions φ connecting x and G_0 such that they do not leave the domain D.

2. The motion of the wave front is the result of the interaction of two factors — random motion of particles and multiplication (or killing) of particles. The multiplication (killing) is described by the nonlinear term in the equation. The motion of the particles is not necessarily a Markov process. We consider as an example the case when the motion of particles is described by a process which is a component of a Markov process. Let $Y_t^{\varepsilon,y}$ be the solution of the equation

$$\dot{Y}_t^{\varepsilon,y} = b(X_{t/\varepsilon}, Y_t^{\varepsilon,y}), \quad Y_0^{\varepsilon,y} = y, \quad 0 < \varepsilon \ll 1, \tag{7.5}$$

where X_t is the Wiener process in $[-1,1]$ with reflection at the ends of the interval. Then we have the following initial-boundary problem for the concentration $u^{\varepsilon}(t,x,y)$:

$$\frac{\partial u^{\varepsilon}(t,x,y)}{\partial t} = \frac{1}{2\varepsilon}\frac{\partial^2 u^{\varepsilon}}{\partial x^2} + b(x,y)\frac{\partial u^{\varepsilon}}{\partial y} + \frac{1}{\varepsilon}f(x,y,u^{\varepsilon}),$$

$$t > 0, \quad |x| < 1, \quad y \in R^1, \quad \frac{\partial u^{\varepsilon}(t,x,y)}{\partial x}\Big|_{|x|=1} = 0, \tag{7.6}$$

$$u^{\varepsilon}(0,x,y) = g(y) \geq 0.$$

We assume for brevity that the initial function depends only on y, $f(x,y,\cdot) \in \mathcal{F}_1$ for $|x| \leq 1$, $y \in R^1$, and $f(x,y,u) = c(x,y,u)u$; $c(x,y) = c(x,y,0)$. The averaging principle implies (see [12], Ch. 7), that for any $T > 0$, $\delta > 0$

$$\lim_{\varepsilon \downarrow 0} P_y\left\{\sup_{0 \leq t \leq T} |y_t^{\varepsilon} - \bar{y}_t| > \delta\right\} = 0,$$

where $\bar{y}_t$ is the trajectory of the averaged equation

$$\dot{\bar{y}}_t = \bar{b}(\bar{y}_t), \quad \bar{y}_0 = y, \quad \bar{b}(y) = \frac{1}{2}\int_{-1}^{1} b(x,y)dx.$$

Let $z_t^\varepsilon = \int_0^t c(X^\varepsilon_{s/\varepsilon}, y^\varepsilon_s)ds$. The action functional for the family of processes $(Y_t^{\varepsilon,y}, Z_t^\varepsilon)$ is as follows (see §2)

$$\varepsilon^{-1}S_{0t}(\varphi^1,\varphi^2) = \varepsilon^{-1}\int_0^t L(\varphi_s^1, \dot{\varphi}_s^1, \dot{\varphi}_s^2)ds,$$

where $L(y,\alpha^1,\alpha^2)$ is the Legendre transformation of $\lambda(y,\beta_1,\beta_2)$ in (β_1,β_2) and $\lambda = \lambda(y,\beta_1,\beta_2)$ is the eigenvalue of the problem

$$\frac{1}{2}\frac{\partial^2 v(x)}{\partial x^2} + (\beta_1 b(x,y) + \beta_2 c(x,y))v(x) = \lambda v(x), \quad |x| < 1$$

$$v'(x)\big|_{x=\pm 1} = 0, \quad y \in R^1; \quad \beta_1,\beta_2 \in R^1,$$

which corresponds to the positive eigenfunction.

Assume for brevity that $\bar{b}(y) \equiv 0$. Then, if a condition similar to condition (N) is fulfilled, the position of the wave front at time t for equation (7.6) is defined by the equation $V(t,y) = 0$, where

$$V(t,y) = \sup\left\{\varphi_t^2 - \int_0^t L(\varphi_s^1, \dot{\varphi}_s^1, \dot{\varphi}_s^2)ds \; : \; \varphi_0^1 = y, \quad \varphi_t^1 \in G_0, \quad \varphi_0^2 = 0\right\}.$$

The assumption $\bar{b}(y) \equiv 0$ is analogous to assuming that there is not drift.

Consider another example. Let $v^\varepsilon(t,x,y)$ be the solution of the problem

$$\frac{\partial v^\varepsilon}{\partial t} = \frac{1}{2\varepsilon}\frac{\partial^2 v^\varepsilon}{\partial x^2} + \frac{\varepsilon}{2}a(x)\frac{\partial^2 v^\varepsilon}{\partial y^2} + \frac{1}{\varepsilon}c(v^\varepsilon)v^\varepsilon,$$

(7.7) $$t > 0, \quad |x| < a, \quad y \in R^1, \quad v^\varepsilon(0,x,y) = g(x,y) \ge 0,$$

$$\frac{\partial v^\varepsilon(t,x,y)}{\partial x}\Big|_{x=\pm a} = 0.$$

We assume that $c(v)v \in \mathcal{F}_1$ and denote $G_0 = \{y : \max_{|x|\le a} g(x,y) > 0\}$, and $c = c(0)$. If we denote by X_t the Markov process in $[-a,a]$ corresponding to the operator $\frac{1}{2}\frac{d^2}{dx^2}$ with reflection in the ends of the interval, then the y-component of the process corresponding to (7.7) can be written in the form

$$y_t^\varepsilon = y_0^\varepsilon + \sqrt{\varepsilon}W\left[\int_0^t a(X_{s/\varepsilon})ds\right], \tag{7.8}$$

where W_t is the Wiener process in R^1. One can see from (7.8) that the process y_t^ε, $t \in [0,T]$, is a continuous transformation of the processes $\sqrt{\varepsilon}W_t$ and $\int_0^t a(X_{s/\varepsilon})ds$, $t \in [0,T_1]$ for some $T_1 < \infty$. Using this remark, it is not difficult to calculate the action functional for the family y_t^ε in C_{0T} as $\varepsilon \downarrow 0$. This action functional is equal to $\varepsilon^{-1}S^y(\varphi)$, where

$$S_{0T}^y(\varphi) = \inf\left\{\int_0^T L(\dot{\psi}_s)ds + \frac{1}{2}\int_0^T \frac{\dot{\varphi}_s^2}{\dot{\psi}_s}ds \,:\, \psi_s \in C_{0T},\right.$$

$$\left.\varphi_s \text{ is absolutely continuous, } \dot{\psi}_s \ge 0\right\},$$

if $\varphi \in C_{0T}$ is absolutely continuous, and $S_{0T}^y(\varphi) = +\infty$ for the rest of C_{0T}. Here $L(\alpha)$ is the Legendre transformation of the function $\lambda(\beta)$, which is the first eigenvalue of the problem

$$\frac{1}{2}\varphi''(x) + \beta a(x)\varphi(x) = \lambda(\beta)\varphi(x), \quad |x| < a, \quad \varphi'(\pm a) = 0.$$

Let us introduce the function $V(t,y)$ as follows:

$$V(t,y) = ct - \inf\{S_{0t}^y(\varphi) : \varphi_0 = y, \ \varphi_t \in G_0\}.$$

Then in the same way as Theorem 3.1 one can prove that the equation $V(t,y) = 0$ defines the position $y^*(t)$ of the wave front at time t.

3. Consider the Cauchy problem for an equation without small parameter

$$\frac{\partial u(t,x)}{\partial t} = \frac{1}{2}\sum_{i,j=1}^r a^{ij}(x)\frac{\partial^2 u}{\partial x^i \partial x^j} + f(x,u), \quad x \in R^r, \quad t > 0,$$
$$u(0,x) = g(x) \ge 0. \tag{7.9}$$

Here $f(x,\cdot) \in \mathcal{F}_1$ for any $x \in R^r$, $G_0 = \text{supp } g$. If the coefficients a^{ij} and nonlinear term f are independent of x, one can introduce the asymptotic (for $t \to \infty$) wave front velocity $v(e)$, where e is a unit vector in R^r; $v(e)$ is the velocity in the direction e. If the coefficients $a^{ij}(x)$ and $f(x,u)$ depend on x in an arbitrary way, one cannot hope that any constant velocity of the front propagation will be established for $t \to \infty$. We can, however, expect a velocity to be established only in the case when $a^{ij}(x)$ and $f(x,u)$ are in one or another sense are homogeneous in x. For example, we can expect an asymptotic velocity $v(e)$ to appear if $a^{ij}(x)$ and $f(x,u)$ are periodic in $x \in R^r$, or if these functions are random fields which are homogeneous in space. Both these cases were considered in [7] Ch. 7. In the case of random coefficients and a nonlinear term, satisfactory results were proved only in the one-dimensional case. The many-dimensional case is still an open problem. These problems belong to a popular class of so-called homogenization problems. The results in this area roughly speaking, can be divided into three parts: results of law of large numbers type; results of type of central limit theorem; and results of limit theorems for large deviations. The wave front propagation problems are connected with the limit theorems for large deviations. There are many open problems in this area even for the periodic media, for example, the problem of wave front propagation in periodic media with "holes." On the boundary of the holes, different boundary conditions can be considered such as $u = 0$ or nonlinear conditions (compare with the next paragraph).

4. We have considered so far nonlinear terms in the equation. I want to consider now a linear equation with nonlinear boundary conditions. As we shall see in this case, the wave front can propagate along the boundary.

Given a domain $D \subset R^r$ with smooth boundary ∂D, consider the problem

$$\frac{\partial u^\varepsilon(t,x)}{\partial t} = \frac{\varepsilon}{2} \sum_{i,j=1}^{r} \frac{\partial}{\partial x^i}\left(a^{ij}(x)\frac{\partial u^\varepsilon}{\partial x^j}\right) = \varepsilon L u^\varepsilon, \quad t > 0, \quad x \in D,$$

(7.10)

$$u^\varepsilon(0,x) = g(x), \quad \frac{\partial u^\varepsilon(t,x)}{\partial \ell(x)} + \varepsilon^{-1} f(x, u^\varepsilon(t,x))\Big|_{x\in\partial D,\, t>0} = 0.$$

We assume that the coefficients $a^{ij}(x)$ and the function $f(x,u)$ are bounded and smooth enough; the field $\ell(x)$, $x \in \partial D$ is supposed to be smooth and non-tangent to ∂D. We assume that $f(x,\cdot) \in \mathcal{F}_1$ for any $x \in \partial D$.

The Feynman-Kac formula in this case gives the following equation for $u^\varepsilon(t,x)$

$$(7.11)\qquad u^\varepsilon(t,x) = E_x g(X_t^\varepsilon)\exp\left\{\frac{1}{\varepsilon}\int_0^t c(X_s^\varepsilon, u^\varepsilon(t-s, X_s^\varepsilon))d\xi_s^\varepsilon\right\},$$

where X_t^ε, ξ_t are the processes with reflection along the field $\ell(x)$ and corresponding local time (see the beginning of this section where the stochastic differential equations for $(X_t^\varepsilon, \xi_t^\varepsilon)$ were written). Formula (7.3) (after replacing $n^i(x)$ by $\ell^i(x)$, $\ell(x) = (\ell^1(x),\dots,\ell^r(x)))$ gives the action functional for the family $(X_t^\varepsilon, \xi_t^\varepsilon)$ as $\varepsilon\downarrow 0$. Using this action functional and equation (7.11), one can calculate where $u^\varepsilon(t,x)$ tends to zero and where it tends to one as $\varepsilon\downarrow 0$. It turns out that in this case the front propagates only along the boundary. Inside the domain D, $\lim_{\varepsilon\downarrow 0} u^\varepsilon(t,x)$ is not equal to 0 only for points of the set $[G_0]$. The wave front in this problem can have jumps even in the case when $\lim_{u\downarrow 0} u^{-1}f(x,u) = c = \text{const.}$ (see [7], §6.6).

A similar problem arises when we consider nonlinear gluing conditions. Let, for example a smooth curve φ_t in R^2 be given, and let $n^+(\varphi_t)$ and $n^-(\varphi_t)$ be the fields of normals to φ directed in opposite directions. Consider the problem

$$(7.12)\qquad \begin{aligned} &\frac{\partial u^\varepsilon(t,x)}{\partial t} = \frac{\varepsilon}{2}\Delta u^\varepsilon(t,x), \quad t>0, \quad x\in R^r\setminus\varphi \\ &u^\varepsilon(0,x) = g(x) \ge 0, \quad \left.\frac{\partial u^\varepsilon(t,x)}{\partial n^+} - \frac{\partial u^\varepsilon(t,x)}{\partial n^-}\right|_{x\in\varphi,\, t>0} = \frac{1}{\varepsilon}f(x,u^\varepsilon), \end{aligned}$$

where $f(x,\cdot)\in \mathcal{F}_1$, $x\in\varphi$. Under some additional minor conditions the wave front will propagate along φ.

Some other generalizations are considered in [6] and [7]. There are some other references there.

§8. Weakly Coupled Reaction-Diffusion Equations.

Suppose we have two equations of the KPP type

$$\frac{\partial u_i(t,x)}{\partial t} = \frac{D_i}{2}\frac{\partial^2 u_i}{\partial x^2} + f_i(u_i), \quad t>0, \quad x\in R^1, \quad i = 1,2, \quad f_1, f_2 \in \mathcal{F}_1.$$

Let us consider the coupling of such equations

$$\frac{\partial u_1(t,x)}{\partial t} = \frac{D_1}{2}\frac{\partial^2 u_1}{\partial x^2} + f_1(u_1) + \varepsilon d_1(u_2-u_1) \tag{8.1}$$

$$\frac{\partial u_2(t,x)}{\partial t} = \frac{D_2}{2}\frac{\partial^2 u_2}{\partial x^2} + f_2(u_2) + \varepsilon d_2(u_1-u_2).$$

Here d_1, d_2 are positive constants, $\varepsilon > 0$ is a parameter characterizing the strength of the coupling. The physical sense of the last terms in (8.1) is as follows. For $d_1 = d_2 = 0$ the equations (8.1) describe diffusion and multiplication (or killing) of the particles of the first and the second types. The particles of the different types have no interaction. The new terms describe transmutation from first to second type and vice-versa. The constant εd_1 is the intensity of the transition from the first to the second type, and εd_2 is the same characteristic for transition from the second to the first type.

If we consider (8.1) on a fixed time interval independent of ε, and take $\varepsilon \downarrow 0$, the functions $u_k(t,x)$, $k = 1,2$, tend to the solutions of the equations (8.1) for $\varepsilon = 0$. For $\varepsilon = 0$ equations (8.1) are independent and have, in general, different velocities of the wave fronts. But in the large time interval, growing together with ε^{-1} when $\varepsilon \downarrow 0$, one can expect that, due to interaction, some velocity of the front common for both component will be established. If the rate of transmutations is small $(\varepsilon \ll 1)$ the establishing of the common velocity takes large time. The position of the wave front at time t also tends to infinity when $t \to \infty$ (at least in the case of initial functions with compact support). To detect the front we should rescale not only the time, but the space also.

As we will see later, the proper scaling is $t \to t/\varepsilon$, $x \to x/\varepsilon$. Put $u_k^\varepsilon(t,x) = u_k(t\varepsilon^{-1}, x\varepsilon^{-1})$, $k = 1,2$, where $u_k(t,x)$ is the solution of the equations (8.1). Then we have the following equations for u_1^ε and u_2^ε:

$$\begin{cases} \dfrac{\partial u_1^\varepsilon(t,x)}{\partial t} = \dfrac{\varepsilon D_1}{2}\dfrac{\partial^2 u_1^\varepsilon}{\partial x^2} + \dfrac{1}{\varepsilon} f_1(u_1^\varepsilon) + d_1(u_2^\varepsilon - u_1^\varepsilon) \\[2ex] \dfrac{\partial u_2^\varepsilon(t,x)}{\partial t} = \dfrac{\varepsilon D_2}{2}\dfrac{\partial^2 u_2^\varepsilon}{\partial x^2} + \dfrac{1}{\varepsilon} f_2(u_2^\varepsilon) + d_2(u_1^\varepsilon - u_2^\varepsilon). \end{cases} \tag{8.2}$$

The generalization of the problem (8.2) for many-dimensional non-homogeneous and non-isotropic in space medium has the form

$$(8.3)\quad \begin{cases} \dfrac{\partial u_k^\varepsilon(t,x)}{\partial t} = \varepsilon L_k u_k^\varepsilon(t,x) + \dfrac{1}{\varepsilon} f_k(x,u_k^\varepsilon(t,x)) + \sum\limits_{j=1}^{r} d_{kj}(u_j^\varepsilon(t,x)-u_k^\varepsilon(t,x)) \\ u_k(0,x) = g_k(x), \quad k=1,\dots,n, \quad t>0, \quad x\in R^r. \end{cases}$$

We assume that $L_k = \frac{1}{2}\sum\limits_{i,j=1}^{r} a_k^{ij}(x)\frac{\partial^2}{\partial x^i \partial x^j}$, $k = 1,\dots,n$, are uniformly elliptic operators with bounded smooth coefficients (say $a_k^{ij} \in C^3$), $d_{kj} > 0$. The nonlinear terms $f_k(x,\cdot)$ are elements of $\mathcal{F}_1$ for any $x \in R^r$ and $k = 1,\dots,n$. Assumptions on the initial functions g_k are the same as in the case of single equation. We denote by G_0 the support of the function $\sum\limits_{k=1}^{n} g_k(x)$, since $g_k(x) \ge 0$, G_0 equal to the union of the supports of g_k.

A Markov process (X_t^ε,ν_t) in the phase space $R^r\times\{1,\dots,n\}$ can be connected with the system (8.3). The component ν_t of this process is the right continuous Markov process with n states such that $P\{\nu_{t+\Delta} = j|\nu_t = i\} = d_{ij}\cdot\Delta + o(\Delta)$, $\Delta\downarrow 0$, $i \ne j$. The first component X_t^ε is defined by the stochastic differential equation:

$$dX_t^\varepsilon = \sqrt{\varepsilon}\,\sigma_{\nu_t}(X_t^\varepsilon)dW_t, \quad \sigma_k(x),\sigma_k^*(x) = (a_k^{ij}(x)),$$

where W_t is an r-dimensional Wiener process. As we explained in §1, the generator $\mathcal{A}^\varepsilon$ of the process (X_t^ε,ν_t) on functions $f(x,k)$, $x \in R^r$, $k = 1,\dots,n$, having uniformly continuous and bounded second derivatives in x, has the form:

$$\mathcal{A}^\varepsilon f(x,k) = \varepsilon L_k f(x,k) + \sum_{j=1}^{r} d_{kj}(f(x,j)-f(x,k)).$$

Taking this into account one can write down the probabilistic representation for the solution of the problem (8.3) in the linear case when $f_k = \tilde{c}_k(t,x)u_k$, $k = 1,2,\dots,n$. In particular, the generalized Feynman-Kac formula for the solution of the problem (8.3) in this case has the form:

$$(8.4)\quad u_k^\varepsilon(t,x) = E_{x,k}\, g_{\nu_t}(X_t^\varepsilon)\exp\left\{\frac{1}{\varepsilon}\int_0^t \tilde{c}_{\nu_s}(t-s,X_s^\varepsilon)ds\right\}.$$

Using (8.4) we get the following integral equation for the solution of the problem (8.3) in the nonlinear case $f_k = c_k(x,u_k)u_k$:

$$(8.5)\qquad u_k^\varepsilon(t,x) = E_{x,k} g_{\nu_t}(X_t^\varepsilon) \exp\left\{\frac{1}{\varepsilon}\int_0^t c_{\nu_s}(X_x^\varepsilon, u_{\nu_s}^\varepsilon(t-s, X_s^\varepsilon))ds\right\},$$

$$x \in R^r,\quad t \ge 0,\quad k = 1,\dots,n.$$

Using the strong Markov property of the process (X_t^ε, ν_t), we can write down the following equation

$$(8.6)\quad u_k^\varepsilon(t,x) = E_{x,k} u_{\nu_{\tau\wedge t}}^\varepsilon\left[t-(\tau\wedge t), X_{\tau\wedge t}^\varepsilon\right] \exp\left\{\frac{1}{\varepsilon}\int_0^{\tau\wedge t} c_{\nu_s}(X_s^\varepsilon, u_{\nu_s}^\varepsilon(t-s, X_s^\varepsilon))ds\right\},$$

where τ is any arbitrary Markov time with respect to the filtration $\{\mathcal{F}_t, t \ge 0\}$, $\mathcal{F}_t = \sigma(X_s^\varepsilon, \nu_s^\varepsilon, 0 \le s \le t)$.

<u>Lemma 8.1</u>. The following properties of the solutions of the system (8.3) hold:

1) $0 \le u_k^\varepsilon(t,x) \le 1 \vee \sup_{x,k} g_k(x)$;

2) $\overline{\lim_{\varepsilon\downarrow 0}}\, u_k^\varepsilon(t,x) \le 1$ for $t > 0$, $x \in R^r$, $k = 1,2,\dots,n$;

3) Let $\{u_k'(t,x)\}$ be the solution of the problem (8.3) for an initial function $g'(x) = (g_1'(x),\dots,g_n'(x))$, and let $\{u_k''(t,x)\}$ be the solution with an initial function $g''(x) = (g_1''(x),\dots,g_n''(x))$. Suppose that $g_k'(x) \ge g_k''(x)$ for $x \in R^r$, $k = 1,\dots,n$. Then $u_k'(t,x) \ge u_k''(t,x)$ for all $t > 0$, $x \in R^r$, $k = 1,\dots,n$.

4) Let $\{u_k'(t,x)\}$ and $\{u_k''(t,x)\}$ be the solutions of problem (8.3) with $f_k = f_k'(x,u)$ and $f_k = f_k''(x,u)$, $k = 1,\dots,n$. Suppose that $f_k'(x,u) \ge f_k''(x,u)$ for all $x \in R^r$, $k = 1,\dots,n$. Then $u_k'(t,x) \ge u_k''(t,x)$ for $x \in R^r$, $t > 0$, $k = 1,\dots,n$.

<u>Proof</u>. 1) Suppose that the set $G = \{(s,y,i) : u_i^\varepsilon(s,y) > 1 \vee \sup_{x,\ell} g_\ell(x)\}$ contains a point (t,x,k), and put $\zeta_1 = \inf\{s : (t-s, X_s^\varepsilon, \nu_s) \notin G\}$. Since ζ_1 is a Markov time and $\zeta_1 < t$ with probability 1 we have from (8.6):

$$(8.7)\qquad u_k^\varepsilon(t,x) = E_{x,k} u_{\nu_{\zeta_1}}^\varepsilon(t-\zeta_1, X_{\zeta_1}^\varepsilon) \exp\left\{\frac{1}{\varepsilon}\int_0^{\zeta_1} c_{\nu_s}(X_s^\varepsilon, u_{\nu_s}^\varepsilon(t-s, X_s^\varepsilon))ds\right\}.$$

According to the definition of ζ_1,

$$(8.8)\qquad u^{\varepsilon}_{\nu_{\zeta_1}}(t-\zeta_1, X^{\varepsilon}_{\zeta_1}) \le 1 \vee \sup_{l,x} g_l(x).$$

Since

$$\inf_{0\le s\le \zeta_1} u^{\varepsilon}_{\nu_s}(t-s, X^{\varepsilon}_s) \ge 1 \quad \text{for trajectories} \quad (X^{\varepsilon}_s, \nu_s) \quad \text{starting from} \quad (x,k),$$

and $c_k(x,u) \le 0$ for $u \ge 1$, the integrand in the exponent of (8.7) is non-positive. Thus from (8.7) and (8.8) we have that $u^{\varepsilon}_k(t,x) \le 1 \vee \sup_{l,x} g_l(x)$, and the set G must be empty.

2) Assume that for some $t > 0$, $x \in R^r$, $k \in \{1,\dots,n\}$, there is a sequence $\varepsilon' \downarrow 0$ such that $\lim_{\varepsilon'\downarrow 0} u^{\varepsilon'}_k(t,x) = 1+2\alpha > 1$. Denote $D = D^{\alpha}_{\varepsilon} = \{(s,y,i) : s < t, u^{\varepsilon}_i(t-s,y) > 1+\alpha\}$ and $\zeta_2 = \inf\{s : s \le t, u^{\varepsilon}_{\nu_s}(t-s, X^{\varepsilon}_s) = 1+\alpha\}$. Let K be a compact in R^r. Denote $Q = \{(i,x,u) : i \in \{1,\dots,n\}$, $x \in K$, $1+\alpha \le u \le 1 + \sup_{l,x} g_l(x)$. We get by replacing ζ_1 by ζ_2 in (8.7), and taking into account that $\max_{(i,x,u)\in Q} c_i(x,u) < -\beta < 0$,

$$(8.9)\qquad u^{\varepsilon}_k(t,x) \le E_{x,k} u_{\nu_{\zeta_2}}(t-\zeta_2, X^{\varepsilon}_{\zeta_2})\chi_{\{\zeta_2<t\}} + E_{x,k} g_{\nu_t}(X^{\varepsilon}_s) e^{\frac{1}{\varepsilon}\int_0^t c_{\nu_s}(X^{\varepsilon}_s)ds} \chi_{\{\zeta_2=t\}}$$

$$\le (1+\alpha)P_{x,k}\{\zeta_2 < t\} + P_{x,k}\{\zeta_2 = t\},$$

if ε is so small that $\sup_{y,\ell} g_{\ell}(y)\cdot e^{-\beta t/\varepsilon} < 1$. From (8.9) one can see that $u^{\varepsilon}_k(t,x) \le 1+\alpha$. This contradiction proves the second statement.

3) The difference $v_k(t,x) = u'_k(t,x) - u''_k(t,x)$ satisfies the following system:

$$\begin{cases} \dfrac{\partial v_k}{\partial t} = \varepsilon L_k v_k + \dfrac{1}{\varepsilon}\hat{c}_k(x, u'_k, u''_k)v_k + \sum d_{kj}\cdot(v_j - v_k) \\ v_k(0,x) = g'_k(x) - g''_k(x) = \delta_k(x) \ge 0, \quad k = 1,\dots,n. \end{cases}$$

where $\hat{c}_k(x,u'_k,u''_k) = \dfrac{f_k(x,u_k)-f_k(x,u''_k)}{u'_k-u''_k}$. Then for $v_k(t,x)$ the following equality holds:

$$(8.10)\qquad v_k(t,x) = E_{x,k}\delta_{\nu_t}(X^{\varepsilon}_s)\exp\left\{\frac{1}{\varepsilon}\int_0^t \hat{c}_{\nu_s}(X^{\varepsilon}_s, u'_{\nu_s}(t-s,X^{\varepsilon}_s), u''_{\nu_s}(t-s,X^{\varepsilon}_s))ds\right\};$$

since $\delta_k(x) = g_k'(x) - g_k''(x) \geq 0$ we derive from (8.10) that $v_k(t,x) \geq 0$.

4) The functions $W_k(t,x) = u_k'(t,x) - u_k''(t,x)$ satisfy the system

$$\begin{cases} \dfrac{\partial W_k(t,x)}{\partial t} = \varepsilon L_k W_k + h(t,x) + \hat{c}_k(t,x)W_k + \sum_{j=1}^{n} d_{kj}(W_j - W_k) \\ W_k(0,x) = 0, \quad k = 1,\dots,n, \end{cases}$$

where

$$h_k(t,x) = f_k'(x,u_k'(t,x)) - f_k''(x,u_k'(t,x)), \quad \hat{c}(t,x) = \frac{f_k''(x,u_k'(t,x)) - f_k''(x,u_k''(t,x))}{u_k'(t,x) - u_k''(t,x)}.$$

One can write down the following equation for $W_k(t,x)$:

$$W_k(t,x) = E_{x,k} \int_0^t h_{\nu_s}(t-s, X_s^\varepsilon) \exp\left\{\frac{1}{\varepsilon} \int_0^s \hat{c}_k(t-s_1, X_{s_1}^\varepsilon) ds_1\right\} ds.$$

Since $h_k(t,x) \geq 0$ for all t, x and k, we have from this equality that $W_k(t,x) \geq 0$. □

Denote by H_{0t} the set of all right continuous step-functions on $[0,t]$, $t < \infty$, with values from the set $\{1,\dots,n\}$ and having a finite number of jumps. For any $\alpha \in H_{0t}$, define the functional

$$S_{0t}^\alpha(\varphi) = \frac{1}{2} \int_0^t \sum_{i,j=1}^{r} a_{ij}^{\alpha_s}(\varphi_s)\dot{\varphi}_s^i \dot{\varphi}_s^j \, ds$$

for absolutely continuous $\varphi \in C_{0t}$, and $S_{0t}^\alpha(\varphi) = +\infty$ for the rest of C_{0t}. It is known that $S_{0t}^\alpha(\varphi)$ is the action functional for the family of processes $X_t^{\alpha,\varepsilon}$, defined by the stochastic differential equation

$$dX_t^{\alpha,\varepsilon} = \sqrt{\varepsilon}\sigma_{\alpha_t}(X_t^{\alpha,\varepsilon})dW_t, \quad \sigma_k(x) = (a_k^{ij}(x))^{1/2},$$

with W_t a Wiener process in R^r (see [12], [22], where it is proved for continuous coefficients, but in the case of finite number of discontinuities the proof preserves). But we need bounds of the probabilities of large deviations which are uniform in $\alpha \in H_{0t}$. Such bounds are given by the following lemma.

<u>Lemma 8.2</u>. For any $\varphi \in C_{0t}$, $\varphi_0 = x$, and any $\gamma, \delta > 0$, one can find $\varepsilon_0 > 0$ such that for all $\alpha \in H_{0t}$

$$P_x\left\{\sup_{0\le s\le t}|X_s^{\alpha,\varepsilon}-\varphi_s|<\delta\right\}>\exp\left\{-\frac{1}{\varepsilon}[S_{0t}^{\alpha}(\varphi)+\gamma]\right\},$$

provided $\varepsilon\in(0,\varepsilon_0]$.

For any $s,\delta,\gamma>0$, one can find $\varepsilon_0>0$ such that for all $\alpha\in H_{0t}$ and $\varepsilon\in(0,\varepsilon_0]$

$$P_x\{\rho_{0t}(X_{\cdot}^{\alpha,\varepsilon},\phi_s^{\alpha})\ge\delta\}\le\exp\{-\frac{1}{\varepsilon}(S-\gamma)\},$$

where $\phi_s^{\alpha}=\{\varphi\in C_{0t},\ \varphi_0=x,\ S_{0t}^{\alpha}(\varphi)\le s\}$ and ρ_{0t} is the uniform metric in C_{0t}.

Proof. The first statement one can get from the standard proof of the lower bound (see [12], [22]).

To prove the upper bound usually the process $X_{\delta_1}^{\alpha,\varepsilon}(t)$ is considered: $dX_{\delta_1}^{\alpha,\varepsilon}(t)=\sqrt{\varepsilon}\sigma_{\alpha_t}(X_{\delta_1}^{\alpha,\varepsilon}(\pi_\delta(t)))dW_t$, $\pi_{\delta_1}(t)=[t/\delta_1]\delta_1$, $0<\delta_1\ll 1$, (see [22], Section 6). For any $\alpha\in H_{0t}$ process $X_{\delta_1}^{\alpha,\varepsilon}(t)$ is a continuous transformation $\Gamma=\Gamma_{\alpha,\delta_1}$ of the process $\sqrt{\varepsilon}\,W_s$, $0\le s\le t$. Thus the action functional for $X_{\delta_1}^{\alpha,\varepsilon}(t)$ can be calculated using the contraction principle. For small enough δ_1 one can have good bounds for $\rho_{o,t}(X_{\cdot}^{\alpha,\varepsilon},X_{\delta_1}^{\alpha,\varepsilon})$. Combining the bounds for $X_{\delta_1}^{\alpha,\varepsilon}$ and for $\rho_{o,t}(X^{\alpha,\varepsilon},X_{\delta_1}^{\alpha,\varepsilon})$, the upper bound included in the large deviation principle can be proved. But the transformation Γ_{α,δ_1} is not uniformly continuous with respect to the number of jumps of the function α_s, $s\in[0,t]$. To have uniform bounds, let us introduce the family of processes $\bar{X}_{\delta_1}^{\alpha,\varepsilon}(s)$, which are defined as follows. Denote $h_i=h_i[\pi_{\delta_1}(t),\pi_{\delta_1}(t)+\delta_1]$ the time which the function α_s spent in the state i during the time interval $[\pi_{\delta_1}(t),\pi_{\delta_1}(t)+\delta]$. Put $h_0=0$,

$$\tilde{\sigma}(t,x)=\sigma_k(x)\quad\text{for}\quad t\in[\pi_{\delta_1}(t)+h_0+\ldots+h_{k-1},\ \pi_{\delta_1}(t)+h_0+\ldots+h_k),$$
$$k=1,\ldots,n.$$

The number of discontinuities of the function $\tilde{\sigma}(s,x)$ on time interval $[0,t]$ is not bigger than $t\delta^{-1}n$. The process $\bar{X}_{\delta_1}^{\alpha,\varepsilon}(s)$ is defined as the solution of the equation

$$d\bar{X}_{\delta_1}^{\alpha,\varepsilon}(s)=\sqrt{\varepsilon}\,\tilde{\sigma}(s,\bar{X}_{\delta_1}^{\alpha,\varepsilon}(\pi_{\delta_1}(s)))dW_s.$$

This equation defines a map $\sqrt{\varepsilon}W\longrightarrow\bar{X}_{\delta_1}^{\alpha,\varepsilon}$ in C_{0t}, which is uniformly continuous in $\alpha \in H_{0t}$, and the contraction principle gives us bounds which are uniform with respect to $\alpha \in H_{0t}$. At the same time, since $\bar{X}^{\alpha,\varepsilon}_{\delta_1}(s) = X^{\alpha,\varepsilon}_{\delta_1}(s)$ for $s = k\delta_1$, k an integer, the distance $\rho_{0,t}(X^{\alpha,\varepsilon}_{\delta_1}, \bar{X}^{\alpha,\varepsilon}_{\delta_1})$ can be properly bounded for small enough δ_1. Together with the bound of $\rho_{0,t}(X^{\alpha,\varepsilon}, X^{\alpha,\varepsilon}_{\delta_1})$ from Section 6 of [22], we have the uniform upper bound for

$$P_x\{\rho_{0,t}(X^{\alpha,\varepsilon}, \phi^s) \geq \delta\}. \qquad \square$$

Define the functional $R_{0t}(\varphi,\alpha)$, $\varphi \in C_{0t}$, $\alpha \in H_{0t}$ by the formula

$$R_{0t}(\varphi,\alpha) = \int_0^t c_{\alpha_s}(\varphi_s)ds - S^{\alpha}_{0t}(\varphi).$$

For any fixed $\alpha \in H_{0t}$ the functional $R_{0t}(\varphi,\alpha)$ is semi-continuous from above in $\varphi \in C_{0t}$.

Introduce the function $V(t,x)$, $t \geq 0$, $x \in R^r$,

$$V(t,x) = \sup\{R_{0t}(\varphi,\alpha) : \varphi \in C_{0t},\ \ \varphi_0 = x,\ \ \varphi_t \in G_0,\ \ \alpha \in H_{0t}\},$$

where G_0 is the support of $\sum_{k=1}^{n} g_k(x)$ and $\{g_k(x)\}$ are the initial functions in problem (8.3). It is easy to check that the function $V(t,x)$ is continuous. Denote $\mathcal{E}_- = \{(t,x) : t > 0,\ \ x \in R^r,\ \ V(t,x) \leq 0\}$.

We say that condition (N) is fulfilled for problem (8.3), if for any $(t,x) \in \mathcal{E}_-$

$$V(t,x) = \sup\{R_{0t}(\varphi,\alpha) : \varphi \in C_{0t},\ \varphi_0 = x,\ \ \varphi_t \in G_0,\ \ \alpha \in H_{0t},$$
$$V(t-s,\varphi_s) < 0 \ \text{ for } \ 0 < s < t\}.$$

<u>Theorem 8.1</u>. The following statements hold for the solution $\{u^{\varepsilon}_k(t,x)\}$ of problem (8.3):

1) Let F_1 be a compact subset of the set $\{(t,x) : t > 0,\ x \in R^r,$ $V(t,x) < 0\}$. Then $\lim_{\varepsilon\downarrow 0} u^{\varepsilon}_k(t,x) = 0$ uniformly in $(t,x) \in F_1$ and $k = 1,\dots,n$.

2) Assume that condition (N) is fulfilled. Let F_2 be a compact subset of the set $\{(t,x) : t > ,\ x \in R^r,\ V(t,x) > 0\}$. Then $\lim_{\varepsilon\downarrow 0} u^{\varepsilon}_k(t,x) = 1$ uniformly in $(t,x) \in F_2$ and $k = 1,\dots,n$.

Proof. It is similar to the proof of Theorem 3.1, and we merely point out the differences.

Since $c_k(x) = c_k(x,0) = \max_{0\le u} c_k(x,u)$, we have from (8.5):

$$u_k^\varepsilon(t,x) \le E_{x,k} g_{\nu_t}(X_t^\varepsilon) \exp\left\{\frac{1}{\varepsilon}\int_0^t c_{\nu_s}(X_s^\varepsilon)ds\right\} \tag{8.11}$$

$$= E_{x,k}E_{x,k}[g_{\nu_t}(X_t^\varepsilon) \exp\left\{\frac{1}{\varepsilon}\int_0^t c_{\nu_s}(X_s^\varepsilon)ds\right\}|\nu_s,\quad 0 \le s \le t].$$

We can calculate the asymptotics of the conditional expectation in (8.11) under the condition $\nu_s = \alpha_s$, using the large deviation principle for the family $\{X_s^{\alpha,\varepsilon}\}$:

$$\lim_{\varepsilon\downarrow 0} \varepsilon \ln E_x g_{\nu_s}(X_t^{\alpha,s}) \exp\left\{\frac{1}{\varepsilon}\int_0^t c_{\nu_s}(X_s^{\alpha,\varepsilon})ds\right\} \tag{8.12}$$

$$= \sup\{R_{0t}^\alpha(\varphi) : \varphi \in C_{0t},\ \varphi_0 = x,\ \varphi_t \in G_{\alpha_t}\},$$

where G_{α_t} is the support of the function $g_{\alpha_t}(x)$. The convergence in (8.12) is uniform with respect to x from any compact set $F \subset R^r$, and, due to Lemma 8.2, with respect to any $\alpha \in H_{0t}$. From (8.11) and (8.12) the first statement of Theorem 8.1 follows.

The proof of the second statement uses the same arguments as the proof of the similar statement in Theorem 3.1: Using condition (N), the first statement and the lower bound from Lemma 8.2, one can prove that for every $\delta > 0$ and any compact set $F \subset \{(t,x) : V(t,x) = 0\}$ there exists $\varepsilon_0 > 0$ such that for $0 < \varepsilon < \varepsilon_0$, $(t,x) \in F$ and $k = 1,\dots,n$

$$u_k^\varepsilon(t,x) \ge \exp\{-\frac{\delta}{\varepsilon}\}. \tag{8.13}$$

Then the second statement of Theorem 8.1 can be proved with the help of equation (8.5) and the strong Markov property as it was done in the case of one equation. □

Now we consider an example where condition (N) is fulfilled and Theorem 8.1 can be used. This example will be helpful also in general situation, when condition (N) is not true.

Let $c_k(x) = c$ be independent of $x \in R^r$ and k. In this case

(8.14)

$$V(t,x) = ct - \frac{1}{2}\inf\left\{\int_0^t \sum_{i,j=1}^r a_{ij}^{\alpha_s}(\varphi_s)\dot\varphi_s^i\,\dot\varphi_s^j\,ds : \varphi_0 = x,\quad \varphi_t \in G_0,\quad \alpha \in H_{0t}\right\}.$$

Calculate

$$D(t,x,y) = \inf\left\{\int_0^t \sum_{i,j=1}^r a_{ij}^{\alpha_s}(\varphi_s)\dot\varphi_s^i\,\dot\varphi_s^j\,ds : \varphi_0 = x,\quad \varphi_t = y,\quad \alpha \in H_{0t}\right\}$$

$$= \inf_{\substack{N \\ i_0, i_1, \dots, i_N \in \{1,\dots,n\}}}\ \inf_{\substack{z_1,\dots,z_{N-1}\in R^r \\ z_0 = x,\ z_N = y}}\ \inf_{\substack{0 < t_0, \dots, t_{N-1} \\ \sum_0^{N-1} t_k = t}}$$

(8.15)

$$\inf_{\substack{\varphi_0 = x,\ \varphi_{t_0} = z_1, \dots, \varphi_{t_0+\dots+t_{N-1}} = z_N \\ \alpha\in H_{0t}:\ \alpha_s = \alpha_{i_k}\ \text{for} \\ s\in[t_0+\dots+t_k,\ t_0+\dots+t_k+t_{k+1})}} \int_0^t \sum_{i,j=1}^r a_{ij}^{\alpha_s}(\varphi_s)\dot\varphi_s^i\,\dot\varphi_s^j\,ds.$$

Note that

$$\inf\left\{\int_0^t \sum_{i,j=1}^r a_{ij}(\varphi_s)\dot\varphi_s^i\,\dot\varphi_s^j\,ds : \varphi_0 = a,\quad \varphi_t = b\right\} = \frac{1}{t}\rho^2(a,b),$$

where ρ is the Riemannian metric corresponding to the form $ds^2 = \sum_{i,j=1}^r a_{ij}(x)dx^i, dx^j$. Then the inner inf in (8.15) equal to

$$D_1 = D_1(t,x;z_1,\dots,z_{N-1};y;t_0,t_1,\dots,t_{N-1},t;\ i_0,i_1,\dots,i_{N-1})$$

$$= \sum_{k=0}^{N-1} \frac{\rho_{i_k}^2(z_k, z_{k+1})}{t_k}$$

where ρ_k is the Riemannian metric corresponding to $(a_{ij}^k(x)) = (a_k^{ij}(x))^{-1}$.

Consider now $\inf_{t_0,\dots,t_{N-1}} D_1$. Using the Cauchy inequality, we have:

$$D_1 = \sum_{k=0}^{N-1} \frac{\rho_{i_k}^2(z_k,z_{k+1})}{t_k} = \sum_{k=0}^{N-1}\left[\frac{\rho_{i_k}(z_k,z_{k+1})}{\sqrt{t_k}}\right]^2 \sum_{k=0}^{N-1}\left[\frac{\sqrt{t_k}}{\sqrt{t}}\right]^2$$

$$\geq \left[\sum_{k=0}^{N-1} \frac{\rho_{i_k}(z_k,z_{k+1})}{\sqrt{t}}\right]^2 = \frac{1}{t}\left[\sum_0^{N-1}\rho_{i_k}(z_k,z_{k+1})\right]^2 .$$

On the other hand, if $t_k = t\cdot\rho_{i_k}(z_k,z_{k+1})\left[\sum_0^{N-1}\rho_{i_k}(z_k,z_{k+1})\right]^{-1}$ then

$$D_1 = \frac{1}{t}\left[\sum_0^{N-1}\rho_{i_k}(z_k,z_{k+1})\right]^2 .$$

Thus

$$\inf_{\substack{t_0+\ldots+t_{N-1}=t\\ t_i>0}} D_1 = \frac{1}{t}\left[\sum_0^{N-1}\rho_{i_k}(z_k,z_{k+1})\right]^2 .$$

Now assume for a moment that all operators L_k have coefficients independent of x. Then we can unite all time intervals on which the same metric ρ_{i_k} is considered. Using the triangle axiom for metrics, we get that

$$D(t,x,y) = \frac{1}{t} \inf_{\substack{z_1,\ldots,z_{n-1}\\ z_0=x,\ z_n=y}} \left[\sum_{k=1}^{N}\rho_k(z_{k-1},z_k)\right]^2 ,$$

where n is the number of equations in our system. Taking in account that in the space-homogeneous case a Riemannian metric ρ has the property that $\rho(\alpha x,\alpha y) = |\alpha|\rho(x,y)$ for any real α, it is easy to check that the set

$$\left\{x : \inf_{\substack{z_1,\ldots,z_{n-1}\in R^r\\ z_0=x,\ z_n=y}} \left[\sum_{k=1}^{N}\rho_k(z_{k-1},z_k)\right] \leq d\right\}$$

is the convex envelope of Riemannian spheres

$$S_k = \{x : \rho_k(x,y) \leq d\},\quad k = 1,2,\ldots,n.$$

To describe $D(t,x,y)$ in the case of space non-homogeneous coefficients, let us define the function $d(x,y)$, $x,y \in R^r$, by the following conditions:

1) $d(x,\alpha y) = |\alpha| d(x,y)$ for any real α;

2) $d(x,y) = 1$ on the boundary of the convex envelope of the sets $\{y \in R^r : \sum_{i,j=1}^{r} a_{ij}^k(x)y^i y^j \le 1\} = S_k$, $k = 1,2,\dots,n$.

Put

$$\bar{\rho}(x,y) = \inf\left\{\int_0^t d(\varphi_s,\dot{\varphi}_2)ds \ :\ \varphi_0 = x,\quad \varphi_t = y\right\}.$$

It is easy to check that this infimum is independent of the parameter t and defines a metric $\bar{\rho}$ in R^r (Finsler metric).

One can calculate that

$$D(t,x,y) = \frac{1}{t}\bar{\rho}(x,y). \tag{8.16}$$

In the space-homogeneous case it was proved above. The general case can be considered by the approximation of the metric by piecewise space-homogeneous ones.

From (8.14), (8.15), and (8.16) we derive

$$V(t,x) = ct - \frac{\bar{\rho}^2(x,G_0)}{2t}.$$

Now we can check that condition (N) is fulfilled: In our case $\mathcal{E}_- = \{(t,x) : t > 0,\ V(t,x) \le 0\} = \{x : \rho(x,G_0) \ge t\sqrt{2c}\}$. For a point $(t,x) \in \mathcal{E}_-$, choose a small $h > 0$ and consider the function φ_s^h such that $\varphi_s^h = x$ for $s \in [0,h]$ and $\varphi_s^h = \hat{\varphi}(s-h)$ for $s \in [h,t]$, where $\hat{\varphi}(s)$ is the minimal geodesic of the metric $\bar{\rho}$, connecting x and G_0, with the parameterization proportional to the length, $\hat{\varphi}(0) = x$, $\hat{\varphi}_{t-h} \in G_0$. The point $(t-s,\varphi_s^h)$ for all $s \in (0,t)$ belongs to the set $\{(s,y),\ V(s,y) < 0\}$, and

$$\sup\{R_{0t}(\varphi^h,\alpha) : \alpha \in H_{0t}\} = ct - \frac{\bar{\rho}^2(x,G_0)}{t-h}.$$

Therefore, $V(t,x) = \lim_{h\downarrow 0}\sup\{R_{0t}(\varphi^h,\alpha),\ \ \alpha \in H_{0t}\}$ and condition (N) is fulfilled. From Theorem 8.1 we have the following result:

<u>Theorem 8.2</u>. Let $c_k(x,0) = c$ for all $x \in R^r$, $k \in \{1,2,\dots,n\}$. Let $\bar{\rho}(x,y)$, $x,y \in R^r$, denote the Finsler metric corresponding to the kernel $d(x,y)$, which is defined by conditions 1 and 2 above. Then

$$\lim_{\varepsilon\downarrow 0} u_k^{\varepsilon}(t,x) = \begin{cases} 1, & \text{for } \bar{\rho}(x,G_0) < t\sqrt{2c}, \\ 0, & \text{for } \bar{\rho}(x,G\ ) > t\sqrt{2c}\ . \end{cases}$$

The convergence is uniform in (t,x) for (t,x) in any compact set F such that $F \cap \{(t,x) : \bar{\rho}(x,G_0) = t\sqrt{2c}\} = \phi$.

The statement of the Theorem 8.2 means that the propagation of the wave front is governed by the Huygens principle. The corresponding velocity field is homogeneous and isotropic in the Finsler metric $\bar{\rho}$.

Lemma 8.3. Suppose that for some $x_0 \in R^r$ and $k_0 \in \{1,\dots,n\}$ and for any $\delta_1, \delta_2 > 0$, there exists $\varepsilon_0 > 0$ such that

$$(8.17) \qquad g_{k_0}(x) = g_k^\varepsilon(x) \ge e^{-\frac{\delta_1}{\varepsilon}} \quad \text{for } 0 < \varepsilon \le \varepsilon_0, \ |x - x_0| \le e^{-\frac{\delta_2}{\varepsilon}}.$$

Then a constant $A > 0$ exists such that $\lim_{\varepsilon\downarrow 0} u_k^\varepsilon(t,x) = 1$ for $t > 0$, $|x - x_0| < At$ and any $k = 1,\dots,n$. The convergence is uniform in any compact subset of the cone $\{(t,x) : t > 0, |x - x_0| < At\}$.

Proof. Because of Lemma 8.1 it is sufficient to consider the case when $g_k(x) \equiv 0$ for $k \ne k_0$, and $\tilde{g}^\varepsilon_{k_0}(x) = g^\varepsilon_{k_0}(x)$ for $|x - x_0| \le \exp\{-\frac{\delta_2}{\varepsilon}\}$ and $\tilde{g}^\varepsilon_{k_0}(x) = 0$ for $|x - x_0| > \exp\{-\frac{\delta_2}{\varepsilon}\}$. Moreover, we can confine ourselves to the case $c(x,u) = c(u)$.

From Theorem 8.2 one can derive that $\lim_{\varepsilon\downarrow 0} u_k^\varepsilon(t,x) = 0$ for $(t,x) \in \{(s,y) : s > 0, \ \bar{\rho}(x,x_0) > s\sqrt{2c(0)}\}$. Here $\bar{\rho}$ is the corresponding Finsler metric. As it was explained when we proved Theorem 8.2, for every point (t,x), $t > 0$, $\bar{\rho}(x,x_0) > t\sqrt{2c(0)}$, and any $\delta > 0$, there exists a function φ_s^δ, $0 \le s \le t$, $\varphi_0^\delta = x$, $\varphi_t^\delta = x_0$, $\rho(x_0,\varphi_s^\delta) > s\sqrt{2c(0)}$ for $0 < s < t$, and $\alpha^\delta \in H_{0t}$ such that

$$(8.18) \qquad \int_0^t \left[c_{\alpha_s^\delta}(\varphi_s^\delta) - \frac{1}{2} \sum_{i,j=1}^r a_{ij}^{\alpha_s^\delta}(\varphi_s^\delta)\dot{\varphi}_s^{\delta,i}\,\dot{\varphi}_s^{\delta,j} \right] ds > -\frac{\delta}{2}.$$

Without loss of generality we can assume that $\varphi_s^\delta = x_0$ for $s \in [t-h,t]$ for $h > 0$ small enough. Taking in account (8.18), the lower bound from Lemma 8.2, condition (8.17), and the bound (5.8) for the transition density of the process corresponding to the operator L_{k_0}, we get that

$$(8.19) \qquad \lim_{\varepsilon\downarrow 0} \varepsilon \ln u_k^\varepsilon(t,x) = 0$$

for $t > 0$, $\rho(x,x_0) \le t\sqrt{2c(0)}$, and any $k = 1,\dots,n$. From (8.19) it follows that $\lim_{\varepsilon\downarrow 0} u_k^\varepsilon(t,x) = 1$ in the interior points of the set $\{(t,x),\ t > 0,\ \bar\rho(x,x_0) > t\sqrt{2c(0)}\}$. The proof of the last statement one can carry out in the same way as the proof of the Theorem 3.1.

To get the statement of the Lemma 8.3, note that

$$\{(t,x),\ t > 0,\ |x - x_0| < At\} \subset \{(t,x),\ t > 0,\ \bar\rho(x,x_0) > t\sqrt{2c(0)}\}$$

for some $A > 0$. □

Let $\tau = \tau(t,\varphi)$, $t \in (-\infty,\infty)$, $\varphi \in C_{0\infty}$, be the Markov functional introduced in §5, and θ be the set of all Markov functionals. Denote

$$V^*(t,x) = \inf_{\tau\in\theta} \sup \left\{ \int_0^{t\wedge\tau} \left[c_{\alpha_s}(\varphi_s) - \frac{1}{2} \sum_{i,j=1}^{r} a_{ij}^{\alpha_s}(\varphi_s)\dot\varphi_s^i \dot\varphi_s^j \right] ds :\right.$$

$$\left. \varphi \in C_{0t},\ \varphi_0 = x,\ \varphi_t \in G_0,\ \alpha \in H_{0t} \right\}.$$

<u>Lemma 8.4</u>. If $V^*(t,x) < 0$, then $\lim_{\varepsilon\downarrow 0} u_k^\varepsilon(t,x) = 0$ for any $k = 1,\dots,n$. The convergence is uniform in (t,x) for (t,x) in any compact subset of the set $\{(s,y) : V^*(s,y) < 0\}$.

Taking into account equation (8.6), Lemma 8.1, and the upper bound from Lemma 8.2, the proof of this lemma is similar to the proof of Lemma 5.1 and we omit it.

The following lemma is a corollary of the Lemma 8.3.

<u>Lemma 8.5</u>. Let $\varepsilon' \downarrow 0$, $\varepsilon' > 0$, $\mathcal{E}^{(\varepsilon')} = \{(t,x) : \lim_{\varepsilon'\downarrow 0} u_{k_0}^{\varepsilon'}(t,x) = 0\}$ for some $k_0 \in \{1,\dots,n\}$, and $(t_0,x_0) \in \mathcal{E}^{(\varepsilon')}$. Then there exists $A > 0$ such that $\overline{\lim}_{\varepsilon'\downarrow 0} \varepsilon' \ln u_k^{\varepsilon'}(t,x) < 0$ for any $k = 1,\dots,n$ and $(t,x) \in D_{t_0,x_0}^A$, where

$$D_{t_0,x_0}^A = \{(s,y) : 0 < s < t_0,\ |x_0 - y| \le A(t_0 - s)\}.$$

<u>Lemma 8.6</u>. Let F be a compact subset of the interior (M) of the set $M = \{(t,x) : V^*(t,x) = 0\}$. Then $\lim_{\varepsilon\downarrow 0} \varepsilon \ln u_k^\varepsilon(t,x) = 0$ for $k = 1,\dots,n$ uniformly in $(t,x) \in F$.

<u>Proof</u>. The proof of this Lemma is similar to the proof of Lemma 5.4, and we omit it.

<u>Theorem 8.3</u>. Let $(u_1^\varepsilon(t,x),\dots,u_n^\varepsilon(t,x))$ be the solution of problem (8.5).

For any compact subset F_1 of the set $\{(s,y) : s > 0\ V^*(s,y) < 0\}$, $\lim_{\varepsilon\downarrow 0} u_k^\varepsilon(t,x) = 0$ for $k = 1,\dots,n$ uniformly in $(t,x) \in F_1$.

For any compact subset F_2 of the interior of the set $\{(s,y) : s > 0,\ V^*(s,y) = 0\}$, $\lim_{\varepsilon\downarrow 0} u_k^\varepsilon(t,x) = 1$ for $k = 1,\dots,n$ uniformly in $(t,x) \in F_2$.

<u>Proof</u>. It follows from Lemmas 8.1 - 8.5.

In Theorem 8.2, we considered the case were $c_k(x) = c$ independent of k and x. Now we consider as an example problem (8.5) when $L_k = L$ the same for all k

$$(8.20)\qquad \begin{cases} \dfrac{\partial u_k^\varepsilon(t,x)}{\partial t} = Lu_k^\varepsilon + c_k(x,u_k^\varepsilon)u_k^\varepsilon + \sum_{j=1}^{n} d_{kj}\cdot(u_j^\varepsilon - u_k^\varepsilon), \\[2ex] u_k^\varepsilon(0,x) = g_k(x), \quad k = 1,\dots,n. \end{cases}$$

The function $V^*(t,x)$ for the system (8.20) has the form

$$V^*(t,x) = \inf_{\tau\in\Theta} \sup\left\{\int_0^{t\wedge\tau} [c_{\alpha_s}(\varphi_s) - \frac{1}{2}\sum_{i,j=1}^{r} a_{ij}^{\alpha_s}(\varphi_s)\dot\varphi_s^i\dot\varphi_s^j]ds : \varphi_0 = x,\ \ \varphi_t \in G_0,\ \ \alpha \in H_{0t}\right\}$$

$$= \inf_{\tau\in\Theta}\ \sup_{\substack{\varphi_0 = x\\ \varphi_t\in G_0}} \left\{\int_0^{t\wedge\tau} \left[\max_{1\le k\le n} c_k(\varphi_s) - \frac{1}{2}\sum_{i,j=1}^{r} a_{ij}(\varphi_s)\dot\varphi_s^i\,\dot\varphi_s^j\right]ds\right\}.$$

Thus the function $V^*(t,x)$ and the law of the wave front propagation will be the same as in the case of single equation

$$(8.21)\qquad \frac{\partial u^\varepsilon(t,x)}{\partial t} = \varepsilon Lu^\varepsilon + \hat c(x)u^\varepsilon(1-u^\varepsilon),\quad u^\varepsilon(0,x) = \sum_{k=1}^{n} g_k(x),$$

where $\hat c(x) = \max_k c_k(x)$.

In particular, let $r = 1$, $a^{11}(x) \equiv 1$, $G_0 = \{x \le 0\}$, and the function $\hat c(x)$ decreases when x increases for $x > 0$. Denote by ψ_s the solution of the equation $\dot\psi_s = \sqrt{2\hat c(\psi_s)}$, $\psi_0 = 0$, for $s \ge 0$. It follows from Example 4 of §4 that

$$\lim_{\varepsilon\downarrow 0} u_k^\varepsilon(t,x) = \lim_{\varepsilon\downarrow 0} u^\varepsilon(t,x) = \begin{cases} 1, & x < \psi_t,\ t > 0, \\ 0, & x > \psi_t,\ t > 0. \end{cases}$$

One can check that in this case condition (N) is not fulfilled, and the infimum in the definition of the function $V^*(t,x)$ is reached on the Markov functional $\tau^*(t,\varphi) = \min\{s : \psi_{t-s} = \varphi_s\}$.

As in the case of the single equation, the wave front may have jumps. For example, if in the case under consideration an interval $(\alpha,\beta) \subset R^+$ exists where the function $\hat{c}(x)$ increases fast enough, then the front will have jumps.

Now we formulate some results on the upper and the lower bounds for the domains where $u_k^\varepsilon(t,x)$ are close to 0 or to 1.

Denote by $V^{k^*}(t,x)$ the V^*-function defined for the single equation $\frac{\partial u}{\partial t} = \varepsilon L_k u + \frac{1}{\varepsilon} c_k(u)u$ with the initial function $u(0,x) = \sum_{k=1}^n g_k(x)$.

The following inclusions are simple implications of the definition of the V^*-function:

$$\{(t,x) : t > 0,\ V^*(t,x) = 0\} \supseteq \bigcup_{k=1}^n \{(t,x),\ \ t > 0,\ \ V^{k^*}(t,x) = 0\}.$$

If for some $k_0 \in \{1,\dots,n\}$

$$\sum_{i,j=1}^r a_{ij}^{k_0}(x)\lambda^i\lambda^j \le \sum_{i,j=1}^r a_{ij}^k(x)\lambda^i\lambda^j \quad \text{for any} \quad \lambda, x \in R^r, \quad k = 1,\dots,n,$$

and $c_{k_0}(x) \ge c_k(x)$ for $x \in R^r$, $k = 1,\dots,n$, then $V^*(t,x) = V^{k_0^*}(t,x)$.

One can have more explicit bounds from below for the set $\{(t,x) : t > 0,\ V^*(t,x) = 0\}$ in the following way. Denote by $h(x,y)$, $x,y \in R^r$, the function defined by the properties: $h(x,ty) = |t|h(x,y)$ for any real t; $h(x,y) = 1$ on the boundary of the convex envelope of the ellipsoids

$$S_k^x = \{y : \sum a_{ij}^k(x)(2c_k(x))^{-1}y^iy^j \le 1\},\ k = 1,\dots,n.$$

Then

$$\{(t,x) : t > 0,\ \ V^*(t,x) = 0\} \supseteq \{(t,x),\ \ t > 0,\ \ \bar{\rho}(x,G_0) < t\},$$

where $\bar{\rho}(x,y) = \inf\left\{\int_0^t h(\varphi_s,\dot{\varphi}_s)ds : \varphi \in C_{0t},\ \ \varphi_0 = x,\ \ \varphi_t = y\right\}$ is the Finsler metric with the kernel $h(x,y)$.

We now give an explicit description of the wave front motion for the space-homogeneous isotropic system. We consider for brevity the case of two equations:

$$\frac{\partial u_1^\varepsilon(t,x)}{\partial t} = \frac{\varepsilon a_1}{2}\Delta u_1^\varepsilon + \frac{1}{\varepsilon}c_1(u_1^\varepsilon)u_1^\varepsilon + d_1\cdot(u_2^\varepsilon - u_1^\varepsilon)$$

$$\frac{\partial u_2^\varepsilon(t,x)}{\partial t} = \frac{\varepsilon a_2}{2}\Delta u_2^\varepsilon + \frac{1}{\varepsilon}c_2(u_2^\varepsilon)u_2^\varepsilon + d_2\cdot(u_1^\varepsilon - u_2^\varepsilon) \qquad (8.22)$$

$$x \in R^r, \quad t > 0, \quad u_1^\varepsilon(0,x) = g_1(x), \quad u_2^\varepsilon(0,x) = g_2(x).$$

We make the usual assumptions on the nonlinear terms and the initial functions, $c_k = c_k(0)$, $G_0 = \mathrm{supp}(g_1+g_2)$. Without loss of generality we assume that $c_1 \ge c_2$. Otherwise we change the indexation.

<u>Theorem 8.4</u>. Let $\rho(\cdot,\cdot)$ be the Euclidian distance in R^r and $c_1 \ge c_2$. Then for $k = 1,2$, $t > 0$, $x \in R^r$,

$$\lim_{\varepsilon\downarrow 0} u_k^\varepsilon(t,x) = \begin{cases} 1, & \text{if } \rho(x,G_0) < vt \\ 0, & \text{if } \rho(x,G_0) > vt \end{cases}$$

uniformly in (t,x) for (t,x) in any compact $F \subset \{(t,x) : t > 0,\ x \in R^r\}$ such that

$$F \cap \{(t,x) : t > 0,\ x \in R^r,\ \rho(x,G_0) = vt\} = \phi.$$

The speed v is given by following formulas:

$$v = \begin{cases} \sqrt{2a_1c_1}, & \text{if } a_1 \ge a_2, \\ \sqrt{2a_1c_1}, & \text{if } a_1 < a_2,\ 2a_1c_1 \ge c_1a_2 + a_1c_2 \\ \dfrac{c_1a_2 - c_2a_1}{\sqrt{2(a_2-a_1)(c_1-c_2)}}, & \text{if } a_1 < a_2,\ 2a_1c_1 \vee 2a_2c_2 < c_1a_2 + a_1c_2 \\ \sqrt{2a_2c_2}\,, & \text{if } a_1 < a_2,\ 2a_2c_2 \ge c_1a_2 + a_1c_2 \end{cases} \qquad (8.23)$$

<u>Proof</u>. If $a_1 \ge a_2$ then, taking in to account our assumption $c_1 \ge c_2$, we have $V(t,x) = c_1t + \dfrac{\rho^2(x,G_0)}{2a_1t}$, where ρ is the Euclidean distance. It is obvious that condition (N) is fulfilled in this case, and our statement follows from Theorem 8.1.

Consider now the case $a_1 < a_2$. Denote

$$V_0(t,x) = \sup\left\{\int_0^t \left[c_{\alpha_s} - \frac{|\dot{\varphi}_s|^2}{2a_{\alpha_s}}\right]ds : \varphi \in C_{0t},\ \ \varphi_0 = x,\ \ \varphi_t = 0,\ \ \alpha \in H_{0t}\right\}.$$

Because of the homogeneity in space

$$V(t,x) = \sup_{y \in G_0} V_0[t, x-y]. \tag{8.24}$$

It is easy to check that

$$V_0(t,x) = \max_{0 \le p \le 1} \left[c_1 pt + c_2(1-p)t - \frac{1}{2} \min_z \left(\frac{|z|^2}{pta_1} + \frac{|x-z|^2}{(1-p)ta_2}\right)\right].$$

Taking into account that

$$\min_z \left(\frac{|z|^2}{pta_1} + \frac{|x-z|^2}{(1-p)ta_2}\right) = \frac{|x|^2}{t(pa_1+(1-p)a_2)}$$

we have from (8.24)

$$V_0(t,x) = t \cdot \max_{0 \le p \le 1} \left[c_1 p + c_2(1-p) - \frac{|x|^2}{2t^2(pa_1+(1-p)a_2)}\right]. \tag{8.25}$$

Denote by $f(P,M)$ the function under the max sign in (8.25), $M = \frac{|x|}{t\sqrt{2}}$. Solving the equation $\frac{df}{dP} = 0$, and taking into account that the smallest of the roots corresponds to the maximum, we see that $\max_{0 \le P \le 1} f(P,M)$ is reached at the point

$$P_0 = \frac{a_2}{a_2 - a_1} - \frac{M}{\sqrt{(c_1-c_2)(a_2-a_1)}},$$

if $P_0 \in [0,1]$. If $P_0 > 1$, the maximum is reached at the point $P = 1$, and $\max_{0 \le P \le 1} f(P,M) = f(0,M)$ in the case $P_0 < 0$.

Thus we get the following expression for $\bar{f}(M) = \max_{0 \le P \le 1} f(P,M)$:

$$\bar{f}(M) = \begin{cases} c_1 - \frac{M^2}{a_1}, & M \le a_1\sqrt{\frac{c_1-c_2}{a_2-a_1}} \\ \frac{c_1a_2-c_2a_1}{a_2-a_1} - 2M\sqrt{\frac{c_1-c_2}{a_2-a_1}}, & a_1\sqrt{\frac{c_1-c_2}{a_2-a_1}} < M < a_2\sqrt{\frac{c_1-c_2}{a_2-a_1}} \\ c_2 - \frac{M^2}{a_2}, & M \ge a_2\sqrt{\frac{c_1-c_2}{a_2-a_1}} \end{cases}$$

From (8.24) and (8.25) we get: $V(t,x) = t\bar{f}(\frac{\rho(x,G_0)}{t\sqrt{2}})$.

Since condition (N) is fulfilled for the function $V(t,x)$, we get from Theorem 8.1 that the position of the wave front at time t is defined by the equation $\bar{f}(\frac{\rho(x,G_0)}{t\sqrt{2}}) = 0$. Solving this equation, we find that $\rho(x,G_0) = tv$, where v is defined by the formulas (8.23) □

Consider the case where a_2 and c_1 are fixed, and $a_1, c_2 \to 0$. Then Theorem 8.4 gives the following expression for the speed v

$$v = \sqrt{\frac{c_1a_2}{2}} + o(1), \quad a_1, c_2 \downarrow 0.$$

The speeds in separated equations, when $d_1 = d_2 = 0$, will be $\sqrt{2a_1c_1}$ and $\sqrt{2a_2c_2}$. We see that in this case the speed of the front in the coupled system is bigger than in separated equations.

In the case $a_1 = c_2 = 0$, system (8.22) has the form:

$$\text{(8.26)} \quad \begin{cases} \frac{\partial u_1^\varepsilon(t,x)}{\partial t} = \frac{1}{\varepsilon} c_1(u_1^\varepsilon) + d_1 \cdot (u_2^\varepsilon - u_1^\varepsilon), & u_1^\varepsilon(0,x) = g_1(x) \\ \frac{\partial u_2^\varepsilon(t,x)}{\partial t} = \frac{\varepsilon a_2}{2} \Delta u_2^\varepsilon + d_2 \cdot (u_1^\varepsilon - u_2^\varepsilon), & u_2^\varepsilon(0,x) = g_2(x) \end{cases}$$

The system (8.26) is formally excluded from our considerations because of the degeneration of the first equation. But equation (8.5) is fulfilled and only minor changes should be made in the proof to show that

$$\lim_{\varepsilon\downarrow 0} u_k^\varepsilon(t,x) = \begin{cases} 1, & \text{if } \rho(x,G_0) < t\sqrt{\frac{c_1a_2}{2}}, \\ 0, & \text{if } \rho(x,G_0) > t\sqrt{\frac{c_1a_2}{2}}. \end{cases}$$

§9. RDE systems of KPP Type.

The nonlinear term in the KPP equation defines a dynamical system in R^1 which has two equilibrium points: a stable one at the point $u = 1$ and unstable equilibrium at $u = 0$. Interplay between diffusion and this dynamical system leads to the wave front propagation. General RDE systems also describe interaction of mutual transmutations and multiplication of the particles and their diffusion in the space. The transmutations are governed by nonlinear terms in the RDE system.

Let us start with the simplest situation, where the nonlinear terms are independent of x and the differential operators are the same in all equations

$$
(9.1)\qquad \begin{cases} \dfrac{\partial u_k(t,x)}{\partial t} = \varepsilon L\, u_k^{\varepsilon}(t,x) + \dfrac{1}{\varepsilon} f_k(u_1,\dots,u_n), \quad x \in R^r, \quad t > 0, \\ u_k(0,x) = g_k(x) \ge 0. \end{cases}
$$

We assume that $L = \frac{1}{2} \sum_{i,j=1}^{r} a^{ij}(x) \frac{\partial^2}{\partial x^i \partial x^j}$ is an elliptic operator with bounded C^3-coefficients. Let, for simplicity,

$$
f_k(u) = c_{kk}(u) u_k + \sum_{j:\, j \ne k} c_{kj} u_k, \quad k = 1,\dots,n,
$$

where c_{kj}, $k \ne j$, are positive constants. Denote

$$
c_k(u) = c_{kk}(u) + \sum_{j:\, j \ne k} c_{kj}, \quad c_{kk} = c_{kk}(0), \quad c_k = c_k(0).
$$

As in the case of the KPP-equation, we assume that the vector field $f(u) = (f_1(u),\dots,f_n(u))$, $u \in R^n$, has in $R^n_+ = \{u \in R^n : u_1 \ge 0,\dots,u_n \ge 0\}$ two equilibrium points: an unstable one at the point $0 = (0,\dots,0)$ and an asymptotically stable one at a point $a = (a_1,\dots,a_n)$. Assume that all the integral curves in the region $R^n_+ \setminus \{0\}$ do not leave R^n_+ and are attracted to the point a (Figure 11). Moreover, we will assume that for some $\alpha > 0$

$$
f_1(u) > 0,\dots,f_n(u) > 0 \quad \text{for} \quad u \in \{u \in R^n_+,\ u \ne 0,\ \sum_1^n u_i < \alpha\},
$$

and in the domain $B_{\alpha/2} = \{u \in R^n_+ : \sum_1^n u_i > \alpha/2\}$ a convex function $V(u)$

(a Lyapunov function) is defined such that $V(u) > 0$ for $u \in \{B_{\alpha/2} \setminus \{a\}\}$, $V(a) = 0$, and $(\nabla V(u), f(u)) < 0$ for $u \in B_{\alpha/2} \setminus \{a\}$.

We assume finally that for $k = 1,2,\ldots,n$,

(9.2)
$$c_{kk} = c_{kk}(0) = \max\{c_{kk}(u),\ \ u = (u_1,\ldots,u_n),\ \ 0 \le u_i \le a_i,\ \ i = 1,\ldots,n\}.$$

Consider problem (9.1) with the initial conditions

$$u_k(0,x) = g_k(x) \ge 0,\ k = 1,\ldots,n. \tag{9.3}$$

For brevity we assume that the functions g_k are bounded and have common support G_0. As usual, we suppose that $[G_0] = [(G_0)]$.

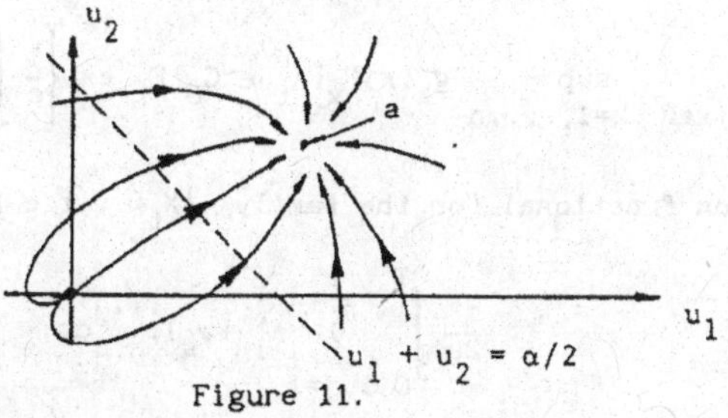

Figure 11.

Denote by λ the eigenvalue of the matrix (c_{ij}) with the largest real part. By the Frobenius theorem such a λ is real, simple and corresponds to the eigenvector with positive components.

Let $d(\cdot,\cdot)$ be the Riemannian metric corresponding to the form $ds^2 = \sum_{i,j=1}^{r} a_{ij}(x)dx^i dx^j$, $(a_{ij}(x)) = (a^{ij}(x))^{-1}$.

<u>Theorem 9.1</u>. Suppose that the above conditions are fulfilled. Then for the solution $(u_1^\varepsilon(t,x),\ldots,u_n^\varepsilon(t,x))$ of problems (9.1) - (9.3)

$$\lim_{\varepsilon\downarrow 0} u_k^\varepsilon(t,x) = \begin{cases} a_k, & \text{if } d(x,G_0) < t\sqrt{2\lambda}, \\ 0, & \text{if } d(x,G_0) > t\sqrt{2\lambda}. \end{cases}$$

<u>Proof</u>. Let ν_t, $t \ge 0$, be the Markov process with a finite number of states $\{1,\ldots,n\}$ such that

$$P\{\nu_{t+\Delta} = j \mid \nu_t = i\} = c_{ij}\cdot\Delta + o(\Delta),\ \ \Delta\downarrow 0,\ \ i \ne j,$$

and let the process $X_t^{\varepsilon,x}$ be defined by the equation

$$dX_t^{\varepsilon,x} = \sqrt{\varepsilon}\,\sigma(X_t^{\varepsilon,x})dW_t, \quad X_0^{\varepsilon,x} = x \in R^r, \quad \sigma(x)\sigma^*(x) = (a^{ij}(x)).$$

As was explained in §1, the functions $u_k^\varepsilon(t,x)$ satisfy the equation

$$(9.4) \qquad u_k^\varepsilon(t,x) = E_{x,k} g_{\nu(t/\varepsilon)}(X_t^\varepsilon)\exp\left\{\frac{1}{\varepsilon}\int_0^t c_{\nu(s/\varepsilon)}(u^\varepsilon(t-s,X_s^\varepsilon))ds\right\}$$

$$k = 1,\dots,n, \quad u = (u_1,\dots,u_n).$$

From (9.4) we have

$$(9.5) \qquad 0 \le u_k^\varepsilon(t,x) \le E_{x,k} g_{\nu(t/\varepsilon)}(X_t^\varepsilon)\exp\left\{\frac{1}{\varepsilon}\int_0^t c_{\nu(s/\varepsilon)}ds\right\}$$

$$\le \sup_{x\in R^r, k=1,\dots,n} g_k(x)P_x\{X_t^\varepsilon \in G_0\}E_k \exp\left\{\frac{1}{\varepsilon}\int_0^t c_{\nu(s/\varepsilon)}ds\right\}.$$

The action functional for the family $\{X_t^\varepsilon\}$, $t \in [0,T]$ is equal to

$$\frac{1}{2\varepsilon}\int_0^T \sum_{i,j=1}^r a_{ij}(\varphi_s)\dot\varphi_s^i\dot\varphi_s^j ds$$

for absolutely continuous $\varphi \in C_{0T}$, and equal to $+\infty$ for the rest of C_{0T}. Therefore, taking into account our assumptions about G_0,

(9.6)

$$\lim_{\varepsilon\downarrow 0} \varepsilon \ln P_x\{X_t^\varepsilon \in G_0\} = -\inf\left\{\frac{1}{2}\int_0^t \sum a_{ij}(\varphi_s)\dot\varphi_s^i\dot\varphi_s^j : \varphi \in C_{0T}, \ \varphi_0 = x, \ \varphi_t \in G_0\right\}.$$

As we proved in Lemma 4.1, the infimum in (9.6) is equal to $\dfrac{d^2(x,G_0)}{2t}$.

To calculate the last factor in the right hand side of (9.5) we can use the following result.

<u>Lemma 9.1</u>. The following equality holds:

$$\lim_{t\to\infty} \frac{1}{T} \ln E_k \exp\left\{\int_0^T c_{\nu(s)}ds\right\} = \lambda.$$

<u>Proof</u>. The family of operators $T_t : (T_t f)(i) = E_i f(\nu(t))e^{\int_0^t c_{\nu(s)}ds}$ is a

semi-group, and the statement of Lemma 9.1 can be written as follows

(9.7) $$\lim_{t\to\infty} \frac{1}{T} \ln(T_t\mathbf{1})(i) = \lambda,$$

where $\mathbf{1}$ is the vector with components equal to one. All components of the eigenvector of the matrix (c_{ij}) corresponding to λ are positive: $e = (e_1,\dots,e_n)$, $\sum_1^n e_k = 1$, $0 < \underline{c} < e_k < \max_k e_k \le 1$. Using the positivity of the semi-group T_t, we conclude that

$$\underline{c}(T_t\mathbf{1})(i) < (T_te)(i) = e^{\lambda t}e_i \le (T_t\mathbf{1})(i), \quad i = 1,\dots,n.$$

Taking the logarithm of this relation and dividing by t

$$\frac{1}{t}\ln \underline{c} + \frac{1}{t}\ln(T_t\mathbf{1})(i) < \frac{1}{t}\ln(T_te)(i) = \lambda + \frac{1}{t}\ln e_i \le \frac{1}{t}\ln(T_t\mathbf{1})(i).$$

letting t tend to ∞ in this chain of inequalities, we obtain (9.7).

From (9.5), (9.6) and Lemma 9.1, we conclude that

$$\overline{\lim_{\varepsilon\downarrow 0}}\, \varepsilon \ln u_k^\varepsilon(t,x) \le \lambda t - \frac{d^2(x,G_0)}{2t},$$

which implies that

$$\lim_{\varepsilon\downarrow 0} u_k^\varepsilon(t,x) = 0 \quad \text{if} \quad d(x,G_0) > t\sqrt{2\lambda}. \qquad \square$$

The proof that $\lim_{\varepsilon\downarrow 0} u_k^\varepsilon(t,x) = a_k$ for $d(x,G_0) < t\sqrt{2\lambda}$ can be divided into two parts. First, one proves that $u^\varepsilon(t,x) = (u_1^\varepsilon(t,x),\dots,u_n^\varepsilon(t,x)) \in B_{\alpha/2}$ for sufficiently small ε and $d(x,G_0) < t\sqrt{2\lambda}$. Then one checks that the solution of the boundary problem for equation (9.1) in the domain $\mathcal{E} = \{(s,x) : s > 0,\ x \in R^r,\ d(x,G_0) < s\sqrt{2\lambda}\}$ with boundary values lying in $B_{\alpha/2}$ for any $(s,x) \in \partial\mathcal{E}$, tends to the equilibrium point $a = (a_1,\dots,a_n) \in R_+^n$ as $\varepsilon\downarrow 0$. The proof of the first of these statements is similar to that of the final part of Theorem 3.1, and we omit it.

To prove the second statement, it is sufficient to verify that $V(u^\varepsilon(t,x)) \to 0$ as $\varepsilon\downarrow 0$ if $(t,x) \in \mathcal{E}$, where $V(u)$ is the Lyapunov function for the point $a \in R_+^n$. Denote $\tau = \tau^\varepsilon$ the exit time of the "heat" process $(t-s, X_s^\varepsilon)$ from the region $\mathcal{E}$: $\tau^\varepsilon = \inf\{s : (t-s, X_s^\varepsilon) \notin \mathcal{E}\}$. One can deduce from (9.4) and the strong Markov property of the process $(\nu_t^\varepsilon, X_t^\varepsilon)$ that

$$u_k^\varepsilon(t,x) = E_{x,k}\, u^\varepsilon_{\nu(\tau/\varepsilon)}(t-\tau, X_\tau^\varepsilon)\exp\left\{\frac{1}{\varepsilon}\int_0^\tau c_{\nu(s/\varepsilon)}(u^\varepsilon(t-s, X_s^\varepsilon))ds\right\},$$

$$(t,x) \in \mathcal{E}, \quad k = 1,\dots,n.$$

Since we assume that the Lyapunov function is convex, taking into account the independence ν_t^ε and X_t^ε, we have

$$(9.8) \qquad V(u^\varepsilon(t,x)) = V(E_{x,1}\zeta_0^\tau,\dots,E_{x,n}\zeta_0^\tau) \le E_x V(E_1\zeta_0^\tau,\dots,E_n\zeta_0^\tau),$$

where

$$\zeta_0^\tau = u_{\nu(\tau/\varepsilon)}(t-\tau, X_\tau^\varepsilon)\exp\left\{\frac{1}{\varepsilon}\int_0^\tau c_{\nu(s/\varepsilon)}(u^\varepsilon(t-s, X_s^\varepsilon))ds\right\}.$$

We denote by $E_k\zeta_0^\tau$ the expectation of ζ_0^τ for a fixed trajectory X^ε under the assumption that $\nu_0 = k$, E_x being the expectation with respect to the measure which corresponds to the process X_t^ε, $X_0^\varepsilon = x$, $E_{xk}\zeta_0^\tau = E_x(E_k\zeta_0^\tau)$.

Note that the mapping

$$M_t : z = (z_1,\dots,z_n) \longrightarrow M_t(z) = (z_1(t),\dots,z_n(t))$$

$$z_k(t) = E_k z_{\nu(t)} \exp\left\{\int_0^t c_{\nu(s)}(z(t-s))ds\right\}$$

is a shift along the trajectories of the dynamical system $\dot{z}_t = f(z_t)$, $f(z) = (f_1(z),\dots,f_n(z))$, $f_k(z) = \sum_{i=1}^n c_{ki}(z)z_i$. Taking into account that $V(z)$ is a Lyapunov function for this system, one can deduce from (9.7) that $V(u^\varepsilon(t,x)) \longrightarrow 0$ as $\varepsilon \downarrow 0$.

This implies that $\lim_{\varepsilon\downarrow 0} u^\varepsilon(t,x) = a$ provided $d(x,G_0) < t\sqrt{2\lambda}$. □

Thus in the case of identical differential operators in all equations of the RDE system, the wave front propagates according to the Huygens principle with constant velocity $\sqrt{2\lambda}$ in the Riemannian metric corresponding to the diffusion matrix. How should the Huygens principle be formulated in the case of different operators? To understand the situation in this case, let us consider coupled RDE's with diffusion coefficients and nonlinear terms which are independent of x:

$$\frac{\partial u_k^\varepsilon(t,x)}{\partial t} = \frac{\varepsilon}{2}\sum_{i,j=1}^{r} a_k^{ij}\frac{\partial^2 u^\varepsilon}{\partial x^i \partial x^j} + \frac{1}{\varepsilon}\Big[f_k(u_k^\varepsilon) + \sum_{j=1}^{n} d_{kj}(u_j^\varepsilon - u_k^\varepsilon)\Big],$$

(9.9)

$$t > 0,\quad x \in R^r,\quad k = 1,\dots,n;\quad u_k^\varepsilon(0,x) = g_k(x) \ge 0.$$

Let, for brevity, $n = 2$, and the initial functions $g_1(x)$ and $g_2(x)$ have common support G_0, $[G_0] = [(G_0)]$. We assume that $f_k(u) = c_k(u)u \in \mathcal{F}_1$, $d_{kj} > 0$, and let, for simplicity, $c_k(0) = c = \text{const.}$ independent of k. Note that the difference between (9.9) and weakly coupled equations considered in §8 is in the large factor ε^{-1} in the term responsible for the jumps. In Figure 12, the unit Riemannian spheres corresponding to matrices (a_1^{ij}) and (a_2^{ij}) are drawn. In the case of weakly coupled systems, the Huygens principle is formulated in the metric where the unit sphere is the convex envelope of these two Riemannian spheres. In the case of system (9.9), the wave front propagation will be slower. To calculate it, let us consider the Markov process $(X_t^\varepsilon, \nu(t/\varepsilon))$ in $R^r\times\{1,2\}$, corresponding to system (9.9). Then we can write

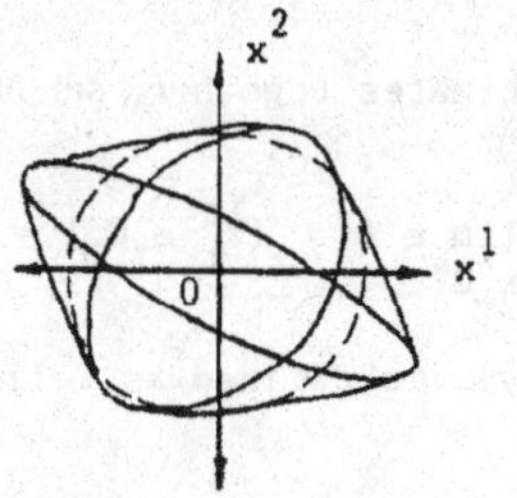

Figure 12.

(9.10) $$u_k^\varepsilon(t,x) = E_{x,k}g_{\nu(t/\varepsilon)}(X_t^\varepsilon)\exp\left\{\left[\frac{1}{\varepsilon}\right]\int_0^\tau c_{\nu(s/\varepsilon)}(u^\varepsilon_{\nu(s/\varepsilon}(t-s,X_s^\varepsilon))ds\right\}$$

$$\le \max_{\substack{x\in R^r\\ k=1,2}} g_k(x)\exp\{\frac{ct}{\varepsilon}\}P_{x,k}\{X_t^\varepsilon \in G_0\}..$$

Denote $\sigma_k = (a_k^{ij})^{1/2}$. Then X_t^ε satisfies the equation

(9.11) $$dX_t^\varepsilon = \sqrt{\varepsilon}\,\sigma_{\nu(t/\delta)}dW_t,\quad X_0^\varepsilon = x.$$

Let $T_\varepsilon(t)$ be the time which $\nu(s/\varepsilon)$ was equal to 1 during the time interval

$[0,t]$. From (9.11) we conclude that X_t^ε has the same distribution as

$$x + \sqrt{\varepsilon}\,\sigma_1 W_{T_\varepsilon(t)} + \sqrt{\varepsilon}\,\sigma_2 (W_t - W_{T_\varepsilon(t)}).$$

Thus, under condition $T_\varepsilon(t) = T$, the vector X_t^ε has Gaussian distribution with mean zero and covariance matrix $\varepsilon[T_\varepsilon(a_1^{ij}) + (t-T_\varepsilon)(a_2^{ij})] =$ $t\varepsilon\left[\frac{T_\varepsilon}{t}(a_1^{ij}) + (1-\frac{T_\varepsilon}{t})(a_2^{ij})\right]$.

The ratio T_ε/t tends to the invariant measure z_0 of the state 1 for the Markov proces $\nu_{t/\varepsilon}$ as $t\varepsilon^{-1} \to \infty$. Large deviations of T_ε/t from z_0 are described by a large deviation principle

$$\lim_{\varepsilon\downarrow 0} \frac{\varepsilon}{t} \ln P\left\{\left|\frac{T_\varepsilon(t)}{t} - z\right| < \delta\right\} = L(z).$$

The action function $L(z)$ is the Legendre transformation of the first eigenvalue $\lambda(\beta)$ of the matrix

$$\begin{bmatrix} -d_{12}+\beta & d_{12} \\ d_{21} & -d_{21} \end{bmatrix}; \quad L(z) = \sup_{\beta\in R^1} (\bar{\beta} z - \lambda(\beta)), \quad L(z_0) = 0.$$

Gathering all these estimates together, we have that

$$\lim_{\varepsilon\downarrow 0} \varepsilon \ln P_x\{X_t^\varepsilon \in G_0\} = -\frac{\rho^2(x,G_0)}{2t}, \tag{9.12}$$

where the metric $\rho(x,y)$ is defined as follows:

$$\rho(x,y) = \rho(0,x-y), \quad \rho^2(0,x) = \min_{0\le z\le 1} [((za_1+(1-z)a_2)^{-1}x,x) + L(z)].$$

From (9.10) and (9.12) we conclude that

$$\overline{\lim}\, \varepsilon \ln u_k^\varepsilon(t,x) \le ct - \frac{\rho^2(x,G_0)}{2t},$$

and thus $\lim_{\varepsilon\downarrow 0} u_k^\varepsilon(t,x) = 0$ if $\rho(x,G_0) > t\sqrt{2c}$. One can prove also that $\lim_{\varepsilon\downarrow 0} u_k^\varepsilon(t,x) = 1$ if $\rho(x,G_0) < t\sqrt{2c}$. This means that the wave front propagates with constant speed $\sqrt{2c}$ in the Finsler metric with unit sphere (with center at the origin 0)

$$S_1 = \{x : \rho(0,x) = \min_{0\le z\le 1} [((za_1+(1-z)a_2)^{-1}x,x) + L(z)] = 1\}.$$

In the case of weakly coupled equations $L(z) \equiv 0$, this sphere is the convex envelope of the unit spheres corresponding to the diffusion matrices a_1 and a_2. In the case under consideration, the sphere will be a subset of the convex envelope (see Figure 12).

To consider the general case with diffusion coefficients and nonlinear term depending on x and k, we should calculate the action functional for the family $(X_t^\varepsilon, \int_0^t h(\nu_s^\varepsilon)ds)$, where $(X_t^\varepsilon, \nu_t^\varepsilon)$ is the process corresponding to the system, and $h(k)$ is a function on $\{1,\dots,n\}$. Then we can use our approach. It was done in [11].

In conclusion of this section, I want to underline that the asymptotic behavior of the solutions of RDE systems is much richer than for the solutions of single equations. Besides the wave fronts, we can have as the asymptotic behavior running impulses, solutions which are periodic in space and in time, spiral waves and even chaotic behavior.

§10. Random Perturbations of RDE's: Perturbed Boundary Conditions.

If we have an RDE system with good enough coefficients we can solve the Cauchy problem or initial-boundary problem for any continuous bounded initial function g. The solution defines the trajectory of a semi-flow in the space of continuous functions, corresponding to this RDE system and starting at the point g. If the equations or the boundary conditions are subjected to random perturbations the solution will be a random process in the space of functions.

As in the case of finite-dimensional dynamical systems, we may be interested in the deviations of this random process from the semi-flow. It can be the results of law-of-large-numbers type or of central-limit-theorem type or results on large deviations.

We start with a problem where the boundary conditions are subjected to perturbations [13]. Consider the following problem

$$(10.1) \qquad \frac{\partial u^\varepsilon(t,x)}{\partial t} = \frac{D}{2}\frac{\partial^2 u^\varepsilon(t,x)}{\partial x^2} + f(x,u^\varepsilon), \quad |x| < 1, \quad t > 0,$$

$$\left.\frac{\partial u^\varepsilon(t,x)}{\partial x}\right|_{x=\pm 1} = \pm\xi_\pm(t/\varepsilon), \quad u^\varepsilon(0,x) = g(x).$$

We assume that the function f is Lipschitz continuous, has continuous derivative in u, and $|f(x,u)| < A + L|u|$ for some constants A and L. The process $(\xi_+(t), \xi_-(t))$ we suppose to be stationary with mean zero and correlation matrix

$$\begin{pmatrix} K_{++}(\tau) & K_{+-}(\tau) \\ K_{-+}(\tau) & K_{--}(\tau) \end{pmatrix}.$$

We assume that this process satisfies some mixing properties, at least that

(10.2)
$$\lim_{T\to\infty} \frac{1}{T}\int_t^{t+T} \xi_\pm(s)ds = 0$$

for any $t \ge 0$, the limit being in probability. This is true if $K_{++}(t)$, $K_{--}(t) \to 0$ when $t \to \infty$. Later on we shall introduce stronger assumptions on the mixing properties.

If (10.2) is fulfilled, one can consider problem (10.1) as a small, in the average sense, perturbation of the problem

(10.3)
$$\frac{\partial u(t,x)}{\partial t} = \frac{D}{2}\frac{\partial^2 u(t,x)}{\partial x^2} + f(x,u), \quad t > 0, \quad |x| < 1,$$
$$\left.\frac{\partial u(t,x)}{\partial x}\right|_{|x|=1} = 0, \quad u(0,x) = g(x).$$

We assume for simplicity that there exists a constant $C < \infty$ such that $P\{|\xi_\pm(t)| \le C\} = 1$. First of all we formulate our main results, and then we shall say several words about the proofs. Detailed proofs can be found in [13].

<u>Theorem 10.1</u>. For any $T > 0$, $\delta > 0$,

$$\lim_{\varepsilon\downarrow 0} P\left\{\sup_{0\le t\le T} |u^\varepsilon(t,x) - u(t,x)| > \delta\right\} = 0.$$

Now we want to consider the difference $u^\varepsilon(t,x) - u(t,x)$. It tends to zero when $\varepsilon \downarrow 0$ uniformly in $0 \le t \le T$, $|x| \le 1$, but we can expect that after deviding by $\sqrt{\varepsilon}$ it will be asymptotically Gaussian. We have the following initial boundary problem for $v^\varepsilon(t,x) = 1/\sqrt{\varepsilon}\,(u^\varepsilon(t,x)-u(t,x))$

(10.4)
$$\frac{\partial v^{\varepsilon}}{\partial t} = \frac{D}{2}\frac{\partial^2 v^{\varepsilon}}{\partial x^2} + f_2'(x,\tilde{u}^{\varepsilon}(t,x))v^{\varepsilon}, \quad t > 0, \quad |x| < 1$$
$$v^{\varepsilon}(0,x) = 0, \quad \frac{\partial v^{\varepsilon}}{\partial x}(t,\pm 1) = \pm\frac{1}{\sqrt{\varepsilon}}\,\xi_{\pm}(t/\varepsilon).$$

Here $\tilde{u}^{\varepsilon}(t,x) \to u(t,x)$ when $\varepsilon \downarrow 0$, $f_2'(x,u) = \frac{\partial f}{\partial u}(x,u)$. Denote $\zeta_{\pm}^{\varepsilon}(t) = \int_0^t \xi_{\pm}(s/\varepsilon)ds$, $\zeta^{\varepsilon}(t) = (\zeta_{+}^{\varepsilon}(t),\zeta_{-}^{\varepsilon}(t))$. It is well known that under some assumptions about the mixing properties of the process $\xi(t) = (\xi_{+}(t),\xi_{-}(t))$ the process $1/\sqrt{\varepsilon}\,\zeta_t^{\varepsilon}$ converges weakly in the space of continuous functions C_{0T} to a Gaussian process, namely, the Brownian motion $(W_{+}(t),W_{-}(t))$ with covariance matrix $(a_{\lambda\mu})$, where

$$a_{\lambda\mu} = \int_{-\infty}^{\infty} K_{\lambda\mu}(\tau)d\tau, \; K_{\lambda\mu}(\tau) = E\xi_{\lambda}(t)\xi_{\mu}(t+\tau), \quad \lambda,\mu \in \{+,-\}.$$

Taking all these things into account we can expect that $v^{\varepsilon}(t,x)$ converges in one or another sense to the solution $v(t,x)$ of the following linear initial boundary problem

(10.5)
$$\frac{\partial v}{\partial t}(t,x) = \frac{D}{2}\frac{\partial^2 v(t,x)}{\partial x^2} + f_2'(x,u(t,x))v(t,x), \quad t > 0, \quad |x| < 1,$$
$$\frac{\partial v}{\partial x}(t,\pm 1) = \pm\dot{W}_{\pm}(t), \quad v(0,x) = 0.$$

We will see that it is actually true, but we should overcome several obstacles.

First of all we should introduce a generalized solution of the problem (10.5). Denote by $p(\tau,x,y)$ the transition density for the process corresponding to the operator $\frac{D}{2}\frac{d^2}{dx^2}$ in the interval $[-1,1]$, and having reflection at the ends of the interval:

$$p(\tau,x,y) = \frac{1}{\sqrt{2\pi D\tau}}\left[\sum_{k=-\infty}^{\infty} e^{-\frac{(y-x-4k)^2}{2D\tau}} + \sum_{k=-\infty}^{\infty} e^{-\frac{(y+x+4k+2)^2}{2D\tau}}\right], \tag{10.6}$$

$$z^0(t,x) = \sum_{+,-}\int_0^t p(t-s,x,\pm 1)dW_{\pm}(s), \quad |x| < 1, \quad t \ge 0. \tag{10.7}$$

A function $v(t,x)$ we call the generalized solution of problem (10.5) if it satisfies the linear equation

$$v(t,x) = z^0(t,x) + \int_0^t ds \int_{-1}^{1} dy\, p(t-s,x,y) f_2'(y,u(s,y)) v(s,y). \tag{10.8}$$

One can see from (10.7) that $z^0(t,x)$ is continuous in $t \ge 0$, $|x| < 1$, with probability 1. Consider the function

$$h(t,x) = \int_0^t z^0(s,x)ds.$$

We deduce from (10.7) by integrating by parts that

$$h(t,x) = \sum_{+,-} \int_0^t dt_1 \int_0^{t_1} p_1'(t_1-s,x,\pm 1) W_\pm(s)ds,$$

where $p_1'(t,x,\pm 1) = \dfrac{\partial p(t,x,\pm 1)}{\partial t}$. For any $\delta > 0$ we can write

$$\begin{aligned} h(t,x) &= \sum_{+,-} \int_0^{t-\delta} dt_1 \int_0^{t_1} p_1'(t_1-s,x,\pm 1) W_\pm(s)ds \\ &+ \sum_{+,-} \int_{t-\delta}^{t} dt_1 \int_0^{t_1} p_1'(t_1-s,x,\pm 1) W_\pm(s)ds. \end{aligned} \tag{10.9}$$

One can see from (10.6) that $p_1'(\tau,x,y) < \dfrac{\text{const}}{\tau^{3/2}}$. Using the last inequality we can bound the second term in (10.9):

$$\begin{aligned} &\left|\sum_{+,-} \int_{t-\delta}^{t} dt_1 \int_0^{t_1} p_1'(t_1-s,x,\pm 1) W_\pm(s)ds\right| \\ &\le \sum_{+,-} \int_{t-\delta}^{t} dt_1 \int_0^{t_1} \frac{C_1}{(t_1-s)^{3/2}} ds \cdot \max_{0 \le s \le t} |W_\pm(s)| < C_2\sqrt{\delta} \sum_{+,-} \max_{0 \le s \le t} |W_\pm(s)|. \end{aligned}$$

We have from this that for almost any trajectory $(W_+(s), W_-(s))$, $0 \le s \le t$, the second term in (10.9) can be made less than any $\lambda > 0$ if δ is small enough. For any fixed $\delta > 0$ the first term in (10.9) is continuous in x.

This implies the existence of $\lim_{x\to 1} h(t,x)$ and $\lim_{x\to -1} h(t,x)$ uniformly in $t \in [0,T]$ It is not difficult to check using the explicit formula for $p(t,x,y)$, that for any $\gamma > 0$

$$\lim_{x\to\pm 1} (1-x^2)^{\gamma} Z^0(t,x) = 0 \quad \text{with probability } 1.$$

Let $a(x)$ be a continuous function in $[-1,1]$, $a(\pm 1) = 0$, $a(x) > 0$ for $|x| < 1$, $\int_{-1}^{1} a^{-1}(x)dx < \infty$, $\lim_{x\to\pm 1} a(x)(1-x^2)^{\gamma} = 0$ for some $\gamma < 0$. The typical example of such a function is $a(x) = (1-x^2)^{\alpha}$, $0 < \alpha < 1$.

Denote by $C_a = C_a([0,T]\times(-1,1))$ the space of all continuous functions $u(t,x)$, $t \in [0,T]$, $|x| < 1$, such that $\lim_{x\to\pm 1} a(x)u(t,x) = 0$ uniformly in $t \in [0,T]$. The norm in this space is defined as follows

$$\|u\|_a = \sup_{\substack{0\le t\le T \\ |x|<1}} a(x)|u(t,x)|.$$

We denote by $\hat{C}_a$ the subspace of C_a, consisting of $u(t,x)$ such that $h(t,x) = \int_0^t u(s,x)ds$ has uniform limit in $t \in [0,T]$ when $x \to 1$ and when $x \to -1$, provided with the norm $|||u||| = \|u\|_a + \sup_{\substack{1\le t\le T \\ |x|<1}} \left|\int_0^t u(s,x)ds\right|$.

As it was explained above, the function $Z^0(t,x)$ defined by (10.7) with probability 1 belongs to the space $\hat{C}_a$.

Taking into account that $Z^0 \in \hat{C}_a$, one can prove that there exists a generalized solution $v(t,x)$ of the problem (10.5) belonging to $\hat{C}_a$. Such a solution is unique. □

So the field $v(t,x)$, which as we expect is the limit field for the normalized differences $v^{\varepsilon}(t,x)$, does not belong to the space $C_{[0,T]\times[-1,1]}$. It belongs only to $\hat{C}_a$, and we shall prove that v^{ε} weakly converges to v in the space $\hat{C}_a$.

The solution $u^{\varepsilon}(t,x)$ of the problem (10.1) can be obtained from $\zeta^{\varepsilon}(t) = (\zeta_+^{\varepsilon}(t),\zeta_-^{\varepsilon}(t))$, $\zeta_{\pm}^{\varepsilon}(t) = \int_0^t \xi_{\pm}(s/\varepsilon)ds$, as the product of two mappings: the linear mapping $\zeta \to w$ and the mapping $w \to u$. The mappings can be expressed in the integral form

$$(10.10)\qquad \zeta \longrightarrow w(t,x) = \int_{-1}^{1} p(t,x,y)g(y)dy + \sum_{+,-}\int_0^t p(t-s,x,\pm 1)d\zeta_\pm(s),$$

where $p(t,x,y)$ is defined by (10.6) and

$$(10.11)\qquad w \longrightarrow u(t,x) = w(t,x) + \int_0^t ds \int_{-1}^{1} dy\, p(t-s,x,y)f(y,u(s,y)).$$

One can prove that the mapping $w \longrightarrow u - w : C_a \longrightarrow C = C([0,T]\times[-1,1])$ is Lipschitz continuous and Fréchet differentiable as well as the mappings $w \longrightarrow u : C_a \longrightarrow C_a$ and $w \longrightarrow u : C \longrightarrow C$.

Let $\mathbb{C} = \mathbb{C}_{0T}$ be the space of all pairs $\zeta(t) = (\zeta_+(t), \zeta_-(t))$ of continuous functions on $[0,T]$, $\zeta_\pm(0) = 0$. Denoted $C^\gamma = C^\gamma_{0T}$ the subspace of $\mathbb{C}$ consisting of Hölder continuous functions with the norm

$$\|\zeta\|_\gamma = \max_{+,-} \sup_{0\le t_1 < t_2 \le T} \frac{|\zeta_\pm(t_1) - \zeta_\pm(t_2)|}{(t_2-t_1)^\gamma}, \qquad 0 < \gamma < 1.$$

Let $a(x) = (1-x^2)^\alpha$, $\alpha > 1 - 2\gamma$, $0 < \gamma < 1/2$. Then one can check that the mapping $\zeta \longrightarrow w$ defined by (10.10) is continuous as a mapping from C^γ to C_a. The mapping $\zeta \longrightarrow \int_0^t w(s,x)ds$ is continuous as a mapping $\mathbb{C}_{0T} \longrightarrow C_{[0,T]\times[-1,1]}$.

We have the linear continuous mapping $\zeta^\varepsilon \longrightarrow w^\varepsilon : C^\gamma \longrightarrow C_a$ and the Fréchet differentiable mapping $w^\varepsilon \longrightarrow u^\varepsilon : C_a \longrightarrow C_a$. Thus the composition of these two mappings of $\zeta^\varepsilon \longrightarrow u^\varepsilon$ is a Fréchet differentiable mapping from C^γ to C_a. Then, to prove the weak convergence of $1/\sqrt{\varepsilon}\,(u^\varepsilon - u)$ to a Gaussian field we should first of all prove that $\zeta^\varepsilon = (\zeta^\varepsilon_+, \zeta^\varepsilon_-)$ is asymptotically Gaussian. Recall that

$$\zeta^\varepsilon_\pm(t) = \int_0^t \xi_\pm(s/\varepsilon)ds,$$

where $\xi_\pm(s)$ are stationary processes, $E\xi_\pm(s) \equiv 0$, and $P\{|\xi_\pm| < C\} = 1$ for some non-random $C < \infty$. There are some results about asymptotic normality of such processes ζ^ε.

Let $\alpha^*(\tau)$ be the strong mixing coefficient for the process $(\xi_+(s), \xi_-(s)) = \xi(s)$:

$$\alpha^*(\tau) = \sup|E\xi\eta - E\xi E\eta|,$$

where supremum is calculated over all random variables ξ, η such that $|\xi| \le 1$, $|\eta| \le 1$, and ξ is measurable with respect to the σ-field $\mathcal{F}_{\le t}$ generated by the process ξ_s for $s \le t$, and η is measurable with respect to the σ-field $\mathcal{F}_{\ge t+\tau}$ generated by ξ_s for $s \ge t+\tau$.

The weak convergence of $1/\sqrt{\varepsilon}\ \zeta_t^\varepsilon$ in the space of continuous functions on $[0,T]$ was proved under the condition that the strong mixing coefficient $\alpha^*(\tau)$ decreases fast enough. But we need a stronger statement: $1/\sqrt{\varepsilon}\ \zeta_t^\varepsilon$ converges weakly in the space C^γ with Hölder norm. It turns out that this slightly stronger statement can be proven by a slight modification of the proof of the standard result.

Theorem 10.2. Let $\int_0^\infty \tau^{k-1}\alpha^*(\tau)d\tau < \infty$ for some $k > 1$. Then for any $\gamma \in (0, \frac{k-1}{2k})$, $T > 0$, the family of processes $1/\sqrt{\varepsilon}\ \zeta_t^\varepsilon$ converges weakly in the space C_{0T}^γ to the two-dimensional Brownian motion with covariance matrix $(a_{\lambda\mu})$, where

$$a_{\lambda\mu} = \int_{-\infty}^{\infty} K_{\lambda\mu}(\tau)d\tau, \quad K_{\lambda\mu}(\tau) = E\xi_\lambda(s)\xi_\mu(s+\tau), \quad \lambda,\mu \in (+,-).$$

Theorem 10.2 and the smoothness of the mapping $\zeta \longrightarrow u$ yield the following result.

Theorem 10.3. Suppose that the process $\xi(t) = (\xi_+(t), \xi_-(t))$ has a mixing coefficient $\alpha^*(\tau)$ such that $\int_0^\infty \tau^{k-1}\alpha^*(\tau)d\tau < \infty$ for some $k > 1$. Let $\alpha \in (\frac{1}{k}, 1)$ and $a(x) = (1-x^2)^\alpha$. Then $v^\varepsilon(t,x) = 1/\sqrt{\varepsilon}\ (u^\varepsilon(t,x) - u(t,x))$ converges when $\varepsilon\downarrow 0$ weakly in the space $\hat{C}_a$ to the Gaussian random field $v(t,x)$, $0 \le t \le T$, $|x| < 1$, which is the generalized solution of problem (10.5).

The deviations of order $\varepsilon^{1/2}$, $\varepsilon\downarrow 0$, of $u^\varepsilon(t,x)$ from $u(t,x)$ are described by Theorem 10.3. Now we turn to the deviations of order ε^κ, $0 < \kappa < \frac{1}{2}$. The probabilities of such deviations tend to zero as $\varepsilon\downarrow 0$. We formulate a result on the logarithmic asymptotics of these probabilities. Such asymptotics are interesting, for example, when we study the exit problem from a neighborhood of size of order ε^κ, $0 < \kappa < \frac{1}{2}$, of a stable equilibrium of equation (10.3).

We say that the family η^ε of random processes with trajectories in the space C^{γ} is $\varepsilon^{-\mu}$-exponentially bounded in C^{γ} if for any $c > 0$ there exists $K > 0$ such that

$$P\{\|\eta^{\varepsilon}\|_{\gamma} \ge K\} \le \exp\{-\frac{c}{\varepsilon^{\mu}}\}.$$

Introduce the functional $S_{OT}(\varphi)$ on C_{OT}

$$S_{OT}(\varphi) = \begin{cases} \frac{1}{2}\int_0^T \sum_{\lambda,\mu} a^{\lambda,\mu}\, \dot{\varphi}_{\lambda}(s)\dot{\varphi}_{\mu}(s)ds, & \text{for } \varphi_+,\varphi_- \text{ absolutely continuous;} \\ +\infty & \text{for the rest of } C_{OT}. \end{cases}$$

Here $(a^{\lambda\mu}) = (a_{\lambda\mu})^{-1}$, $a_{\lambda\mu} = \int_{-\infty}^{\infty} K_{\lambda\mu}(\tau)d\tau$, $K_{\lambda\mu}(\tau) = E\xi_{\lambda}(t)\xi_{\mu}(t+\tau)$, $\lambda,\mu \in \{+,-\}$. We assume that the matrix $(a_{\lambda\mu})$ is not degenerate.

Theorem 10.4. Assume that the functional $\varepsilon^{2\kappa-1}S_{OT}(\varphi)$ is the action functional for the family $\eta^{\varepsilon} = \varepsilon^{-\kappa}\zeta^{\varepsilon} = \varepsilon^{-\kappa}\int_0^t \xi(s/\varepsilon)ds$ in the space C_{OT} when $\varepsilon\downarrow 0$, and that the family η^{ε} is $\varepsilon^{2\kappa-1}$-exponentially bounded in C^{γ_0} for some $\gamma_0 \in (0,\frac{1}{2})$. Then the family of random fields

$$v_{\kappa}^{\varepsilon}(t,x) = \varepsilon^{-\kappa}(u^{\varepsilon}(t,x)-u(t,x)), \quad \varepsilon\downarrow 0,$$

has the action functional $\varepsilon^{2\kappa-1}\cdot S^{\kappa}(g)$ in the space $\hat{C}_a$ for $a(x) = (1-x^2)^{\alpha}$, $1 > \alpha > 1-2\gamma_0$, where

$$S^{\kappa}(g) = \begin{cases} S_{OT}(\frac{\partial g}{\partial x}(s,1),-\frac{\partial g}{\partial x}(s,-1)), & \text{if } \frac{\partial g}{\partial t} = \frac{D}{2}\frac{\partial^2 g}{\partial x^2} \\ \quad + f_u'(x,u(t,x))g(t,x) & \text{for } 0 < t \le T,\ |x| < 1 \\ \quad \text{and } \frac{\partial g}{\partial x}(s,\pm 1) & \text{are absolutely continuous;} \\ +\infty, & \text{for the rest of } \hat{C}_a \end{cases}$$

The following lemma gives sufficient conditions for exponential boundedness.

Lemma 10.1. Assume that there exist C_1, C_2 such that

$$E\exp\left\{z\varepsilon^{-1/2}\int_t^{t+\tau}\xi_{\pm}(s/\varepsilon)ds\right\} < \exp\left\{\frac{C_2 z^2\tau}{2}\right\}$$

for $|z| \le \frac{C_2}{\sqrt{\varepsilon}}$, $\tau > \varepsilon$. Then the family $\eta_t^\varepsilon = \varepsilon^{-\kappa}\int_0^t \xi_{s/\varepsilon} ds$, $\varepsilon\downarrow 0$ is $\varepsilon^{2\kappa-1}$-exponentially bounded in the norm C^γ for any $\gamma \in (0,\frac{1}{2})$.

We shall consider an example later in this section. Now let us formulate a result concerning the deviations of $u^\varepsilon(t,x)$ from $u(t,x)$ of order 1 as $\varepsilon\downarrow 0$.

Theorem 10.5. Assume that the family $\zeta_t^\varepsilon = \left[\int_0^t \xi_+(s/\varepsilon)ds, \int_0^t \xi_-(s/\varepsilon)ds\right]$, $\varepsilon\downarrow 0$, in the space $\mathbb{C}_{0T}$ has the action functional $\varepsilon^{-1}S^\zeta(\varphi_+,\varphi_-)$. The action functional for the field $u^\varepsilon(t,x)$ in the space $C_{[0,T]\times[-1,1]}$ when $\varepsilon\downarrow 0$ is equal to $\varepsilon^{-1}S^u(g)$, $g \in C_{[0,T]\times[-1,1]}$, where

$$S^u(g) = \begin{cases} S^\zeta(\frac{\partial g}{\partial x}(s,1), -\frac{\partial g}{\partial x}(s,-1)), \quad \text{if } \frac{\partial g}{\partial t} = \frac{D}{2}\frac{\partial^2 g}{\partial x^2} + f(x,g) \\ \quad \text{for } 0 < t \le T,\ |x| < 1, \text{ and the functions} \\ \quad \frac{\partial g}{\partial x}(t,\pm 1) \text{ are absolutely continuous;} \\ +\infty, \quad \text{for the rest of } C_{[0,T]\times[-1,1]}. \end{cases}$$

Example. Let $\xi_t = (\xi_+(t),\xi_-(t))$ be the two-dimensional diffusion process in a domain $D \subset R^2$, corresponding to a second order elliptic operator L with reflection in the co-normal on the boundary ∂D of D. We assume that the coefficients of the operator L are smooth enough and that the domain D is bounded and has smooth boundary. Denote by $p(t,x^1,x^2)$, $x^i = (x_+^i, x_-^i)$, the transition density, and by $m(x)$, $x \in D \cup \partial D$, the invariant density of the process ξ_t. We take $m(x)$ as the initial density of ξ_t and then the process ξ_t is stationary. Assume that $E\xi_t = \int xm(x)dx = 0$. The correlation matrix is given as follows:

$$K_{\lambda\mu}(\tau) = E\xi_\lambda(t)\xi_\mu(t+\tau) = \int_D\int_D x_\lambda^1 x_\mu^2 p(\tau,x^1,x^2)m(x^1)dx^1dx^2, \quad \lambda,\mu \in \{+,-\}.$$

It is easy to check that the strong mixing coefficient $\alpha^*(\tau)$ for the process ξ_t decreases exponentially fast and thus

$$\int_0^\infty \tau^{k-1}\alpha^*(\tau)d\tau < \infty$$

for any $k > 1$. Then we conclude from Theorem 10.1 that the solution $u^\varepsilon(t,x)$

of problem (10.1) with $\xi_\pm(t)$ described above converges in $C_{[0,T]\times[-1,1]}$ in probability to the solution $u(t,x)$ of problem (10.3) when $\varepsilon\downarrow 0$.

The normalized difference $v^\varepsilon(t,x) = 1/\sqrt{\varepsilon}\,(u^\varepsilon(t,x)-u(t,x))$ converges weakly in the space $\hat{C}_a$ with $a(x) = (1-x^2)^\alpha$ for any $\alpha \in (0,1)$ to the solution of problem (10.5), where $a_{\lambda\mu} = 2\int_0^\infty K_{\lambda\mu}(\tau)d\tau$.

To describe the large deviations for the field $u^\varepsilon(t,x)$ we need first of all to recall results on large deviations for the family $\zeta_t^\varepsilon = \int_0^t \xi_{s/\varepsilon}ds$, $\varepsilon\downarrow 0$ (see §2 or [12], Ch. 7]). Consider the eigenvalue problem

$$L\varphi(x) + (\beta,x)\varphi(x) = \lambda\varphi(x), \quad x \in D, \quad \frac{\partial\varphi}{\partial n}\Big|_{\partial D} = 0.$$

Here $\beta = (\beta_1,\beta_2)$ is a parameter and $n = n(x)$ is the co-normal corresponding to the operator L. Let $\lambda = \lambda(\beta)$ be the eigenvalue corresponding to the positive eigenfunction. Such an eigenvalue is simple, real and continuously differentiable in β (we assume that the coefficients of L and the boundary of the domain D are smooth enough and that D is bounded). One can check that $\lambda(\beta)$ is convex. Denote $L(\alpha)$ the Legendre transformation of $\lambda(\beta)$: $L(\alpha) = \sup_\beta((\alpha,\beta)-\lambda(\beta))$, $\alpha \in R^2$. The action functional for the family ζ_t^ε, $t \in [0,T]$, $\varepsilon\downarrow 0$, in $\mathbb{C}_{0T}$ is equal to $\varepsilon^{-1}\int_0^T L(\dot{\varphi}_+(s),\dot{\varphi}_-(s))ds$ for absolutely continuous φ, and equal to $+\infty$ for the rest of $\mathbb{C}_{0T}$.

The action functional for the process $\varepsilon^{-\kappa}\zeta_t^\varepsilon$, $0 < \kappa < \frac{1}{2}$, in the space $\mathbb{C}_{0T}$ is $\varepsilon^{2\kappa-1}S_{0T}^\kappa(\varphi)$, where $S_{0T}^\kappa(\varphi) = \frac{1}{2}\int_0^T \sum_{\lambda,\mu} a^{\lambda\mu}\dot{\varphi}_\lambda(s)\dot{\varphi}_\mu(s)ds$ for $\varphi_s = (\varphi_+(s),\varphi_-(s))$ absolutely continuous, and $S_{0T}^\kappa(\varphi) = +\infty$ for the rest of $\mathbb{C}_{0T}$; $a^{\lambda\mu} = (a_{\lambda\mu})^{-1}$, $\lambda,\mu \in (+,-)$.

Now we can describe large deviations for the field $u^\varepsilon(t,x)$. The action functional for the family $u^\varepsilon(t,x)$, $\varepsilon\downarrow 0$, in the space $C_{[0,T]\times[-1,1]}$, according to Theorem 10.5, has the form $\varepsilon^{-1}S^u(g)$, $g \in C_{[0,T]\times[-1,1]}$, where

$$S^u(g) = \begin{cases} \int_0^T L(\frac{\partial g}{\partial x}(s,1),-\frac{\partial g}{\partial x}(s,-1)), & \text{if } \frac{\partial g}{\partial t} = D\,\frac{\partial^2 g}{\partial x^2} + f(x,g) \\ & \text{for } 0 < s \le T,\ |x| < 1, \text{ and the functions} \\ & \frac{\partial g}{\partial x}(s,\pm 1),\ 0 \le s \le T, \text{ are absolutely continuous;} \\ +\infty, & \text{for the rest of } C_{[0,T]\times[-1,1]}. \end{cases}$$

Using this result and taking into account that the couple $(u^{\varepsilon}(t,\cdot),\xi_t)$ is a Markov process in the functional space we can describe transitions between different stable stationary solutions of the non-perturbed problem. We can also consider the exit problems and asymptotic behavior of the invariant measure of the perturbed problem when $\varepsilon\downarrow 0$ in the way similar to the finite-dimensional case (see [12]).

To describe the deviations $u^{\varepsilon}(t,x)$ from $u(t,x)$ of order ε^{κ}, $0<\kappa<\frac{1}{2}$, we should check the conditions of Lemma 10.1. Consider the semigroup P^t_{zf} acting in the space C_D of continuous functions on $D\cup\partial D$ with values in R^1:

$$(P^t_{zf}\varphi)(x) = E_x\varphi(\xi_t)e^{z\int_0^T f(\xi_s)ds}, \quad t\ge 0,$$

where $f(x)$, $x\in D\cup\partial D$, is a smooth enough function such that $\int_D f(x)m(x)dx = 0$. This semigroup has a positive eigenfunction $\varphi_z(x)$, and the corresponding eigenvalue has the form $\exp\{t\mu(zf)\}$, where $\mu(zf)=\mu$ is the first eigenvalue of the problem

$$L\psi(x) + zf(x)\psi(x) = \mu\psi(x), \quad x\in D, \quad \frac{\partial\psi}{\partial n}\Big|_{\partial D} = 0 \tag{10.12}$$

(see for example, [12], Ch. 7). The first eigenvalue μ is real, simple, and has two continuous derivatives in z. The eigenfunction $\varphi_z(x)$ is differentiable in z. It is easy to check that $\mu(0)=0$ and $\varphi_0(x)\equiv 1$. Then, differentiating (10.12) in z, we have

$$L\psi_0'(x) = -f(x) + \mu_0', \quad x\in D, \quad \frac{\partial\psi_0'}{\partial n}\Big|_{\partial D} = 0. \tag{10.13}$$

Problem (10.13) is solvable only if the right side of (10.13) is orthogonal to the invariant density $m(x)$. Since we assumed that $\int_D f(x)m(x)dx = 0$, we get $\mu_0' = 0$. Now taking into account that $\mu(0)=\mu'(0)=0$, we have that

$$\mu(zf) \le \frac{c|z|^2}{z}$$

for some constant C and $|z|<1$. Using this bound we conclude that the conditions of Lemma 10.1 are fulfilled in this example:

$$E\exp\left\{z\int_t^{t+\tau}\frac{f(\xi(s/\varepsilon))}{\sqrt{\varepsilon}}ds\right\} = \int_D m(x)\left[P^{\tau/\varepsilon}_{z\sqrt{\varepsilon}f}\,1\right](x)dx$$

$$\le \text{const}\cdot\exp\left\{\frac{\tau}{\varepsilon}\mu(z\sqrt{\varepsilon})\right\} \le \text{const.}\cdot\exp\left\{\frac{c|z|^2}{2}\tau\right\}$$

for $|z| < 1/\sqrt{\varepsilon}$. As function $f = f(x_+,x_-)$, we should take $f = x_+$ of $f = x_-$.

We conclude from Lemma 10.1 and Theorem 10.4 that the action functional for the family of fields $1/\varepsilon^{\kappa}\,(u^{\varepsilon}(t,x)-u(t,x)) = v^{\varepsilon}_{\kappa}(t,x)$, $0 < \kappa < \frac{1}{2}$, in the space $\hat{C}_a$, $a = (1-x^2)^{\alpha}$, $0 < \alpha < 1$, has the form $\varepsilon^{2\kappa-1}S^{u,\kappa}(g)$, where

$$S^{u,\kappa}(g) = \begin{cases} \frac{1}{2}\int_0^T\sum_{\lambda,\mu} a^{\lambda\mu}\cdot\lambda\mu\,\frac{\partial g(s,\lambda)}{\partial x}\,\frac{\partial g(s,\mu)}{\partial x}ds, & \text{if } \frac{\partial g}{\partial t} = \frac{D\partial^2 g}{\partial x^2} + f'(x,u)g \\ & \text{for } t\in[0,T],\ |x|<1, \text{ and } \frac{\partial g}{\partial x}(t,\pm 1) \text{ are} \\ & \text{absolutely continuous;} \\ +\infty, & \text{for the rest of } \hat{C}_a. \end{cases}$$

§11. Random Perturbations of RDE's: White Noise in the Equation.

In this section we consider an RDE system perturbed by white noise:

$$(11.1)\qquad \begin{cases} \frac{\partial u^{\varepsilon}_k(t,x)}{\partial t} = D_k\,\frac{\partial^2 u^{\varepsilon}_k}{\partial x^2} + f_k(x,u^{\varepsilon}_1,\dots,u^{\varepsilon}_n) + \varepsilon\zeta_k(t,x) \\ t > 0,\quad x\in S^1,\quad k = 1,\dots,n,\quad u^{\varepsilon}_k(0,x) = g_k(x). \end{cases}$$

Here S^1 is the unit circle, D_k are positive constants, g_k are continuous functions on S^1, and ε is a small parameter. Let $\zeta_k(t,x)$, $k = 1,\dots,n$, be independent white noise fields. This means that $\zeta_k(t,x) = \frac{\partial^2 W_k(t,x)}{\partial t\partial x}$, where $W_k(t,x)$ are independent Brownian sheets, i.e., Gaussian random fields with mean zero and the correlation function $EW(s,x)W(t,y) = (s\wedge t)(x\wedge y)$. The theory of such stochastic differential equations was developed by J. Walsh in [23].

It is easy to prove that under some mild assumptions on the nonlinear terms f_k, the solution $u^\varepsilon = (u_1^\varepsilon, \dots, u_n^\varepsilon)$ converges in the space $C_{[0,T]\times S^1}$ in probability to the solution $u(t,x)$ of the non-perturbed problem (i.e., of problem (11.1) with $\varepsilon = 0$) as $\varepsilon \downarrow 0$. It is well known that the non-perturbed problem can have many stable stationary solutions. The corresponding semi-flow in C_{S^1} may have many invariant measures. Equations (11.1) define a Markov process u_t^ε on C_{S^1}. Under some natural assumptions on the nonlinear terms one can prove that the process u_t^ε has only one invariant distribution μ^ε for any $\varepsilon \neq 0$. One can expect that μ^ε tends to an invariant measure of the non-perturbed system of $\varepsilon \downarrow 0$, but to which of them? This question can be answered with the help of the limit theorems for the large deviations for the family $u^\varepsilon(t,x)$. Another problem which can also be solved using the limit theorems for large (of order 1, as $\varepsilon \downarrow 0$) deviations is the so-called exit problem for the process u_t^ε.

Let us start with the case of potential field: assume that a function $\mathcal{F}(x,u)$, $x \in S^1$, $u \in R^n$, exists such that

$$f_k(x,u) = -\frac{\partial \mathcal{F}(x,u)}{\partial u_k} \quad \text{for any} \quad x \in S, \quad u \in R^r, \quad k = 1,\dots,n.$$

Note that in the case of one equation such a function $\mathcal{F}$ always exists.

If $f(x,u) = (f_1(x,u),\dots,f_n(x,u)) = -\nabla\mathcal{F}(x,u)$, then the non-perturbed system is potential. This means that

$$D_k \frac{\partial^2(\varphi_k(x))}{\partial x^2} + f_k(x;\varphi_1,\dots,\varphi_n) = -\frac{\delta U(\varphi)}{\delta \varphi_k},$$

where $\frac{\delta}{\delta\varphi_k}$ is the variational derivative with respect to φ_k, and

$$U(\varphi) = \int_0^{2\pi} \left[\frac{1}{2}\sum_{k=1}^n D_k \left(\frac{d\varphi_k}{dx}\right)^2 + \mathcal{F}(x,\varphi)\right]dx, \quad \varphi = (\varphi_1,\dots,\varphi_n).$$

The potential $U(\varphi)$ is defined on the Sobolev space W_2^1 of functions $\varphi(x) : S^1 \to R^n$, having square integrable generalized first derivative. Note that in the one-dimensional case, $W_2^1 \subset C_{S^1}$.

Thus, system (11.1) can be written in the form

$$\text{(11.2)} \qquad \dot{u}^\varepsilon(t,x) = -\nabla U(u^\varepsilon(t,x)) + \varepsilon\zeta(t,x), \quad u^\varepsilon(0,x) = g(x).$$

Let us consider the finite-dimensional counterpart of system (11.2):

(11.3) $$\dot{u}_t^\varepsilon = -\nabla F(u_t) + \varepsilon \dot{W}_t, \quad u_0 = g \in R^m.$$

Here $u \in R^m$, $\dot{W}_t$ is the m-dimensional white noise. The density $m^\varepsilon(u)$ of the invariant distribution for the m-dimensional diffusion process (11.2) (with respect to the Lebesgue measure in R^m) can be written explicitly:

(11.4) $$m^\varepsilon(u) = c_\varepsilon \exp\left\{-\frac{2F(u)}{\varepsilon^2}\right\},$$

where

$$c_\varepsilon^{-1} = \int_{R^m} e^{-2F(u)/\varepsilon^2}\, du.$$

Finiteness of this integral is the condition for existence of the unique invariant measure.

We can easily deduce from (11.4) that the invariant measure of the process u_t^ε converges as $\varepsilon \downarrow 0$ to the measure concentrated in the absolute minimum of the potential $F(u)$, provided such an absolute minimum exists.

Since our infinite-dimensional system (11.2) is also potential, we would like to write a formula for the density of the invariant measure similar to (11.4) and then use this formula for calculations of the limit as $\varepsilon \downarrow 0$. But with respect to which measure, should we consider the density? There is no such universal measure on C_{S^1} which is invariant with respect to all shifts. One can overcome these obstacles as follows.

Let us introduce a Gaussian measure in the space C_{S^1}. Consider the linear problem

(11.5) $$\frac{\partial v_k^\varepsilon}{\partial t} = D_k \frac{\partial^2 v_k^\varepsilon}{\partial x^2} - \alpha_k v_k + \varepsilon\, \zeta_k(t,x)$$

$$t > 0, \quad x \in S^1, \quad v_k^\varepsilon(0,x) = 0, \quad k = 1,\dots,n.$$

Here $\zeta_k(t,x)$ are the same white noise fields as in (11.1); $\alpha_1,\dots,\alpha_n$ are positive constants. Problem (11.5) can be solved by the Fourier method:

(11.6) $$v_\ell^\varepsilon(t,x) = \frac{\varepsilon}{\sqrt{2\pi}} A_0^\ell(t) + \frac{\varepsilon}{\sqrt{\pi}} \sum_{k=1}^{\infty} (A_k^\ell(t)\sin kx + B_k^\ell(t)\cos kx).$$

In (11.6) $A_k^\ell(t), B_k^\ell(t)$ are independent Ornstein-Uhlenbeck processes, i.e., Gaussian processes with $EA_k(t) = EB_k(t) = 0$, and

$$EA_k^\ell(s)A_k^\ell(t+s) = EB_k^\ell(s)B_k^\ell(t+s) = \frac{1}{2\lambda_k}\,[e^{-\lambda_k^\ell t} - e^{-\lambda_k^\ell (t+s)}],$$

$\lambda_k^\ell = D_\ell k^2 + \alpha_\ell$, $k = 0,1,2,\dots$; $\ell = 1,\dots,n$.

One can see from (11.6) that the Markov Gaussian process in C_{S^1}, corresponding to the linear system (11.5) has invariant measure ν^ε which is the mean zero Gaussian measure in C_{S^1} with diagonal correlation matrix $(B_{ij}(z))$:

$$B_{ii}(z) = \frac{\varepsilon^2}{2\pi}\sum_{k=0}^{\infty}\frac{1}{\lambda_k^i}\cos kz, \quad i = 1,\dots,n.$$

It turns out that in the potential case the density of the invariant measure μ^ε of the process u_t^ε in C_{S^1} with respect to the Gaussian measure ν^ε can be written explicitly:

$$(11.7) \qquad \frac{d\mu^\varepsilon}{d\nu^\varepsilon}(\varphi) = C_\varepsilon^{-1}\exp\left\{-\frac{2}{\varepsilon^2}\int_0^{2\pi}\left[F(x,\varphi(x)) - \frac{1}{2}\sum_{k=1}^{n}\alpha_k\varphi_k^2(x)\right]dx\right\},$$

if $C_\varepsilon = E\exp\{-\frac{2}{\varepsilon^2}\int_0^{2\pi}[F(x,\varphi(x)) - \frac{1}{2}\sum_{k=1}^{n}\alpha_k\varphi_k^2(x)]dx\} < \infty$.

Equality (11.7) is to some extent the counterpart of (11.4). The difference, however, is that the reference measure ν^ε in (11.7) depends on ε itself. Therefore, it is not true that μ^ε tends as $\varepsilon\downarrow 0$ to the measure concentrated in the absolute minimum of the functional in the exponent of (11.7). But taking into account the large deviations principle for the family ν^ε, one can prove the following result [9]:

<u>Theorem 11.1</u>. Assume that the potential $U(\varphi)$ has a unique point $\hat\varphi \in C_{S^1}$ of absolute minimum $U(\varphi) > U(\hat\varphi)$ for $\varphi \neq \hat\varphi$. Then for any $\delta > 0$,

$$\lim_{\varepsilon\downarrow 0}\mu^\varepsilon\{\varphi \in C_{S^1} : \max_{x\in S^1}|\varphi(x) - \hat\varphi(x)| > \delta\} = 0.$$

If the potential has several points of absolute minimum, the limit measure will be distributed between them with weights which one can calculate (see [9], Theorem 5). it is worth noting that extremals of the potential are the stationary points of the semi-flow u_t.

If the field $f(x,u)$ is not potential, we cannot write an explicit formula for the density. To calculate the limit behavior of the invariant measure in the general case and to study exit probabilities from a neighborhood of a stable stationary point of the flow u_t, we need to calculate the action functional.

Theorem 11.2. The action functional for the family of random fields $u^\varepsilon(t,x)$, $0 \le t \le T$, $x \in S^1$, defined by equations (11.1) in $C_{[0,T]\times S^1}$ as $\varepsilon\downarrow 0$, has the form $\varepsilon^{-2}S^u(\varphi)$, with

$$S^u(\varphi) = \begin{cases} \dfrac{1}{2}\displaystyle\int_0^T\int_0^{2\pi}\sum_{k=1}^n \left|\frac{\partial\varphi_k}{\partial t} - D_k\frac{\partial^2\varphi_k}{\partial x^2} - f_k(x,\varphi(x))\right|^2 dtdx, & \varphi \in W_2^{1,2}, \\ +\infty, \quad \text{for} \quad \varphi \in C_{[0,T]\times S^1}\setminus W_2^{1,2}. & \end{cases}$$

where $W_2^{1,2}$ is the Sobolev space of functions $\varphi : [0,T]\times S^1 \longrightarrow R^n$ with square integrable generalized first derivatives in t and second derivatives in x.

Proof. The proof of this theorem can be found in [9].

To describe the limiting behavior of the stationary distribution μ^ε of the process u_t^ε in C_{S^1} as $\varepsilon\downarrow 0$, we need an auxiliary construction. The same one is also employed in the finite-dimensional case [12]. We introduce the function

$$V(g,h) = \inf\{S^u(\varphi) : \varphi \in C_{[0,T]\times S^1},\ \ \varphi(0,x) = g(x),\ \ \varphi(T,x) = h(x),\ \ T > 0\}.$$

Two points $g,h \in C_{S^1}$ are said to be equivalent $(g \sim h)$, whenever $V(g,h) = V(h,g) = 0$.

We say that condition (D) is fulfilled if there is a finite number of compactums $K_1,\dots,K_\ell \subset C_{S^1}$ such that

1) $g \sim h$ for any g and h of one and the same compactum;

2) if $g \in K_i$ and $h \notin K_i$, then $g \nsim h$;

3) every w-limit set of the non-perturbed system belongs to one of K_i, $i = 1,\dots,\ell$.

We put $V_{ij} = V(g,h)$ for $g \in K_i$, $h \in K_j$. It is easy to see that V_{ij} does not depend on the choice of g and h from K_i and K_j, respectively.

Denote $\mathcal{L} = \{1,2,\dots,\ell\}$. By the i-graph over the set $\mathcal{L}$ we mean a graph consisting of arrows $(m\longrightarrow n)$, $m,n \in \mathcal{L}$, in which, from every point $j \in \mathcal{L}$ except the point $i \in \mathcal{L}$, exactly one arrow issues, and which has no closed loops.

The collection of all possible i-graphs over $\mathcal{L}$ will be designated by G_i.

Theorem 11.3. Suppose that condition (D) is fulfilled. We put

$$\sigma(\gamma) = \sum_{(m\to n)\in\gamma} V_{mn}, \quad \gamma \in G_i,$$

(the summation is taken over all arrows $(m\to n)$ involved in the i-graph γ). We assume that there is a unique $i_0 \in \mathcal{L} = \{1,\dots,\ell\}$, such that

$$\min_{\gamma\in G_{i_0}} \sigma(\gamma) < \min_{\gamma\in G_i} \sigma(\gamma) \quad \text{for} \quad i \neq i_0.$$

Then the invariant measure of the process u_t^ε in C_{S^1} is concentrated as $\varepsilon\downarrow 0$ on the compactum K_{i_0}, that is, for every $\delta > 0$,

$$\lim_{\varepsilon\downarrow 0} \mu^\varepsilon\{\varphi \in C_{S^1},\ \rho(\varphi, K_{i_0}) > \delta\} = 0,$$

where $\rho(\cdot,\cdot)$ is the metric in C_{S^1}.

The function $V(g,h)$ is important for the exit problem, too. Suppose that $\varphi_0 \in C_{S^1}$ is an asymptotically stable equilibrium point for the non-perturbed system and let D be a bounded open neighborhood of the point φ_0, which is attracted to φ_0 together with its boundary as $t\to\infty$. Denote $\tau_D^\varepsilon = \tau^\varepsilon$ the exit time from D : $\tau^\varepsilon = \inf\{t : u_t^\varepsilon \notin D\}$, $V_0 = \inf\{V(\varphi_0,\varphi) : \varphi \in \partial D\}$. Assume that there exists a unique point $\hat{\varphi} \in \partial D$ such that $V_0 = V(\varphi_0,\hat{\varphi})$. Then, under some minor assumptions about the domain D, for any starting point $g \in D$ and $\delta > 0$,

$$\lim_{\varepsilon\downarrow 0} \varepsilon^2 \ln E_g \tau^\varepsilon = V_0,$$

$$\lim_{\varepsilon\downarrow 0} P_g\left\{\sup_{x\in S^1} |u^\varepsilon_{\tau^\varepsilon}(x) - \hat{\varphi}(x)| > \delta\right\} = 0.$$

If the field $f(x,u) = (f_1(x,u),\dots,f_n(x,u))$ is potential, then V_0 can be expressed through the potential $U(\varphi)$.

The function $V(g,h)$ and the matrix V_{ij} describe some other characteristics of the behavior of u_t^ε for small ε such as the sequence of transitions between the compacts K_i, and the main terms of the times of transitions.

Let us consider now the case when the intensity of the perturbations depends on the position in the phase space:

$$\frac{\partial u^{\varepsilon}(t,x)}{\partial t} = \frac{D}{2}\frac{\partial^2 u^{\varepsilon}}{\partial x^2} + f(x,u^{\varepsilon}) + \varepsilon\sigma(x,u^{\varepsilon})\frac{\partial^2 W}{\partial t\partial x}, \quad x \in S^1, \tag{11.8}$$

$$u^{\varepsilon}(0,x) = g(x).$$

Here $\sigma(x,u)$ is a positive, bounded and smooth enough function, $W(t,x)$ as before, is the Brownian sheet. We consider for brevity the case of one equation.

One can expect that the action functional for the family $u^{\varepsilon}(t,x)$ in $C_{[0,T]\times S^1}$ is equal to $\varepsilon^{-2}\tilde{S}^u(\varphi)$, where

$$\tilde{S}^u(\varphi) = \begin{cases} \frac{1}{2}\int_0^T\int_0^{2\pi} \sigma^{-2}(x,\varphi(s,x))\left[\frac{\partial\varphi}{\partial t}(s,x) - \frac{D}{2}\frac{\partial^2\varphi}{\partial x^2} - f(x,\varphi(s,x))\right]^2 dsdx, & \varphi \in W_2^{1,2} \\ +\infty, \quad \text{for } \varphi \in C_{[0,T]\times S^1}\setminus W_2^{1,2}. \end{cases}$$

But the proof of this statement is more delicate than in the case $\sigma \equiv 1$. It was done by R. Sowers in [21]. White noise perturbations of RDE's were considered also in [5], [24]. One can consider a fast oscillating noise in the nonlinear term

$$\frac{\partial u^{\varepsilon}}{\partial t}(t,x) = \frac{D}{2}\frac{\partial^2 u^{\varepsilon}}{\partial x^2} - f(x,u,\xi(t/\varepsilon,x)), \quad t > 0, \quad x \in R^1,$$

$$u^{\varepsilon}(0,x) = g(x).$$

Results of the central limit theorem type for such a problem were considered in [19]. Some results on the large deviations are available in [16].

We considered some examples of perturbations of the differential equations and the boundary conditions. Random perturbations of the domain, where the initial boundary problem is considered, are also of interest.

Random perturbations of PDE's is a relatively new field and there are many interesting unsolved problems in this area.

References

1. D.G. Aronson and H.F. Weinberger (1975), Non-linear diffusion in population genetic, combustion and nerve propagation, Proceedings of the Tulane program in PDE's, Lecture Notes in Math., Springer-Verlag, Berlin-Heidelberg-New York.

2. D.G. Aronson and H.F. Weinberger (1978), "Multi-dimensional nonlinear diffusion in population genetics," Advances in Math. 30, 33-76.

3. M.D. Bramson (1978), "Maximal displacement of branching Brownian motion," Comm. Pure Appl. Math. 31, 531-582.

4. L.C. Evans and P.E. Souganidis (1989), "A PDE approach to geometric optics for certain semi-linear parabolic equations," Indiana Univ. Mathematics Journal 83, 1, 141-127.

5. W.G. Faris and G. Jona-Lasinio (1982), "Large deviations for a nonlinear heat equation with noise," Jour. of Physics, A. Math. Gen. 15, 441-459.

6. M.I. Freidlin (1985), "Limit theorems for large deviations and reaction-diffusion equations," The Annals of Probability 13, 3, 639-676.

7. M.I. Freidlin (1985), Functional integration and PDE's, Princeton Univ. Press, 540 pp.

8. M.I. Freidlin (1987), "Tunneling soliton in the equations of reaction-diffusion type," Annals of the New York Academy of Sciences 491, 149-157.

9. M.I. Freidlin (1988), "Random perturbations of RDE's," TAMS 305, 2, 665-697.

10. M.I. Freidlin (1990), "Weakly coupled RDE's," to be published in the Annals of Probability.

11. M.I. Freidlin and T.-Y. Lee (1990), "Wave front propagation in a class of semi-linear systems, preprint.

12. M.I. Freidlin and A.D. Wentzell (1984), Random perturbations of dynamical systems, Springer-Verlag, Berlin-Heidelberg-New York.

13. M.I. Freidlin and A.D. Wentzell (1990), "Reaction-diffusion equations with randomly perturbed boundary conditions," submitted to the Annals of Probability.

14. J. Gartner (1980), "Non-linear diffusion equations and excitable media," Dokl. Acad. Nauk, USSR, 254, 1310-1314.

15. J. Gartner and M.I. Freidlin (1979), "On the propagation of concentration waves in periodic and random media," Dokl. Acad. Nauk, USSR 249, 521-525.

16. D. Ioffe, "On some applicable version of abstract large deviations theorem," submitted to the Annals of Probability.

17. A. Kolmogorov, I. Petrovskii, N. Piskunov (1937), "Étude de l'équation de la diffusion avec croissence de la matière et sone application a un problem biologique," Moscow Univ. Bull. Math. 1, 1-25.

18. H.P. McKean (1975), "Application of Brownian motion to the equation of Kolmogorov-Petrovskii-Piskunov," Comm. Pure Appl. Math. 29:5, 553-564.

19. E. Pardoux (1985), Asymptotic analysis of a semi-linear PDE with wide-band noise disturbance, stochastic space-time models and limit theorems, Reidel Publishing Company, Dordrecht-Boston-Lancaster, 227-243.

20. J. Smoller (1983), Shock waves and RDE's, Springer-Verlag, New York-Heidelberg-Berlin.

21. R. Sowers, "Large deviations for a reaction-diffusion equation with non-Gaussian perturbations," submitted to the Annals of Probability.

22. S.R.S. Varadhan (1984), "Large deviations and applications," SIAM, CMBS, 46.

23. J. Walsh (1984), An introduction to stochastic partial differential equations, École d'Éte de Probabilités de Saint Flour, XIV, Lecture Notes in Mathematics 1180, Springer-Verlag, New York.

24. J. Zabchyk (1988), "On large deviations for a stochastic evolution equation," Proceedings of the 6th IFIP Working Conference on Stochastic Systems and Optimization, Warsaw.

INDEX

SOME PROPERTIES

OF

PLANAR BROWNIAN MOTION

Jean-Francois LE GALL

SOME PROPERTIES OF PLANAR BROWNIAN MOTION

Jean-François LE GALL

Université Pierre et Marie Curie
Laboratoire de Probabilités
4, Place Jussieu
Tour 56

75230 PARIS Cedex 05, France

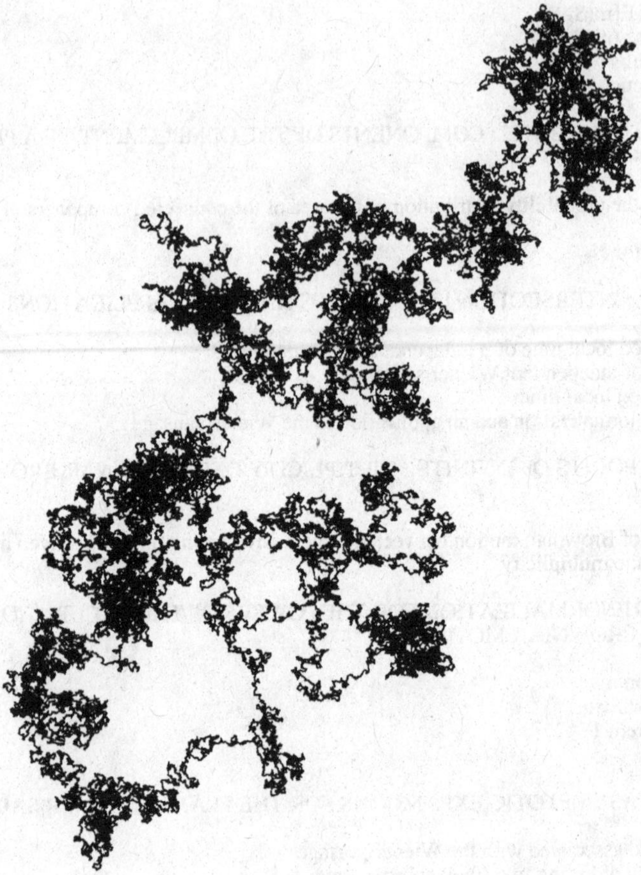

A planar Brownian path.

CHAPTER I

Introduction

1. Historical sketch and program of the course.

The physical Brownian motion attracted the interest of the British botanist Robert Brown in 1828 . The mathematical study of the Brownian motion started in 1900 when the French mathematician Louis Bachelier guessed several important properties of this process, including a weak form of the Markov property, and the Gaussian distribution of Brownian motion at a fixed time. A more rigorous derivation of the Gaussian character of the one-dimensional marginals was provided by Albert Einstein in 1905. The first complete construction of Brownian motion as a continuous stochastic process is due to Norbert Wiener in 1923. Later, in collaboration with Paley and Zygmund, Wiener proved the non-differentiability of the Brownian paths, which had been conjectured by the French physicist Perrin.

Much of what we know about Brownian motion is due to Paul Lévy. Lévy discovered many remarkable sample path properties, as well as several important distributions connected with Brownian motion. Lévy also introduced the local times of linear Brownian motion, which have given rise to many important developments.

Since Lévy's work, linear Brownian motion has been studied extensively, sometimes with the help of Itô's stochastic calculus, which among other applications yields a very simple construction of local times. The books of Knight [Kn], Revuz and Yor [ReY] and Rogers and Williams [RoW] contain much information about properties of one-dimensional Brownian motion.

Multidimensional Brownian was not neglected after Lévy : see in particular Chapter 7 of Itô and Mc Kean [IM]. However several questions raised by Lévy were left aside until very recently.

In the last few years, there has been much interest in properties of planar Brownian motion : e.g. geometric properties of sample path (Burdzy, Mountford, Shimura,...), asymptotic distributions (Pitman, Yor,...) or multiple points and intersection problems (Dynkin, Rosen,...). The purpose of these lectures is to provide a detailed account of a number of these recent developments. We will mainly consider sample path (that is, almost sure) properties, although for instance Chapter II presents a proof of the celebrated Spitzer theorem on the winding number of planar Brownian motion.

We also restrict our attention to the two-dimensional case (with the important exception of Chapter VI). Sometimes the extension to higher dimensions is possible if not straightformward (this is the case for Chapters III, IV), sometimes on the contrary the extension is impossible or has no meaning (this is the case for most of the results concerning multiple points). Generally speaking, planar Brownian motion has several very nice properties, which disappear in higher dimensions. This can be explained by the relationship between Brownian motion and holomorphic functions, and also by the fact that the dimension 2 is critical for Brownian motion, meaning that a planar Brownian path, on the time interval $[0,\infty)$, is dense in the plane although it does not hit a given point.

In Chapter II, we recall the conformal invariance of planar Brownian paths and we use it to derive their first basic properties. The main topics of the next chapters are :

- the existence of the exceptional points of the path called cone points (Chapters III-IV);
- the smoothness of the convex hull of the planar Brownian path, on the time interval $[0,1]$ (Chapter III) ;
- the connected components of the complement of the Brownian path (Chapter VII);
- the shape of the Brownian path near a typical point of the boundary of one such component (Chapter V);
- the area of a tubular neighborhood of the path, and more generally of the so-called Wiener sausage (Chapters VI-VIII);
- the existence of points of finite and infinite multiplicity (Chapters VIII-IX);
- the associated "self-intersection local times" (Chapter VIII);
- the renormalization of self-intersections and its application to asymptotic expansions of the area of the Wiener sausage (Chapters X-XI).

It is worth noting that most of the previous topics are related to questions raised by Lévy. The non-existence of angular points on the boundary the convex hull of a Brownian path was stated without proof in [Lé4, p. 240]. As for the boundary of the complement of the Brownian path, it is interesting to compare the twist points theorem of Chapter V with Lévy's assertion that "la plupart des points de cette frontière ne sont accessibles que par des chemins très compliqués, le long desquels l'angle polaire n'est pas borné" [Lé4, p. 239]. Lévy [Lé4, p. 325-329] also raised many questions about multiple points, and most of them can be answered using the modern notion of intersection local time. See in particular Chapter IX for a rigorous version of Lévy's heuristic assertion that "un point double choisi sur la courbe n'a aucune chance d'être triple" [Lé4, p. 325].

2. Some comments about the proofs.

Generally speaking, we do not assume much from the reader, except for some well-known facts such as the strong Markov property or the Brownian scaling property, which both play an essential role throughout this work. We use stochastic calculus only to derive the conformal invariance of planar Brownian paths in Chapter II. Some classical results of probabilistic potential theory are used in Chapters VI and XI. They are recalled in detail at the beginning of Chapter VI. In Chapter V, we use a rather involved result of complex analysis, namely Mc Millan's theorem. It is presumably possible, and it would have been more satisfactory in a sense, to prove at least part of this result using Brownian motion. This however would have taken us too far, and we simply recall this theorem without proof in Chapter V.

An important concept in this work is the notion of local time. We use the term local time in a very wide sense. A local time is a random measure supported either on the state space $\mathbb{R}^d$ of the process or on $\mathbb{R}_+$, the set of times (or even on $\mathbb{R}_+ \times \mathbb{R}_+$, when we consider double points, etc...). This random measure is supported on a certain class of exceptional points of the path (in the first case) or on a set of exceptional times. For instance, the local time at 0 of a linear Brownian motion B may be viewed as a random measure supported on the zero set of B. Note that the zero set is a set of exceptional times since for every $t > 0$, $B_t \neq 0$ w.p. 1. As is well-known, this local time is very useful when investigating various properties of the zero set. In these lectures, we do not use the local times of linear Brownian motion (except in a remark of Chapter II and, up to some extent, in Chapters X, XI). However, in Chapters IV, VIII, we construct local times associated with certain classes of random sets, and we then apply these local times to various sample path properties. A typical example is provided by Chapter IX, where we use intersection local times (associated with points of finite multiplicity) to get the existence of points of infinite multiplicity.

The previously mentioned topics are related to various problems in probability theory or in other branches of mathematics and physics. The shortest proof of the existence of cone points uses the notion of reflected Brownian motion in a wedge, which has been studied extensively in the last few years. Many properties of planar Brownian motion can be proved from complex analysis via the conformal invariance theorem (Theorem II-1). A typical example is the twist points theorem of Chapter V, whose proof uses both McMillan's theorem and a weak form of Makarov's theorem on the support of harmonic measure. The asymptotics of the volume of the Wiener sausage give information about certain problems connected with the heat equation. The mathematical notion of intersection local time was motivated by the models of polymer physics, as well as by Symanzik's approach to quantum field theory. In the same connection, Dynkin's renormalization for multiple self-intersections of planar Brownian motion was inspired by the renormalization techniques of field theory. It was not possible in these notes to explain all the connections between Brownian motion and various problems of mathematics or physics. It should however be kept in mind that these connections often motivated the proof of the results that are presented below.

Remerciements. Je tiens ici à remercier l'ensemble des participants de l'Ecole d'Eté de Probabilités de St-Flour pour l'intérêt qu'ils ont porté à ce cours et leurs remarques souvent pertinentes. Je remercie tout particulièrement Paul-Louis Hennequin pour l'excellente organisation de l'école d'été. Je veux aussi remercier Chris Burdzy et Jay Rosen pour leurs commentaires sur la première version de ce travail. Enfin, je remercie Nicolas Bouleau pour la simulation de trajectoire brownienne plane, réalisée au CERMA, qui illustre ce cours.

Interconnections between chapters.

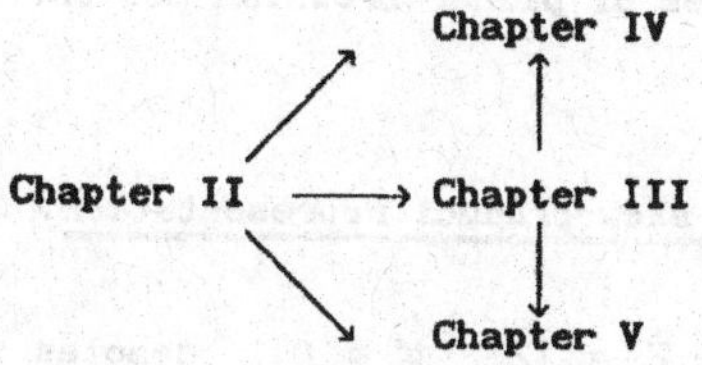

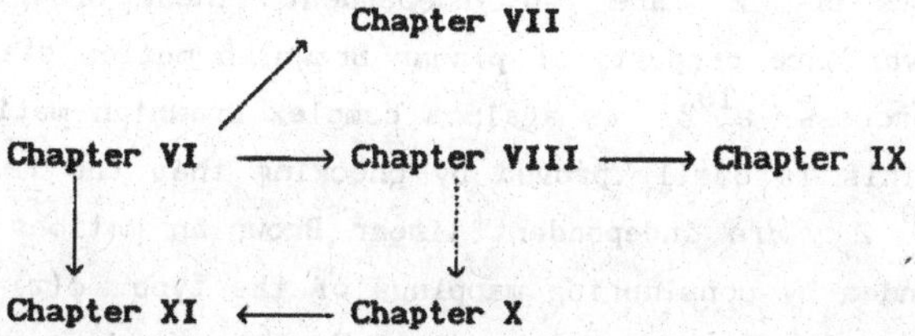

Main notation.

The complex Brownian motion is denoted by $(B_t)_{t\geq 0}$, or $(Z_t)_{t\geq 0}$. As usual, B_0 (or Z_0) = z under the probability P_z.

$B[u,v] = \{ B_s \,;\, u \leq s \leq v \}$.

m denotes the Lebesgue measure on $\mathbb{C} = \mathbb{R}^2$ (or on $\mathbb{R}^d$ in Chapter VI).

The Brownian transition density is denoted by:

$$p_t(x,y) = (2\pi t)^{-1} \exp(-|y-x|^2/2t)$$

For $z \in \mathbb{C}$ and $\varepsilon > 0$,

$$D(z,\varepsilon) = \{ y \in \mathbb{C} \,;\, |z - y| < \varepsilon \} \,, \quad D = D(0,1)$$

$$T_\varepsilon(z) = \inf\{ t \geq 0 \,;\, |B_t - z| \leq \varepsilon \}$$

If K is a subset of $\mathbb{C}$ and $\varepsilon > 0$,

$$\varepsilon K = \{ \varepsilon y \,;\, y \in K \} \,,$$

$$z - K = \{ z - y \,;\, y \in K \} \,,$$

$$T_K = \inf\{ t \geq 0 \,;\, B_t \in K \} \qquad (\inf \emptyset = +\infty)$$

$$T_K(z) = T_{z-K} = \inf\{ t \geq 0 \,;\, B_t \in z - K \}$$

dim A denotes the Hausdorff dimension of a subset A of $\mathbb{R}^d$ and diam A is the diameter of A .

CHAPTER II

Basic properties of planar Brownian motion.

1. Conformal invariance and the skew-product representation.

Throughout this chapter, $Z = (Z_t,\ t \geq 0)$ denotes a complex-valued Brownian motion started at $z_0 \in \mathbb{C}$. This simply means that the real and imaginary parts of Z are two independent linear Brownian motions. The rotational invariance property of planar Brownian motion states that for any $\theta \in \mathbb{R}$, the process $e^{i\theta}Z_t$ is again a complex Brownian motion, which starts at $e^{i\theta}z_0$. This is easily proved by checking that the real and imaginary parts of $e^{i\theta}Z_t$ are independent linear Brownian motions. This result is slightly extended by considering mappings of the type $\phi(z) = az + b$, with $a \neq 0$. Then using the scaling property of Brownian motion, we obtain that:

$$\phi(Z_t) = Z'_{\lambda^2 t}$$

where $\lambda = |a|$ and Z' is a complex Brownian motion started at $\phi(z_0)$. In particular, the image of a Brownian path under ϕ is a (time-changed) Brownian path. A very important theorem of Lévy shows that the latter property still holds if we only assume that ϕ is locally tangent to a mapping of the type $z \longrightarrow az + b$, that is if ϕ is conformal. More precisely, we have the following result.

Theorem 1. *Let* U *be an open subset of* $\mathbb{C}$, *such that* $z_0 \in U$, *and let* $\phi : U \longrightarrow \mathbb{C}$ *be holomorphic. Set*

$$\tau_U = \inf\{\ t \geq 0\ ;\ Z_t \notin U\ \} \leq +\infty$$

Then there exists a complex Brownian motion Z' *such that, for any* $t \in [0,\tau_U)$,

$$\phi(Z_t) = Z'_{C_t}$$

where

$$C_t = \int_0^t |\phi'(Z_s)|^2\, ds\ .$$

Remark. At any point $z_1 \in U$, the holomorphic function ϕ is "locally tangent" to the mapping $z \longrightarrow \phi(z_1) + \phi'(z_1)\,(z - z_1)$. Notice that the derivative of $t \longrightarrow C_t$ is precisely $|\phi'(Z_t)|^2$.

Proof : Set $\phi = g + ih$, so that g and h are harmonic on U. By the Itô formula applied to $g(Z_t^1 + iZ_t^2)$, we get for $t < \tau_U$,

$$g(Z_t) = g(z_0) + \int_0^t \frac{\partial g}{\partial x}(Z_s)\, dZ_s^1 + \int_0^t \frac{\partial g}{\partial y}(Z_s)\, dZ_s^2$$

and similarly,

$$h(Z_t) = h(z_0) + \int_0^t \frac{\partial h}{\partial x}(Z_s)\, dZ_s^1 + \int_0^t \frac{\partial h}{\partial y}(Z_s)\, dZ_s^2 \ .$$

This shows that $M_t = g(Z_t)$, $N_t = h(Z_t)$ are two continuous local martingales on the stochastic interval $[0,\tau_U)$.

By the Cauchy-Riemann equations, $\frac{\partial g}{\partial x} = \frac{\partial h}{\partial y}$, $\frac{\partial g}{\partial y} = -\frac{\partial h}{\partial x}$. It follows that:

$$\langle M\rangle_t = \langle N\rangle_t = \int_0^t |\phi'(Z_s)|^2\, ds = C_t$$

$$\langle M,N\rangle_t = 0 \ .$$

By a standard result of stochastic calculus, the last two properties imply the existence of two *independent* linear Brownian motions such that, for $t \in [0,\tau_U)$

$$M_t = B^1_{C_t} \ , \quad N_t = B^2_{C_t} \ .$$

The desired result follows, with $Z'_t = B^1_t + iB^2_t$. □

Notice that, in the previous proof, the Brownian motion Z' is determined from Z only on the time interval $[0,\tau_U)$. As a matter of fact, when $P[\,\tau_U < \infty\,] > 0$, in order to define Z'_s for $s \geq C_{\tau_U}$, it may be necessary to enlarge the underlying probability space. This fact is unimportant in applications of Theorem 1.

The proof of Theorem 1 also yields the formula:

$$\phi(Z_t) = \phi(z_0) + \int_0^t \phi'(Z_s)\, dZ_s \ .$$

Here the "complex stochastic integral" $\int_0^t \phi'(Z_s)\, dZ_s$ is obviously defined by:

$$\int_0^t \phi'(Z_s) dZ_s = \int_0^t (\mathrm{Re}\ \phi'(Z_s)\, dZ_s^1 - \mathrm{Im}\ \phi'(Z_s)\, dZ_s^2)$$

$$+\ i\int_0^t (\mathrm{Re}\ \phi'(Z_s)\, dZ_s^2 + \mathrm{Im}\ \phi'(Z_s) dZ_s^1) \ .$$

Theorem 1 can be used to interpret (and sometimes to prove) many results of complex analysis in terms of planar Brownian motion. On the other hand, it allows one to prove properties of Brownian motion using holomorphic functions. We shall be interested in this second type of applications. We first use Theorem 1 to establish the polarity of single points for planar Brownian motion.

Corollary 2 : *Let* $z_1 \in \mathbb{C} \setminus \{z_0\}$. *Then,*

$$P[\, Z_t = z_1 \quad \textit{for some} \quad t \geq 0 \,] = 0 .$$

Let m *denote Lebesgue measure on* $\mathbb{C}$. *Then,*

$$m(\{ Z_t \; ; \; t \geq 0 \}) = 0 \quad , \quad a.s.$$

Proof : We may assume that $z_0 = 1$; $z_1 = 0$. Let $\Gamma = (\Gamma_t, \ t \geq 0)$ be a planar Brownian motion started at 0 . By Theorem 1,

$$\exp(\Gamma_t) = Z'_{C_t}$$

where

$$C_t = \int_0^t \exp(2 \operatorname{Re} \Gamma_s)\, ds \ ,$$

and Z' is a complex Brownian motion started at 1 . It is immediate that $\lim_{t\to\infty} C_t = +\infty$, a.s. , so that

$$\{ Z'_t \; ; \; t \geq 0 \} = \{ \exp \Gamma_t \; ; \; t \geq 0 \} \quad \text{a.s.}$$

Obviously, $0 \notin \{ \exp \Gamma_t \; ; \; t \geq 0 \}$, and we get the desired result, with Z replaced by Z' . This however makes no difference since the two processes Z, Z' have the same distribution.

To get the second assertion, write

$$E[\, m(\{ Z_t \; ; \; t \geq 0\}) \,] = E\Big[\int dy\, 1_{(Z_t = z \text{ for some } t \geq 0)} \Big]$$
$$= \int dy\, P[\, Z_t = z \text{ for some } t \geq 0 \,] = 0 \ . \ \square$$

As a second consequence of Theorem 1, we get the skew-product representation of planar Brownian motion.

Theorem 3 : *Suppose that* $z_0 \neq 0$, *and write* $z_0 = \exp(r + i\theta)$, *with* $r \in \mathbb{R}$ *and* $\theta = \arg(z) \in (-\pi,\pi]$. *There exist two independent linear Brownian motions* β, γ , *started respectively at* r, θ , *such that, for every* $t \geq 0$,

$$Z_t = \exp(\, \beta_{H_t} + i\, \gamma_{H_t} \,)$$

where

$$H_t = \int_0^t \frac{ds}{|Z_s|^2} = \inf\{\, u \geq 0 \ , \ \int_0^u \exp(2\beta_v)\, dv > t \,\} \ .$$

Remark. Corollary 2 shows that H_t is well-defined for any $t \geq 0$.

Proof : The "natural" method would be to apply Theorem 1 to $\phi(z) = \mathrm{Log}\, z$, that is to some determination of the complex logarithm. This approach however leads to certain minor technical difficulties (due to the fact that one cannot take $U = \mathbb{C} \setminus \{0\}$!). Therefore we will use another method, similar to the proof of Corollary 2.

We may assume that $z_0 = 1$ and thus $r = \theta = 0$. Let $\Gamma_t = \Gamma_t^1 + i\Gamma_t^2$ be a complex Brownian motion started at 0 . By Theorem 1,

$$\exp \Gamma_t = Z'_{C_t} , \tag{1}$$

where

$$C_t = \int_0^t \exp(2\,\Gamma_s^1)\, ds ;$$

Let $(H_t,\ t \geq 0)$ be the inverse function of C_t :

$$H_s = \int_0^s \exp(-2\,\Gamma^1_{H_u})\, du = \int_0^s \frac{du}{|Z'_u|^2} ,$$

since $\exp(\Gamma^1_{H_u}) = |Z'_u|$. By (1) with $t = H_s$,

$$Z'_s = \exp(\,\Gamma^1_{H_s} + i\,\Gamma^2_{H_s}\,) .$$

This is the desired result, with $\beta = \Gamma^1$, $\gamma = \Gamma^2$, except that we have replaced Z by Z'.

To complete the proof, we argue as follows. Theorem 1 is equivalent to saying that, if

$$\beta_t = (\,\log |Z|\,)_{\inf\{\, s\ ;\ \int_0^s |Z_u|^{-2}\, du > t\,\}} ,$$

$$\gamma_t = (\,\arg Z\,)_{\inf\{\, s\ ;\ \int_0^s |Z_u|^{-2}\, du > t\,\}} ,$$

then β, γ are two independent linear Brownian motions. Observe that β, γ are deterministic functions of the process Z. Therefore their joint distribution depends only on that of Z. □

Another approach to Theorem 3, avoiding the use of Theorem 1, would be to check that:

$$\log |Z_t| = r + \int_0^t \frac{Z_s^1\, dZ_s^1 + Z_s^2\, dZ_s^2}{|Z_s|^2} , \quad \arg Z_t = \theta + \int_0^t \frac{Z_s^1\, dZ_s^2 - Z_s^2\, dZ_s^1}{|Z_s|^2}$$

and then to use the same argument as in the proof of Theorem 1 in order to write $\log |Z_t|$, $\arg Z_t$ as time-changed independent linear Brownian motions.

Notice the intuitive contents of Theorem 3 . When $|Z_t|$ is large, then H_t increases slowly, so that $\arg Z_t$ also varies slowly.

The formula

$$\log|Z_t| = \beta_{\inf\{ u \geq 0 \,,\, \int_0^u \exp(2\beta_v)\, dv > t \}}$$

shows that $|Z|$ is completely determined by the linear Brownian motion β (and conversely). This is related to the fact that $|Z|$ is a Markov process, namely a two-dimensional Bessel process. On the other hand, $\arg Z_t = \gamma_{H_t}$ is a linear Brownian motion time-changed by an independent increasing process. The independence of γ and H_t is especially important in applications of the skew-product representation.

2. Some applications of the skew-product representation.

We start by proving that planar Brownian motion is recurrent.

Theorem 4 : *For any open subset* U *of* $\mathbb{C}$,

$$P[\limsup_{t\to\infty} \{ Z_t \in U \}] = 1 .$$

Proof : We may take $z_0 = 1$, $U = D(0,\varepsilon)$ for $\varepsilon \in (0,1)$. Theorem 3 gives:

$$\log|Z_t| = \beta_{H_t}$$

and the obvious facts: $\lim_{t\to\infty} H_t = +\infty$, $\liminf_{s\to\infty} \beta_s = -\infty$ imply

$$\liminf_{t\to\infty} \log|Z_t| = -\infty \quad \text{a.s.} \ \square$$

From now on, we assume $z_0 \neq 0$. Let $(\theta_t, t \geq 0)$ be the continuous determination of $\arg Z_t$ such that $\theta_0 = \arg z_0 \in (-\pi,\pi]$.

Proposition 5 : *With probability* 1 ,

$$\limsup_{t\to\infty} \theta_t = +\infty \quad , \quad \liminf_{t\to\infty} \theta_t = -\infty .$$

Proof : Similar to that of Theorem 4 , using now $\theta_t = \gamma_{H_t}$. $\square$

Remark. We can use the previous results to prove the conformal invariance of harmonic measure, in a special case that will be used in Chapter V . Let $\hat{\mathbb{C}} = \mathbb{C} \cup \{\infty\}$ denote the Riemann sphere and let V be a simply connected subset of $\hat{\mathbb{C}}$, such that $\hat{\mathbb{C}} \setminus V$ has a nonempty interior. Let D be the open unit disk of $\mathbb{C}$. The Riemann mapping theorem yields a one-to-one conformal mapping f from D onto V . The value of f at 0 may be chosen arbitrarily, so that we can impose $f(0) \neq \infty$. By Fatou's theorem, the radial limits

$$\lim_{r\to 1, r<1} f(re^{i\theta}) =: \tilde{f}(e^{i\theta})$$

exist for $d\theta$-a.a. $\theta \in [0,2\pi)$. We take $U = D$ if $\infty \notin V$, $U = D \setminus \{f^{-1}(\infty)\}$ if $\infty \in V$ and we will apply Theorem 1 to $\phi = f_{|U}$, with $z_0 = 0$. Notice that the distribution of Z_{τ_U} is the uniform distribution $\sigma(d\zeta)$ over ∂D (by rotational invariance, and Corollary 2 if $\infty \in V$).

Let Z' be another Brownian motion started at $z'_0 = f(0)$ and let $\tau'_V = \inf\{ t \,;\, Z'_t \notin V \}$ ($\tau'_V < \infty$ a.s. by Theorem 4). Theorem 1 implies that the limit

$$\lim_{t\to\tau_U} f(Z_t)$$

exists a.s. and is distributed as $Z'_{\tau'_V}$. Furthermore, using the skew-product representation and some well-known properties of linear Brownian motion, it is easy to check that this limit coincides a.s. with $\tilde{f}(Z_{\tau_U})$. We conclude that $\tilde{f}(Z_{\tau_U}) \overset{(d)}{=} Z'_{\tau'_V}$. In other words, the harmonic measure in V relative to z'_0 is the image of $\sigma(d\zeta)$ under $\tilde{f}$.

Theorem 4 and Proposition 5 are straightforward applications of the skew-product representation. The idea of these applications is that many properties of Z can be derived by looking at the independent Brownian motions β, γ and taking account of the time-change H_t. Until now, we only used the simple fact $\lim_{t\to\infty} H_t = +\infty$. For further applications, it is important to control the asymptotic behavior of H_t. We know that H_t has a simple expression in terms of the Brownian motion β. The next lemma relates the asymptotic behavior of H_t to that of an even simpler functional of β. This result is a key ingredient in the proof of several asymptotic theorems for planar Brownian motion.

Lemma 6 : *For every* $\lambda > 0$, *set:*

$$\beta^{(\lambda)}_t = \frac{1}{\lambda}\beta_{\lambda^2 t} \quad (t \ge 0) \;, \quad T^{(\lambda)}_1 = \inf\{ t \ge 0 \,,\, \beta^{(\lambda)}_t = 1 \} .$$

Then,

$$\frac{4}{(\log t)^2} H_t - T_1^{(\frac{1}{2}\log t)} \xrightarrow[t\to\infty]{\text{Probability}} 0 .$$

Remark. For every $\lambda > 0$, $\beta^{(\lambda)}$ is a linear Brownian motion started at $\lambda^{-1}\log|z_0|$. Therefore, Lemma 6 entails in particular that $4\,(\log t)^{-2} H_t$ converges in distribution towards the hitting time of 1 by a linear Brownian motion started at 0.

Proof : By scaling, we may assume $|z_0| = 1$, so that $\beta_0 = 0$. To simplify notation, we write

$$\lambda = \lambda(t) = \frac{1}{2}\log t \quad .$$

Let $\varepsilon > 0$ and $T^{(\lambda)}_{1+\varepsilon} = \inf\{ t \geq 0 , \beta^{(\lambda)}_t = 1 + \varepsilon\}$. We first prove that:

(2) $$P[\lambda^{-2} H_t > T^{(\lambda)}_{1+\varepsilon}] \xrightarrow[t\to\infty]{} 0 \ .$$

Since

$$H_t = \inf\{ u \geq 0 , \int_0^u \exp(2\beta_v)\, dv > t \} .$$

we have

$$\{ \lambda^{-2} H_t > T^{(\lambda)}_{1+\varepsilon} \} = \left\{ \int_0^{\lambda^2 T^{(\lambda)}_{1+\varepsilon}} \exp(2\beta_v)\, dv < t \right\}$$

$$= \left\{ \frac{1}{2\lambda} \log \int_0^{\lambda^2 T^{(\lambda)}_{1+\varepsilon}} \exp(2\beta_v)\, dv < 1 \right\}$$

(recall that $2\lambda = \log t$). However,

$$\frac{1}{2\lambda}\log \int_0^{\lambda^2 T^{(\lambda)}_{1+\varepsilon}} \exp(2\beta_v)\, dv = \frac{\log\lambda}{\lambda} + \frac{1}{2\lambda}\log\int_0^{T^{(\lambda)}_{1+\varepsilon}} \exp(2\lambda\, \beta^{(\lambda)}_v)\, dv$$

$$\overset{(d)}{=} \frac{\log\lambda}{\lambda} + \frac{1}{2\lambda}\log\int_0^{T^{(1)}_{1+\varepsilon}} \exp(2\lambda\, \beta^{(1)}_v)\, dv$$

since the processes $\beta^{(\lambda)}$ are identically distributed. We now use the fact that, for any continuous function $f : \mathbb{R}_+ \longrightarrow \mathbb{R}$, for any $t > 0$,

$$\frac{1}{2\lambda}\log\int_0^t \exp(2\lambda\, f(v))\, dv \xrightarrow[\lambda\to\infty]{} \sup_{[0,t]} f(s) \ .$$

It follows that

$$\frac{1}{2\lambda}\log\int_0^{T^{(1)}_{1+\varepsilon}} \exp(2\lambda\, \beta_v)\, dv \xrightarrow[\lambda\to\infty]{} \sup_{[0,T^{(1)}_{1+\varepsilon}]} \beta_s = 1 + \varepsilon \quad , \quad \text{a.s.}$$

and thus

$$\frac{1}{2\lambda}\log\int_0^{\lambda^2 T^{(\lambda)}_{1+\varepsilon}} \exp(2\lambda\, \beta^{(\lambda)}_v)\, dv \xrightarrow[\lambda\to\infty]{\text{Probability}} 1 + \varepsilon \ .$$

This completes the proof of (2). Exactly the same arguments give:

$$P[\lambda^{-2} H_t < T^{(\lambda)}_{1-\varepsilon}] \xrightarrow[t\to\infty]{} 0$$

which completes the proof of Lemma 6. □

The next theorem, due to Spitzer (1958), gives precise information on the order of θ_t when t is large.

Theorem 7 : *As* $t \to \infty$, $\frac{2}{\log t} \theta_t$ *converges in distribution towards the standard symmetric Cauchy distribution. Equivalently, for any* $x \in \mathbb{R}$,

$$\lim_{t\to\infty} P[\frac{2}{\log t} \theta_t \le x] = \int_{-\infty}^{x} \frac{dy}{\pi(1+y^2)} .$$

Proof : For $\lambda > 0$, write

$$\gamma_t^{(\lambda)} = \frac{1}{\lambda} \gamma_{\lambda^2 t} \quad (t \ge 0) .$$

Take $\lambda = \lambda(t) = \frac{1}{2} \log t$ as previously. Then,

$$\lambda^{-1} \theta_t = \lambda^{-1} \gamma_{H_t} = \gamma^{(\lambda)}_{\lambda^{-2} H_t} .$$

Hence, by Lemma 6,

$$\lambda^{-1} \theta_t - \gamma^{(\lambda)}(T_1^{(\lambda)}) \xrightarrow[\lambda\to\infty]{\text{Probability}} 0 .$$

To complete the proof, note that the variable $\gamma^{(\lambda)}(T_1^{(\lambda)})$ obviously converges in distribution towards $\gamma^{(\infty)}(T_1^{(\infty)})$ where $T_1^{(\infty)} = \inf\{ t \ge 0 , \beta_1^{(\infty)} = 1\}$, and $\beta^{(\infty)}$, $\gamma^{(\infty)}$ are two independent linear Brownian motions started at 0 . It is easy to compute the characteristic function of $\gamma^{(\infty)}(T_1^{(\infty)})$ and to check that it is that of a standard symmetric Cauchy distribution (an alternative method is to observe that the process $(\gamma^{(\infty)}(T_a^{(\infty)}), a \ge 0)$ is a symmetric Lévy process, stable with index 1 , hence must be a symmetric Cauchy process: this is Spitzer's construction of the Cauchy process). □

Lemma 6 can be applied to the proof of other asymptotic theorems for planar Brownian motion. Let us mention the following result of Kallianpur and Robbins, which yields information on the time spent by the Brownian path in domains of the plane. Let $f : \mathbb{C} \longrightarrow \mathbb{R}_+$ be a bounded measurable function with compact support. Then,

$$\frac{2}{\log t} \int_0^t f(Z_s)\, ds \xrightarrow[t\to\infty]{(d)} \left(\frac{1}{\pi} \int f(y)\, dy\right) e$$

where e denotes an exponential variable with parameter 1 . When f is radial, that is $f(z) = f(|z|)$, Lemma 6 yields a simple proof of this convergence. Denote by $L_t^a(\beta)$ the local time of the Brownian motion β at level a , at time t . Then,

$$\frac{2}{\log t}\int_0^t f(Z_s)\,ds = \frac{1}{\lambda}\int_0^t f(\exp\beta_{H_s})\,ds$$

$$= \frac{1}{\lambda}\int_0^{H_t} f(\exp\beta_u)\exp 2\beta_u\,du$$

$$= \lambda\int_0^{\lambda^{-2}H_t} f(\exp\lambda\,\beta_v^{(\lambda)})\exp(2\lambda\,\beta_v^{(\lambda)})\,dv$$

$$= \lambda\int_{\mathbb{R}} f(\exp\lambda a)\exp(2\lambda a)\,L^a_{\lambda^{-2}H_t}(\beta^{(\lambda)})\,da$$

$$= \int_0^\infty r\,f(r)\,L^{\lambda^{-1}\log r}_{\lambda^{-2}H_t}(\beta^{(\lambda)})\,dr$$

$$\xrightarrow[\lambda\to\infty]{(d)} \left(\int_0^\infty r\,f(r)\,dr\right) L^0_{T_1^{(\infty)}}(\beta^{(\infty)})$$

by Lemma 6. To complete the proof, note that $L^0_{T_1^{(\infty)}}(\beta^{(\infty)}) \overset{(d)}{=} 2\,e$. The general case (f non radial) can then be handled using the Chacon-Ornstein ergodic theorem.

Via a scaling argument, we can use Theorem 7 to get information about the behavior of the process Z in small time. Suppose now that $z_0 = 0$. Of course we can no longer define θ_t . However, by Corollary 2 and the Markov property, we know that $Z_t \neq 0$ for every $t > 0$, a.s. Hence, for every $\varepsilon>0$, we may consider $\theta_{[\varepsilon,1]}$, defined as the variation of (a continuous determination of) $\arg Z_t$ between times ε and 1 . By scaling,

$$\theta_{[\varepsilon,1]} \overset{(d)}{=} \theta_{[1,1/\varepsilon]}\,.$$

Therefore, Theorem 7 and the Markov property imply the convergence in distribution of $2|\log\varepsilon|^{-1}\,\theta_{[\varepsilon,1]}$ towards a standard Cauchy distribution. An application of the zero-one law also gives, a.s. for any $\delta > 0$,

$$\limsup_{\varepsilon\to 0}\ \theta_{[\varepsilon,\delta]} = +\infty \quad , \quad \liminf_{\varepsilon\to 0}\ \theta_{[\varepsilon,\delta]} = -\infty$$

Informally, on any interval $[0,\delta]$, the Brownian path performs an infinite number of windings around its starting point.

3. The Hausdorff dimension of the Brownian curve.

In this section, which is independent of the previous two ones, we propose to compute the Hausdorff dimension of the Brownian path. We have already noticed that the Lebesgue measure of the path is zero a.s. Nonetheless we will check that its Hausdorff dimension is 2 , which shows that in a sense the Brownian path is not far from having positive Lebesgue measure (see Lévy [Lé4, p. 242-243] for comments about the area of the planar Brownian curve).

We first recall the definitions of Hausdorff measure and Hausdorff dimension. Let h be a continuous monotone increasing function from $\mathbb{R}_+$ into $\mathbb{R}_+$. For any Borel subset A of $\mathbb{R}^d$, the Hausdorff measure $h\text{-}m(A)$ is defined by:

$$h\text{-}m(A) = \lim_{\substack{\varepsilon\to 0\\ \varepsilon>0}} \left(\inf_{(R_i)\in\mathcal{R}_\varepsilon(A)} \sum_i h(\operatorname{diam} R_i) \right)$$

where $\operatorname{diam}(R_i)$ denotes the diameter of the set R_i, and, for $\varepsilon > 0$, $\mathcal{R}_\varepsilon(A)$ is the collection of all countable coverings of A by subsets of $\mathbb{R}^d$ of diameter less than ε. Notice that the limit exists in $[0,\infty]$ since the infimum is a nonincreasing function of ε.

In what follows, we only consider functions h such that $h(2x) \le C\, h(x)$ for some constant C, and we are interested in knowing whether $h\text{-}m(A) > 0$, or $h\text{-}m(A) < \infty$. To this end, we may restrict our attention to coverings by balls, or cubes, or rectangles (notice for instance that any bounded subset R of $\mathbb{R}^d$ is contained in a ball of diameter $2 \operatorname{diam} R$).

For any $\alpha > 0$, we set $h_\alpha(x) = x^\alpha$. It can be proved that

$$h_d\text{-}m(A) = C_d\, m(A)$$

for some universal constant $C_d > 0$. It is also easy to check that, for any Borel subset A of $\mathbb{R}^d$, there exists a number $\dim A \in [0,d]$ such that:

$$h_\alpha\text{-}m(A) = \begin{cases} +\infty & \text{if } \alpha < \dim A\ , \\ 0 & \text{if } \alpha > \dim A\ . \end{cases}$$

The number $\dim A$ is the Hausdorff dimension of A. If $A = \bigcup_{n\in\mathbb{N}} A_n$, we have

$$\dim A = \sup_n \dim A_n\ .$$

Theorem 8. *With probability* 1 *, for every* $t > 0$,

$$\dim(\{ Z_s\ ;\ 0\le s\le t \}) = 2\ .$$

Proof. We need only check that

$$\dim(\{ Z_s\ ;\ 0\le s\le t \}) \ge 2 - \delta\ ,$$

for any $\delta > 0$. We introduce the random measure

$$\mu(A) = \int_0^t 1_A(Z_s)\, ds\ .$$

Fix $\delta > 0$. We will prove that, w.p. 1 , there exists a constant $\rho(\omega) > 0$ such that, for any subset A of $\mathbb{C}$ with $\operatorname{diam} A < \rho(\omega)$,

(3) $$\mu(A) \le 16\ (\operatorname{diam} A\ )^{2-\delta} .$$

Then, if (R_i) is a countable covering of $\{ Z_s\ ;\ 0\le s\le t \}$ by sets of diameter less than $\rho(\omega)$, we have

$$\sum_i\ (\operatorname{diam} R_i)^{2-\delta} \ge \frac{1}{16} \sum_i \mu(R_i) \ge \frac{1}{16}\mu(\{ Z_s\ ;\ 0\le s\le t \}) = \frac{1}{16}\, t .$$

Therefore, $h_{2-\delta}\text{-}m(\{ Z_s\ ;\ 0\le s\le t \}) > 0$ and $\dim(\{ Z_s\ ;\ 0\le s\le t \}) \ge 2-\delta$.

It remains to prove (3). Suppose first that A is a square of the type $A = [u,u+r] \times [v,v+r]$. For every integer $p \ge 1$, we evaluate

$$E[\mu(A)^p] = E\Big[\Big(\int_0^t 1_A(Z_s)\, ds\Big)^p\Big] = E\Big[\int_{[0,t]^p} ds_1\ldots ds_p\ 1_A(Z_{s_1})\ \ldots\ 1_A(Z_{s_p})\Big]$$

$$= p!\int_{A^p} dy_1\ldots dy_p \int_{0\le s_1\le\ldots\le s_p\le t} ds_1\ldots ds_p\ p_{s_1}(z_0,y_1)p_{s_2-s_1}(y_1,y_2)\ldots p_{s_p-s_{p-1}}(y_{p-1},y_p)$$

$$\le p!\ \Big(\sup_{z\in\mathbb{C}}\int_A dy\int_0^t ds\ p_s(z,y)\Big)^p .$$

At this point, we use the easy bound:

$$\int_0^t ds\ p_s(z,y)\ \le\ C\ \Big(1 + \log_+\frac{1}{|z-y|}\Big)\ e^{-|z-y|}$$

and after integration over A we get

$$E[\mu(A)^p] \le p!\ C'^{\,p}\ \varphi(m(A))$$

where $\varphi(x) = x\ (1 + \log_+ 1/x)$. It follows that, for $\lambda > 0$ small enough, for any square A ,

$$E[\ \exp \lambda\frac{\mu(A)}{\varphi(m(A))}\ ] \ \le\ 2 .$$

Then, by the Tchebicheff inequality, for every $r > 0$,

(4) $$P[\ \mu(A) \ge r\ \varphi(m(A))\ ] \le 2 \exp -\lambda r .$$

The proof of (3) is now easily completed. Denote by $A(n,j,k)$ the dyadic square $[j2^{-n},(j+1)2^{-n}] \times [k^{-n},(k+1)2^{-n}]$. By (4) ,

$$\sum_{n=1}^{\infty}\ \sum_{j=-2^{2n}}^{2^{2n}}\ \sum_{k=-2^{2n}}^{2^{2n}}\ P[\ \mu(A(n,j,k) \ge 2^{-(2-\delta)n}\ ]\ <\ \infty .$$

Therefore, by the Borel-Cantelli lemma, we may w.p. 1 find $n_0(\omega)$ such that:

$$\mu(A(n,j,k)) \le 2^{-(2-\delta)n}$$

for every $n \geq n_0(\omega)$, $j, k \in \{-2^{2n},\dots,2^{2n}\}$. The bound (3) now follows: use the fact that any set A such that $\operatorname{diam} A < 1/2$ is contained in the union of 4 dyadic squares $A(n,j,k)$, with n such that $\operatorname{diam} A \leq 2^{-n} \leq 2 \operatorname{diam} A$.□

The previous proof is certainly not the shortest one (in particular, the connection between Hausdorff measures and capacities can be used to give a very short proof). It is however interesting as it serves as a prototype for the evaluation of the Hausdorff dimension of random sets. The upper bound is usually easy (here it was trivial) because it suffices to construct good coverings. The lower bound requires the introduction of an auxiliary measure (a "local time") which is in a sense uniformly distributed over the random set. See Chapter IV for an application of this technique to cone points and [L9] for an application to multiple points of the Brownian path. In the latter case, the auxiliary measure is provided by the intersection local time introduced in Chapter VIII.

Bibliographical notes. *The conformal invariance of Brownian paths was stated by Lévy (see* [Lé4, p. 254]*), with a heuristic proof. A (succinct) proof using stochastic calculus was provided by McKean* [MK, p. 109] *(see also* [IMK, p. 279-280] *for a different approach). Several results related to Theorem 1, and a detailed proof of the needed arguments of stochastic calculus, may be found in Getoor and Sharpe* [GS] *(see also the Chapter 5 of Revuz and Yor* [ReY]*). Applications of Theorem 1 to complex analysis are given in Davis* [Da] *and Durrett* [Du2]. *Corollary 2 and Theorem 4 are due to Lévy (Lévy's proof of Corollary 2 is elementary, it uses only the scaling properties of Brownian motion, see* [Lé4, p. 240-241]*). The skew-product representation is stated in Itô and McKean* [IMK, p. 265] , *in the more general setting of* d*-dimensional Brownian motion. Theorem 7 was first proved by Spitzer* [Sp1], *using explicit calculations of the Fourier transform of* θ_t . *The basic idea of our proof is due to Durrett* [Du1] *(see also Pitman and Yor* [PY1] *and Le Gall and Yor* [LY], *the latter paper dealing with diffusions more general than Brownian motion). See* [KR] *for the original proof of the Kallianpur-Robbins law. Pitman and Yor* [PY1,PY3] *(see also the Chapter 13 of* [ReY]*) have obtained limit theorems which extend Spitzer's result and the Kallianpur-Robbins law in many respects. A typical example is the determination of the asymptotic joint distribution of the winding numbers around several points of the plane* [PY1]. *Theorem 8 is only a weak form of Taylor's result on the exact Hausdorff measure of the sample path of planar Brownian motion* [T1] . *See Lévy* [Lé2] *and Ciesielski and Taylor* [CT] *for the analogous theorem in higher dimensions, and* [L1] *for a unified approach to these results.*

CHAPTER III

Two-sided cone points and the convex hull of planar Brownian motion.

1. The definition of cone points.

We consider a standard complex-valued Brownian motion $(B_t, t \geq 0)$ started at 0. As was noticed in Chapter II, for every fixed $t > 0$, with probability 1, the curve $(B_{t+s}, 0 < s \leq 1)$ performs an infinite number of windings around the point B_t. The same is true for the curve $(B_{t-s}, 0 < s \leq t)$. These results hold for any fixed t with probability 1. It is natural to ask whether there can be exceptional times (depending on ω) for which these properties fail to hold. A simple geometric argument shows that there must exist such times. Write

$$B_t = B_t^1 + i\, B_t^2$$

and set :

$$T = \inf\{t \geq 0 \; ; \; B_t^1 = \sup_{0 \leq s \leq 1} B_s^1\}.$$

It is very easy to see that $0 < T < 1$ a.s. Furthermore the definition of T shows that both curves $(B_{T-s}, 0 \leq s \leq T)$ and $(B_{T+s}, 0 < s \leq 1-T)$ lie in the hyperplane $\{x \leq B_T^1\}$. Therefore the previous properties cannot hold for $t = T$.

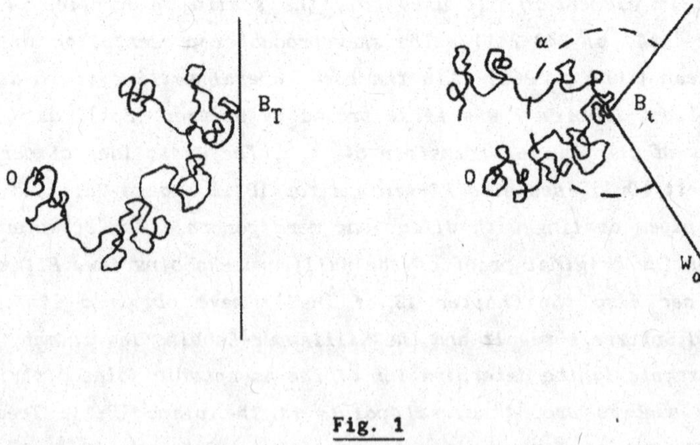

Fig. 1

Definition : *Let* $\alpha \in (0,2\pi)$ *and* $t > 0$. *We say that* B_t *is a two-sided cone point with angle* α *if there exist* $\delta > 0$ *and a closed wedge* W_α *with vertex* B_t *and angle* α, *such that the two curves* $(B_{t+s}, 0 < s \le \delta)$ *and* $(B_{t-s}, 0 < s \le \delta)$ *lie inside the wedge* W_α. *We say that* B_t *is a one-sided cone point with angle* α *if the same property holds for one of the two curves* $(B_{t+s}, 0 \le s \le \delta)$, $(B_{t-s}, 0 \le s \le \delta)$.

The point B_T constructed above is with probability 1 a two-sided cone point with angle π. One may ask whether there exist two-sided cone points with angle less than π. We shall see that the answer is no and that this fact is closely related to the non-existence of "corners" on the boundary of the convex hull of $(B_s, 0 \le s \le 1)$. One-sided cone points will be studied in the next chapter.

2. Estimates for two-sided cone points.

As we have already observed, for any fixed $t > 0$, B_t is w.p. 1 not a cone point. It will therefore be convenient to introduce a weaker notion of "approximate cone point". Fix $A > 0$ and let $z \in \mathbb{C} \setminus \{0\}$. Write the skew-product decomposition of the Brownian motion $z - B_t$:

$$z - B_t = R_t \exp(i\,\theta_t) \qquad (\theta_0 = \arg(z) \in (-\pi\ ;\ \pi]).$$

Set :

$$T_\varepsilon(z) = \inf\{s \ge 0\ ;\ R_s \le \varepsilon\},$$

$$S_\varepsilon(z) = \inf\{s \ge T_\varepsilon(z)\ ;\ R_s \ge A\}.$$

For $\varepsilon < |z|$, we say that z is an ε-approximate (two-sided) cone point with angle α if :

$$\forall\, s \le S_\varepsilon(z), \quad |\theta_s| \le \frac{\alpha}{2}.$$

Note that we do not require z to belong to the Brownian curve. We will discuss later the connection between cone points and approximate cone points. Notice that z is an ε-approximate cone point iff the curve $\{B_s, 0 \le s \le S_\varepsilon(z)\}$ lies inside the wedge $\{z - r\,e^{i\gamma}\ ;\ r \ge 0,\ |\gamma| \le \frac{\alpha}{2}\}$.

We will now get upper bounds on the probability that z is an ε-approximate cone point. Clearly, the only non-trivial case is when $\theta_0 = \arg(z) \in (-\frac{\alpha}{2}, \frac{\alpha}{2})$, which we assume now. The basic idea is to split the interval $[0,\ S_\varepsilon(z)]$ as $[0,\ T_\varepsilon(z)] \cup [T_\varepsilon(z),\ S_\varepsilon(z)]$ and to bound separately the corresponding probabilities, making use of the Markov property at time $T_\varepsilon(z)$.

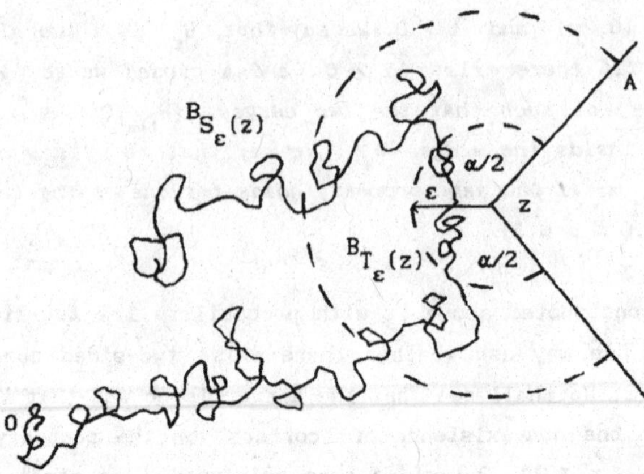

Fig. 2

The skew-product representation gives us

$$\log R_t = \beta_{H_t}, \ \theta_t = \gamma_{H_t}$$

where $H_t = \int_0^t R_s^{-2}\, ds$ and β, γ are two independent linear Brownian motions, with $\beta_0 = \log|z|$ and $\gamma_0 = \arg(z)$. Clearly,

$$H_{T_\varepsilon(z)} = \inf\{u,\ \beta_u \le \log \varepsilon\} =: \sigma_{\log \varepsilon}.$$

Therefore,

$$\{\forall\ s \le T_\varepsilon(z)\ ,\ |\theta_s| \le \frac{\alpha}{2}\} = \{\forall\ u \le \sigma_{\log\, \varepsilon}\ ,\ |\gamma_u| \le \frac{\alpha}{2}\}.$$

The probability of the last event is easy to estimate. We note that $\sigma_{\log\varepsilon}$ and γ are independent and we make use of the following classical lemma.

<u>Lemma 1</u> : *Let* $(W_t, t \ge 0)$ *be a standard linear Brownian motion started at* 0, *and let* $a < 0 < b$. *Then for every* $t > 0$,

$$P[\forall s \le t,\ a \le W_s \le b] = \sum_{k=0}^{\infty} \frac{4}{(2k+1)\pi} \sin\left(\frac{(2k+1)\pi b}{b-a}\right) \exp - \frac{(2k+1)^2\pi^2}{2(b-a)^2}\, t.$$

<u>Proof</u> : (see e.g. Feller (1971), p. 342) The function

$$\varphi(t,x) = P[\forall s \le t,\ a \le x + W_s \le b]$$

solves $\frac{\partial\varphi}{\partial t} = \frac{1}{2}\Delta\varphi$ in $(0,\infty) \times (a,b)$, with Dirichlet boundary conditions and initial value 1. This equation is solved by the usual eigenfunction expansion. □

It follows that :

$$P[\forall\, u \le \sigma_{\log\varepsilon};\ |\gamma_u| \le \tfrac{\alpha}{2}]$$

$$= \sum_{k=0}^{\infty} \frac{4}{(2k+1)\pi} \sin\left(\frac{(2k+1)\pi(\alpha/2-\arg(z))}{\alpha}\right) E\left[\exp - \frac{(2k+1)^2\pi^2}{2\alpha^2}\,\sigma_{\log\,\varepsilon}\right]$$

$$= \sum_{k=0}^{\infty} \frac{4}{(2k+1)\pi} \sin\left(\frac{(2k+1)\pi(\alpha/2-\arg(z))}{\alpha}\right)\left(\frac{\varepsilon}{|z|}\right)^{(2k+1)\pi/\alpha},$$

using the well-known formula for the Laplace transform of hitting times of points for linear Brownian motion:

$$E[\exp - \lambda\, \sigma_{r+\log|z|}\ ] = \exp - |r|\sqrt{2\lambda}\ .$$

In this chapter, we will only need the following simple consequence of the previous explicit formula. There exists a constant C, independent of $z \in \mathbb{C}$, $\varepsilon \in (0,1)$, such that :

$$\text{(1)}\qquad P\left[\forall\, s \le T_\varepsilon(z),\ |\theta_s| \le \frac{\alpha}{2}\right] \le C\,\left(\frac{\varepsilon}{|z|}\right)^{\pi/\alpha}.$$

Formula (1) is trivial when $|z| \le 2\varepsilon$ and follows from the previous expansion when $|z| > 2\varepsilon$.

Let $(\mathcal{F}_t)$ be the canonical filtration of B. Our next goal is to bound

$$P\left[\forall\, s \in [T_\varepsilon(z),S_\varepsilon(z)],\ |\theta_s| \le \frac{\alpha}{2}\ \Big|\ \mathcal{F}_{T_\varepsilon(z)}\right].$$

The Markov property at time $T_\varepsilon(z)$ leads us to consider a Brownian motion started at some point of $D(z,\varepsilon) := \{y, |z-y| \le \varepsilon\}$, and to bound the probability that it exits $D(z,A)$ before exiting the wedge $\{z-re^{iu}\ ;\ r \ge 0, |u| \le \frac{\alpha}{2}\}$. However the previous calculations apply as well to this situation. Therefore we get the bound :

$$\text{(2)}\qquad P\left[\forall\, s \in [T_\varepsilon(z),S_\varepsilon(z)],\ |\theta_s| \le \frac{\alpha}{2}\ |\ \mathcal{F}_{T_\varepsilon(z)}\right] \le C\,\left(\frac{\varepsilon}{A}\right)^{\pi/\alpha}.$$

Let $\Theta_\varepsilon^{\alpha,A}$ denote the set of all ε-approximate two-sided cone points with angle α. The next lemma follows readily from (1) and (2).

<u>**Lemma 2**</u> : *There exists a constant* C_α *such that :*

$$P[z \in \Theta_\varepsilon^{\alpha,A}] \le C_\alpha\ |z|^{-\pi/\alpha}\ A^{-\pi/\alpha}\ \varepsilon^{2\pi/\alpha}.$$

As a simple consequence of Lemma 2, we get that for any compact subset K of $\mathbb{C} \setminus \{0\}$, for $\varepsilon \in (0,1)$,

$$E[m(K \cap \Theta_\varepsilon^{\alpha,A})] = \int_K dz\ P[z \in \Theta_\varepsilon^{\alpha,A}] \le C'_{\alpha,A,K}\ \varepsilon^{2\pi/\alpha}$$

so that, by Fatou's lemma,

(3) $$\liminf_{\varepsilon \to 0} \varepsilon^{-2\pi/\alpha}\, m(K \cap \Theta_{\varepsilon}^{\alpha,A}) < \infty, \text{ a.s.}$$

This fact will be the main ingredient in the proof of the following theorem.

Theorem 3 : *Let* Γ_α *denote the set of all two-sided cone points with angle* α. *Then, with probability* 1,

(i) *if* $\alpha \in (0,\pi)$, $\Gamma_\alpha = \emptyset$;

(ii) *if* $\alpha \in [\pi,2\pi)$, $\dim \Gamma_\alpha \le 2 - \frac{2\pi}{\alpha}$

($\dim \Gamma_\alpha$ *denotes the Hausdorff dimension of* Γ_α).

Remark : In case (ii), it can in fact be proved that $\dim \Gamma_\alpha = 2 - \frac{2\pi}{\alpha}$ (see Evans [Ev1]).

Proof : We set

$$\Theta^{\alpha,A} = \bigcap_{\varepsilon>0} \Theta_{\varepsilon}^{\alpha,A}$$

and

$$\Theta^{\alpha} = \bigcup_{A>0} \Theta^{\alpha,A}.$$

It is easy to check that $z \in \Theta^\alpha$ iff $z = B_t$ for some $t > 0$, and, for some $\delta > 0$, the curves $(B_s, 0 \le s \le t)$, $(B_{t+s}, 0 \le s \le \delta)$ lie inside the wedge

$$W_\alpha(z) = \{y = z - re^{iu}\ ;\ r \ge 0,\ |u| \le \tfrac{\alpha}{2}\}.$$

In particular, Θ^α is contained in Γ^α.

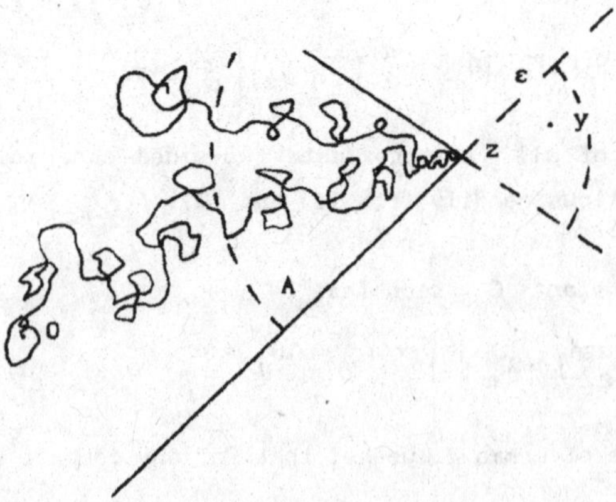

Fig. 3

Consider first the case $\alpha \in (0,\pi)$. We make use of the following simple observation. Fix $A > 0$, then for ε small, if $z \in \Theta^{\alpha,A}$, any point y of the form $y = z + re^{iu}$ with $0 \le r \le \varepsilon$, $|u| < \frac{\alpha}{2}$ belongs to $\Theta_{\varepsilon}^{\alpha,A/2}$ (see fig. 3).

It follows that for any compact subset K of $\mathbb{C} \setminus \{0\}$,

$$(4) \qquad m(\Theta_{\varepsilon}^{\alpha,A/2} \cap K_{\varepsilon}) \ge c_{\alpha}\, \varepsilon^2\, 1_{(\Theta^{\alpha,A} \cap K \neq \emptyset)}$$

for some $c_{\alpha} > 0$ (here K_{ε} denotes the ε-neighborhood of K). Since $2\pi/\alpha > 2$, (3) and (4) give

$$\Theta^{\alpha,A} \cap K = \emptyset \quad , \quad \text{a.s.}$$

Since this is true for any $A > 0$ and any compact subset K we conclude that

$$\Theta^{\alpha} = \emptyset \quad , \quad \text{a.s.}$$

It is then quite easy to show also that $\Gamma^{\alpha} = \emptyset$ a.s. First observe that we may replace B by any of the Brownian motions $B_t^{(p)} = B_{p+t} - B_p$, for all rational p. Then choose $\alpha' \in (\alpha,\pi)$ and notice that we may find a finite number of wedges with vertex 0 and angle α' such that any wedge with vertex 0 and angle α is contained in one of these. From the fact that $\Theta^{\alpha'} = \emptyset$ a.s. and the rotational invariance of Brownian motion it is easy to deduce that $\Gamma^{\alpha} = \emptyset$ a.s.

We now turn to the case $\alpha > \pi$ (we may forget about the case $\alpha = \pi$). It will be enough to show that for any $A > 0$,

$$\dim \Theta^{\alpha,A} \le 2 - \frac{2\pi}{\alpha} \quad \text{a.s.}$$

Indeed the previous arguments then show that, for any $\alpha' > \alpha$, Γ_{α} is contained in a countable union of sets of the type $\Theta^{\alpha',A}$, hence has dimension less than $2 - 2\pi/\alpha'$.

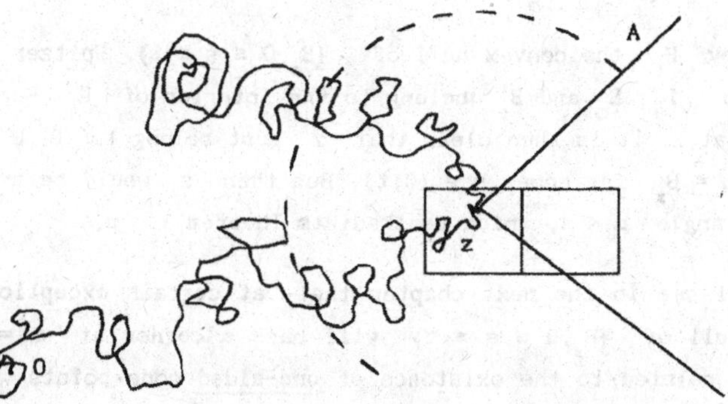

Fig. 4

Let K be a compact subset of $\mathbb{C} \setminus \{0\}$. For every $n \geq 1$, denote by $\mathscr{C}_n$ the collection of all squares $Q^n_{i,j} = [i\,2^{-n},(i+1)2^{-n}] \times [j\,2^{-n},(j+1)2^{-n}]$ for $i,j \in \mathbb{Z}$. Let

$$N_n = \sum_{Q \in \mathscr{C}_n} 1_{(Q \cap K \cap \Theta^{\alpha,A} \neq \emptyset)}$$

be the number of squares in $\mathscr{C}_n$ that intersect $K \cap \Theta^{\alpha,A}$. We observe that for n large, for every square $Q^n_{i,j}$ which intersects $\Theta^{\alpha,A}$, we may find a subset of $Q^n_{i+1,j}$ of measure larger than $C_\alpha 2^{-2n}$, which is contained in $\Theta^{\alpha,A/2}_{4\cdot 2^{-n}}$ (here C_α is some positive constant depending on α). See fig. 4.

This shows that

$$C_\alpha\, 2^{-2n}\, N_n \leq m(\Theta^{\alpha,A/2}_{4\cdot 2^{-n}} \cap K_{4\cdot 2^{-n}}). \tag{5}$$

Then (3) and (5) imply :

$$\liminf_{n\to\infty} 2^{n(2\pi/\alpha-2)}\, N_n < \infty\,, \quad \text{a.s.}$$

From the definition of Hausdorff measures we conclude that

$$\dim(\Theta^{\alpha,A} \cap K) \leq \frac{2\pi}{\alpha}\,, \quad \text{a.s.} \quad \square$$

3. Application to the convex hull of planar Brownian motion.

Let H be a compact convex subset of $\mathbb{C}$. We say that H has a corner at $z \in \partial H$ if H is contained in a wedge with vertex z and opening $\alpha < \pi$.

Theorem 4 : *Let* $t > 0$. *With probability* 1, *the convex hull of* $\{B_s, 0 \leq s \leq t\}$ *has no corners.*

Proof : Denote by H_t the convex hull of $\{B_s, 0 \leq s \leq t\}$. Spitzer's theorem implies that w.p. 1, B_0 and B_t belong to the interior of H_t. Suppose that H_t has a corner at z. It is then clear that z must belong to $\{B_s, 0 \leq s \leq t\}$, and therefore $z = B_s$ for some $s \in (0,t)$. But then z would be a two-sided cone point with angle $\alpha < \pi$, which contradicts Theorem 3. $\square$

Remark : We will see in the next chapter that, at certain exceptional times t, the convex hull of $\{B_s, 0 \leq s \leq t\}$ will have a corner at $z = B_t$. This fact is closely related to the existence of one-sided cone points with angle $\alpha < \pi$.

As a consequence of Theorem 4 we get that the boundary of the convex hull of $\{B_s, 0 \le s \le t\}$, parametrized by the argument, is with probability one a C^1-curve. We also get the following result.

Theorem 5 : *With probability one, the convex hull of* $\{B_s, 0 \le s \le t\}$ *has no isolated extreme points, and the set of all extreme points has dimension* 0.

Proof : Let H_t be as above. It is easy to check that any extreme point of H_t must belong to $\{B_s, 0 \le s \le t\}$ (this is true for the convex hull of any continuous curve). It follows that the set of extreme points is contained in the set of two-sided cone points with angle π, hence has dimension 0 by Theorem 3. Finally, if $z = B_s$ is an isolated extreme point, the set H_t must also be the convex hull of $\{z\} \cup (H_t \setminus D(z,\delta))$ for some $\delta > 0$ (use the Krein - Milman theorem). However this implies that H_t has a corner at z (otherwise z would not be extremal) and so the desired result follows from Theorem 4. □

4. The first intersection of a line with the Brownian path.

For any $y \in \mathbb{R}$ let D_y be the horizontal line $D_y = \{x + iy \; ; \; x \in \mathbb{R}\}$. Fix $t > 0$ and set $B[0,t] = \{B_s, 0 \le s \le t\}$, and

$$x(y) = \sup\{x \; ; \; x + iy \in B[0,t]\}$$

(by convention $\sup \emptyset = -\infty$).

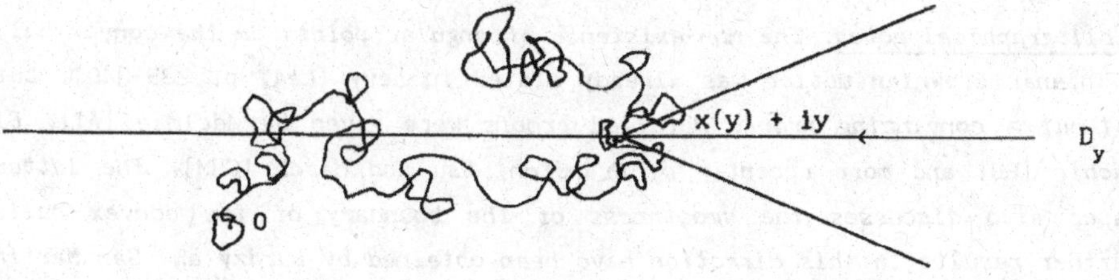

Fig. 5

If we imagine a particle coming from infinity along the line D_y, the point $x(y) + iy$ is the first hitting point of the Brownian path by this particle. One might expect this point to be a two-sided cone point. The next result shows that this is usually not the case.

Theorem 6 : *With probability one, for* dy-*almost all* $y \in \mathbb{R}$, *either* $x(y) = -\infty$ *or, for any* $\theta > 0$

$$\{x(y) + iy + r\,e^{iu} \; ; \; r > 0, \; |u| < \theta\} \cap B[0,t] \neq \emptyset \, .$$

Proof : Theorem 3 shows that for every $\theta > 0$

$$\dim \Gamma_{2\pi-\theta} < 1, \quad \text{a.s.}$$

Let p denote the projection $p(x + iy) = y$. It follows that

$$\dim p(\Gamma_{2\pi-\theta}) < 1 , \quad \text{a.s.}$$

and so

$$m(p(\Gamma_{2\pi-\theta})) = 0 , \quad \text{a.s.}$$

where m denotes Lebesgue measure on $\mathbb{R}$. Taking a sequence (θ_n) decreasing to 0, we get

$$m\left(\bigcup_{\theta>0} p(\Gamma_{2\pi-\theta})\right) = 0 , \quad \text{a.s.}$$

which gives the statement of Theorem 6. □

Remark : The previous proof shows that a statement analogous to Theorem 6 holds simultaneously for all directions, for (almost) all lines of the chosen direction.

The result of Theorem 6 can be stated in a slightly different form as follows. With probability one, for any $\theta > 0$

$$\{x(0) + re^{iu} \; ; \; r > 0, \; |u| < \theta\} \cap B[0,t] \neq \emptyset.$$

To check that this property holds, apply the Markov property at time $\delta > 0$ small, and use the fact that the law of B^2_δ is absolutely continuous w.r.t. Lebesgue measure.

Bibliographical notes. The non-existence of angular points on the convex hull of planar Brownian motion was already stated in Lévy [Lé4, p. 239-240], *but without a convincing proof. Detailed proofs were given by Adelman* [A1], *El Bachir* [EB] *and more recently by Cranston, Hsu and March* [CHM]. *The latter paper also discusses the smoothness of the boundary of the convex hull. Further results in this direction have been obtained by Burdzy and San Martin* [BSM]. *The approach taken here is inspired from* [L7], *although this paper deals with one-sided cone points. Theorem 5 is from Evans* [Ev1], *who has also obtained precise estimates on the Hausdorff dimension of cone points (Theorem 3 is only a very weak form of Evans' results). Finally, Burdzy* [B3] *contains many interesting results along the lines of Theorem 6 and Shimura* [Sh3] *treats a problem closely related to two-sided cone points with angle* π .

CHAPTER IV

One-sided cone points and a two-dimensional version of Lévy's theorem on the Brownian supremum process

1. A local time for one-sided cone points.

In this chapter, $B = (B_t, t \geq 0)$ is again a standard complex-valued Brownian motion started at 0. Let $\alpha \in (0,\pi]$. We shall be interested in a special class of one-sided cone points with angle α. We set

$$W_\alpha = \{z = r\,e^{i\theta}\ ;\ r \geq 0,\ |\theta| \leq \frac{\alpha}{2}\}.$$

Observe that W_α is convex since $\alpha \leq \pi$. Set

$$H_\alpha = \{t \geq 0\ ;\ \forall s \leq t,\ B_t - B_s \in W_\alpha\}$$

$$\Delta^\alpha = \{B_t\ ;\ t \in H_\alpha\}$$

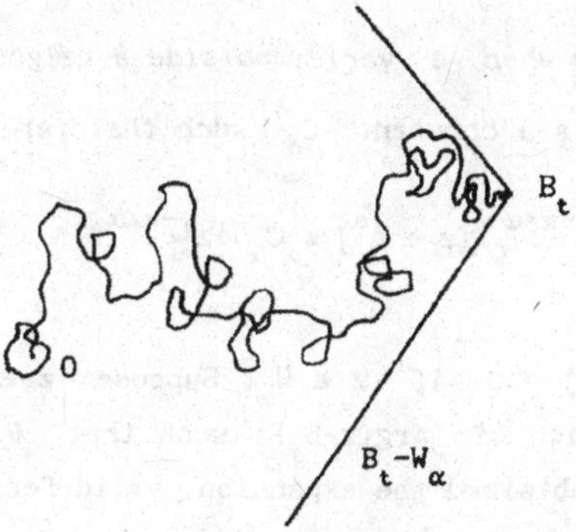

Fig. 1

Notice that $0 \in \Delta^\alpha$. According to the definitions of the previous chapter any $z \in \Delta^\alpha \setminus \{0\}$ is a one-sided cone point with angle α. This gives only a rather special class of one-sided cone points. However it is easy to see that much useful information (such as existence or non-existence, Hausdorff measure properties...) can be derived from the consideration of this special class.

We intend to show that Δ^α (or H_α) $\neq \{0\}$ if $\alpha > \pi/2$. To this end, we will construct a non-trivial measure supported on $\Delta^{\alpha} \setminus \{0\}$. This measure, the so-called local time of cone points, will also be extremely useful when investigating various properties of the cone points. The local time is constructed by approximation from the (suitably normalized) Lebesgue measure on a class of approximate cone points similar to the one used in Chapter III.

For $\varepsilon > 0$ we set

$$\Delta_{\varepsilon}^{\alpha} = \{z \in \mathbb{C} \; ; \; \forall s \le T_{\varepsilon}(z), \; z - B_s \in W_{\alpha}\}$$

where $T_{\varepsilon}(z) = \inf\{s \; ; \; |B_s - z| \le \varepsilon\}$.

Lemma 1 : *(i) For* $z \neq 0$,

$$\lim_{\varepsilon \to 0} \varepsilon^{-\pi/\alpha} P[z \in \Delta_{\varepsilon}^{\alpha}] = h_{\alpha}(z),$$

where

$$h_{\alpha}(re^{i\theta}) = \begin{cases} \frac{4}{\pi} \cos(\frac{\pi\theta}{\alpha}) \, r^{-\pi/\alpha} & \text{if } \theta \in (-\frac{\alpha}{2}, \frac{\alpha}{2}) , \\ 0 & \text{if } \theta \in [-\pi,\pi] \setminus (-\frac{\alpha}{2},\frac{\alpha}{2}) . \end{cases}$$

The convergence is uniform when z *varies outside a neighborhood of* 0.

(ii) There exists a constant C_{α} *such that for any* $z \neq 0$, *for any* $\varepsilon \in (0,1]$,

$$\varepsilon^{-\pi/\alpha} P[z \in \Delta_{\varepsilon}^{\alpha}] \le C_{\alpha} \, |z|^{-\pi/\alpha}.$$

Proof : Clearly $P[z \in \Delta_{\varepsilon}^{\alpha}] = 0$ if $z \notin W_{\alpha}$. Suppose $z \in W_{\alpha}$ and let (θ_s) be the continuous determination of $\arg(z-B_s)$ such that $\theta_0 = \arg(z)$. In the previous chapter we have obtained the expansion, valid for $|z| > \varepsilon$,

$$P[z \in \Delta_{\varepsilon}^{\alpha}] = P[\forall \; s \le T_{\varepsilon}(z) \; , \; |\theta_s| \le \alpha \;]$$

$$= \sum_{k=0}^{\infty} \frac{4}{(2k+1)\pi} \sin\left(\frac{(2k+1)\pi(\frac{\alpha}{2} - \arg(z))}{\alpha}\right) \left(\frac{\varepsilon}{|z|}\right)^{(2k+1)\pi/\alpha} .$$

Both assertions of Lemma 1 are immediate consequences of this formula. □

We shall also need estimates for the probability that two or more given points belong to $\Delta_{\varepsilon}^{\alpha}$.

Lemma 2 : *(i) For* $z, z' \in \mathbb{C} \setminus \{0\}$, $z \neq z'$

$$\lim_{\varepsilon, \varepsilon' \to 0} (\varepsilon\varepsilon')^{-\pi/\alpha} P[z \in \Delta_{\varepsilon}^{\alpha} \; , \; z' \in \Delta_{\varepsilon'}^{\alpha}] = h_{\alpha}(z) \, h_{\alpha}(z'-z) + h_{\alpha}(z') \, h_{\alpha}(z-z').$$

(ii) There exists a constant C'_α *such that for any* $n \geq 1$, $z_1,\dots,z_n$ *distinct points of* $\mathbb{C}\setminus\{0\}$ *and* $\varepsilon_1,\dots,\varepsilon_n \in (0,1]$,

$$(\varepsilon_1\dots\varepsilon_n)^{-\pi/\alpha}\, P[z_1 \in \Delta^\alpha_{\varepsilon_1},\dots,z_n \in \Delta^\alpha_{\varepsilon_n}] \leq (C'_\alpha)^n \sum_{\sigma\in\Sigma_n} \prod_{i=1}^{n} |z_{\sigma(i)} - z_{\sigma(i-1)}|^{-\pi/\alpha}.$$

Here Σ_n *denotes the set of all permutations of* $\{1,\dots,n\}$ *and for* $\sigma \in \Sigma_n$, $z_{\sigma(0)} = 0$ *by convention.*

Proof : (i) We may assume that $z \in W_\alpha$, $z'-z \in W_\alpha$, or $z' \in W_\alpha$, $z-z' \in W_\alpha$. Indeed, if not the case, $P[z \in \Delta^\alpha_\varepsilon, z' \in \Delta^\alpha_{\varepsilon'}]$ will be zero for ε,ε' small enough. Suppose $z \in W_\alpha$, $z'-z \in W_\alpha$. For ε,ε' small, the conditions $z \in \Delta^\alpha_\varepsilon$, $z' \in \Delta^\alpha_{\varepsilon'}$, force $T_\varepsilon(z) < T_{\varepsilon'}(z')$. Also, if $z \in \Delta^\alpha_\varepsilon$, we have automatically $B[0,T_\varepsilon(z)] \subset z - W_\alpha \subset z' - W_\alpha$ (because W_α is a convex cone !) and it is then enough to check that $B[T_\varepsilon(z),T_{\varepsilon'}(z')] \subset z' - W_\alpha$. The desired result follows from Lemma 1 (i) by using the Markov property at time $T_\varepsilon(z)$.

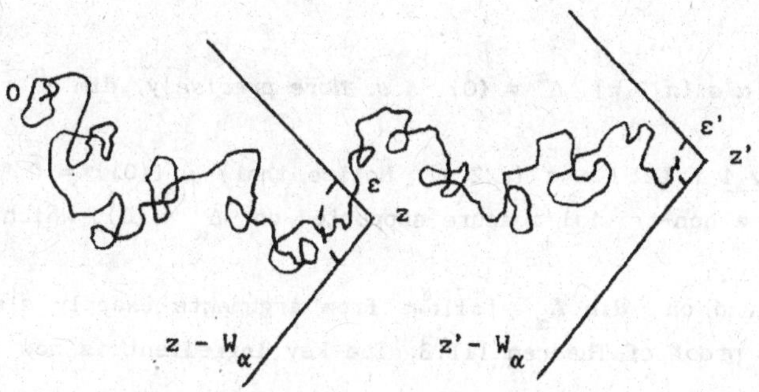

Fig. 2

(ii) We only treat the case $n = 2$. The idea is to deal separately with the cases $T_{\varepsilon_1}(z_1) \leq T_{\varepsilon_2}(z_2)$ and $T_{\varepsilon_1}(z_2) \leq T_{\varepsilon_1}(z_1)$. Suppose first that $|z_2 - z_1| \geq 2(\varepsilon_1 + \varepsilon_2)$. Then the Markov property at time $T_{\varepsilon_1}(z_1)$ and Lemma 1 (ii) give the bound

$$P[z_1\in \Delta^\alpha_{\varepsilon_1},\ z_2\in \Delta^\alpha_{\varepsilon_2},\ T_{\varepsilon_1}(z_1) \leq T_{\varepsilon_2}(z_2)] \leq P[z_1 \in \Delta^\alpha_{\varepsilon_1}]\, C_\alpha\Big(\frac{|z_2-z_1|}{2}\Big)^{-\pi/\alpha} (\varepsilon_2)^{\pi/\alpha}$$

$$\leq C^2_\alpha\, 2^{\pi/\alpha}\, (|z_1||z_2-z_1|)^{-\pi/\alpha}\, (\varepsilon_1\varepsilon_2)^{\pi/\alpha}.$$

If $|z_2 - z_1| \leq 2(\varepsilon_1 + \varepsilon_2)$ we can directly bound $P[z_1 \in \Delta^\alpha_{\varepsilon_1},\ z_2 \in \Delta^\alpha_{\varepsilon_2}]$. Supposing for instance $\varepsilon_1 \leq \varepsilon_2$ we write

$$P[z_1 \in \Delta^\alpha_{\varepsilon_1}, z_2 \in \Delta^\alpha_{\varepsilon_2}] \le P[z_1 \in \Delta^\alpha_{\varepsilon_1}] \le C_\alpha |z_1|^{-\pi/\alpha} \varepsilon_1^{\pi/\alpha}$$
$$\le C_\alpha 4^{\pi/\alpha}(|z_1||z_2-z_1|)^{-\pi/\alpha} (\varepsilon_1\varepsilon_2)^{\pi/\alpha}$$

since in this case $|z_2 - z_1| \le 4\,\varepsilon_2$. □

Theorem 3 : *Suppose* $\alpha \in (\pi/2,\pi]$. *With probability* 1 *there exists a (unique) Radon measure* μ_α *on* $\mathbb{C}$ *such that, for any compact subset* K *of* $\mathbb{C}$,

$$\mu_\alpha(K) = L^2 - \lim_{\varepsilon\to 0} \varepsilon^{-\pi/\alpha}\, m(\Delta^\alpha_\varepsilon \cap K).$$

Moreover, for any $\varepsilon, M > 0$, *there exists w.p.* 1 *a constant* $C_{\varepsilon,M}(\omega)$ *such that, for any square* $[u,u+r] \times [v,v+r]$ *contained in* $[-M,M]^2$,

$$\mu_\alpha([u,u+r] \times [v,v+r]) \le C_{\varepsilon,M}\, r^{2-\frac{\pi}{\alpha}-\varepsilon}. \tag{1}$$

The measure μ_α *is w.p.* 1 *supported on* Δ^α. *Furthermore* $\mu_\alpha(D(0,\varepsilon)) > 0$ *for any* $\varepsilon > 0$, *a.s.*

Corollary 4 : *If* $\alpha \in (\pi/2,\pi]$, $\Delta^\alpha \ne \{0\}$ *a.s. More precisely,* $\dim \Delta^\alpha = 2 - \frac{\pi}{\alpha}$.

Proof of Corollary 4 : Let $\alpha \in (\pi/2,\pi]$. Notice that $\mu_\alpha(\{0\}) = 0$ by (1). Therefore μ_α is a non-trivial measure supported on $\Delta_\alpha \setminus \{0\}$, which implies $\Delta_\alpha \ne \{0\}$.

The upper bound on $\dim \Delta_\alpha$ follows from arguments exactly similar to those used in the proof of Theorem III.3. The key ingredient is now the fact that

$$\liminf_{\varepsilon \to 0} \varepsilon^{-\pi/\alpha}\, m(\Delta^\alpha_\varepsilon \cap K) < \infty \quad , \quad \text{a.s.}$$

The lower bound follows from (1). Let (R_i) be a covering of $\Delta^\alpha \cap [-M,M]^2$ by squares contained in $[-M,M]^2$. Then, a.s.,

$$\sum_i (\mathrm{diam}(R_i))^{2-\frac{\pi}{\alpha}-\varepsilon} \ge (C_{\varepsilon,M})^{-1} \sum_i \mu_\alpha(R_i) \ge (C_{\varepsilon,M})^{-1} \mu_\alpha(\Delta^\alpha \cap [-M,M]^2),$$

since μ_α is supported on Δ^α. Using the last assertion of Theorem 3 we get that $\dim \Delta^\alpha \ge 2 - \frac{\pi}{\alpha} - \varepsilon$ a.s. □

Remark : Since $\Delta^\alpha = \{B_s, s \in H_\alpha\}$, a result of Kaufman [Ka] implies that

$$\dim H_\alpha = 1 - \frac{\pi}{2\alpha} \qquad \text{a.s.}$$

Proof of Theorem 3 : Set $\mu_{\alpha,\varepsilon}(K) = \varepsilon^{-\pi/\alpha}\, m(\Delta^{\alpha}_{\varepsilon} \cap K)$. Then,

$$E[\mu_{\alpha,\varepsilon}(K)\, \mu_{\alpha,\varepsilon'}(K)] = \int_{K\times K} dz\, dz'\ (\varepsilon\varepsilon')^{-\pi/\alpha}\, P[z \in \Delta^{\alpha}_{\varepsilon}, z' \in \Delta^{\alpha}_{\varepsilon'}].$$

Lemma 2 and the dominated convergence theorem imply that

$$\lim_{\varepsilon,\varepsilon'\to 0} E[\mu_{\alpha,\varepsilon}(K)\mu_{\alpha,\varepsilon'}(K)] = 2\int_{K\times K} dz\, dz'\ h_{\alpha}(z)h_{\alpha}(z'-z)$$

(notice that the function $|z|^{-\pi/\alpha}$ is locally integrable since $\alpha > \pi/2$). It follows that $(\mu_{\alpha,\varepsilon}(K))_{\varepsilon>0}$ is Cauchy in L^2, so that we may set :

$$\bar{\mu}_{\alpha}(K) = L^2\text{-}\lim_{\varepsilon\to 0} \mu_{\alpha,\varepsilon}(K).$$

Lemma 2 (ii) and Fatou's lemma give the bound

$$E[(\bar{\mu}_{\alpha}(K))^n] \le n!\,(C'_{\alpha})^n \int_{K^n} dz_1 \ldots dz_n \prod_{i=1}^{n} |z_i - z_{i-1}|^{-\pi/\alpha} \le n!\,(C''_{\alpha})^n\, m(K)^{n(1-\pi/2\alpha)}$$

(notice that, if $m(K)$ is fixed, $\int_K |z-y|^{-\pi/\alpha}\,dz$ is maximal when K is a disk centered at y, with radius $\pi^{-1/2}\, m(K)^{1/2}$). This bound and the multidimensional version of the Kolmogorov lemma imply the existence of a continuous version of the mapping $(a,b,c,d) \longrightarrow \bar{\mu}_{\alpha}([a,b]\times[c,d])$ (for $a\le b$, $c\le d$). Denote by $\mu_{\alpha}([a,b]\times[c,d])$ this continuous version. Obviously $\mu_{\alpha}([a,b]\times[c,d])$ is a nondecreasing function of $[a,b]\times[c,d]$. Standard measure-theoretic arguments show that $\mu_{\alpha}(\cdot)$ can be extended to a Radon measure on $\mathbb{C}$. Furthermore, the monotone class theorem gives $\mu_{\alpha}(K) = \bar{\mu}_{\alpha}(K)$ a.s. for any compact K.

It remains to prove (1). The previous bound on the moments of $\mu_{\alpha}(K)$ and the arguments of the proof of Theorem II-8 give (1) for any dyadic square contained in $[-M,M]^2$. A simple covering argument completes the proof of (1).

Let us check that μ_{α} is a.s. supported on Δ . Let R be a compact rectangle with rational coordinates. We have

$$\mu_{\alpha}(R) \le \liminf_{\varepsilon\to 0} \mu_{\alpha,\varepsilon}(R) \qquad \text{a.s.}$$

Note that $\Delta^{\alpha} = \bigcap_{\varepsilon>0} \Delta^{\alpha}_{\varepsilon}$ and that every $\Delta^{\alpha}_{\varepsilon}$ is closed. It follows that, on $\{R \cap \Delta^{\alpha} = \emptyset\}$ we have for ε small $R \cap \Delta^{\alpha}_{\varepsilon} = \emptyset$ so that $\mu_{\alpha,\varepsilon}(R) = 0$ and $\mu_{\alpha}(R) = 0$.

Finally a scaling argument gives

$$P[\mu_{\alpha}(D(0,1)) > 0] = P[\mu_{\alpha}(D(0,\varepsilon)) > 0] = P[\forall\varepsilon > 0,\ \mu_{\alpha}(D(0,\varepsilon)) > 0].$$

However $P[\mu_{\alpha}(D(0,1) > 0] > 0$ since

$$E[\mu_\alpha(D(0,1))] = \lim_{\varepsilon\to 0} E[\mu_{\alpha,\varepsilon}(D(0,1))] = \int_{D(0,1)} h_\alpha(z)dz$$

by Lemma 1. It is easy to check from the construction of μ_α that the event $\{\forall\varepsilon > 0,\ \mu_\alpha(D(0,\varepsilon)) > 0\}$ is asymptotic. The 0-1 law then gives the desired result. □

Remark : For any $t \in H_\alpha \setminus \{0\}$, the convex hull of $\{B_s, 0 \le s \le t\}$ has a corner at B_t, with opening (less than) α. This comes in contrast to the fact (Theorem III-4) that for a fixed t, w.p. 1 the convex hull of $\{B_s, 0 \le s \le t\}$ has no corners.

2. A stable process embedded in two-dimensional Brownian motion.

At this point, we have proved that, for $\alpha > \pi/2$, $\Delta_\alpha \neq \{0\}$ so that in particular there exist one-sided cone points with angle α. We will prove in the next section that $\Delta_\alpha = \{0\}$ for $\alpha \le \frac{\pi}{2}$. In the present section we will use the local time constructed in Theorem 3 to get certain interesting probabilistic properties of the sets Δ_α and H_α.

Let $(\mathcal{F}_t)$ denote the canonical filtration of B. A random closed subset H of $\mathbb{R}_+$ is called $(\mathcal{F}_t)$-regenerative if $0 \in H$ and :

(i) $\forall t \ge 0$, $\{(s,\omega)\ ;\ s \le t,\ s \in H(\omega)\}$ is $\mathcal{B}_{[0,t]} \otimes \mathcal{F}_t$ measurable ($\mathcal{B}_{[0,t]}$ denotes the Borel σ-field on $[0,t]$)

(ii) For any $(\mathcal{F}_t)$ stopping time T such that $T \in H$ a.s., the set $\{(t-T)_+, t \in H\}$ is independent of $\mathcal{F}_T$ and distributed as H.

With every regenerative set H we can associate its local time process (ℓ_t), defined up to a multiplicative constant. The process (ℓ_t) is càdlàg, non decreasing and $(\mathcal{F}_t)$-adapted. It is characterized (up to a multiplicative constant) by the following two properties :

(i) $\ell_0 = 0$ and ℓ_t increases only on H.

(ii) For any stopping time T such that $T \in H$ a.s., the process $\ell_t^T = \ell_{T+t} - \ell_T$ is independent of $\mathcal{F}_T$ and distributed as (ℓ_t).

Theorem 5 : *Let* $\alpha \in (\frac{\pi}{2}, \pi]$. *The set* Δ^α *is an* $(\mathcal{F}_t)$*-regenerative set. Its local time may be defined by :*

$$\ell_t^\alpha = \mu_\alpha(\{B_s, 0 \le s \le t\}).$$

Set

$$\tau_t^\alpha = \inf\{s,\ \ell_s^\alpha > t\} < \infty \qquad a.s.$$

The process (τ_t^α) *is a stable subordinator with index* $1 - \pi/2\alpha$. *The process* $(B(\tau_t^\alpha))$ *is a two-dimensional stable process with index* $2 - \pi/\alpha$. *In particular,* $(B^1(\tau_t^\alpha))$ *is a stable subordinator and* $(B^2(\tau_t^\alpha))$ *is a symmetric stable process. Finally,* H_α *coincides with the closure of the range of* τ^α *and* Δ^α *coincides with the closure of the range of* $B \circ \tau^\alpha$.

Before proving Theorem 5 let us discuss the limiting case $\alpha = \pi$. In this case it is easy to check that

$$H_\pi = \{t\ ;\ B_t^1 = \sup_{s\le t} B_s^1\},$$

so that H_π coincides with the zero set of the process $\sup_{s\le t} B_s^1 - B_t^1$, which by a famous theorem of Lévy is a (one-dimensional) reflecting Brownian motion. Therefore, H_π is distributed as the zero set of a linear Brownian motion, which is the typical example of a regenerative set. Moreover, ℓ_t^π ($= C \sup_{s\le t} B_s^1$) is distributed as (C times) the local time process at 0 of a linear Brownian motion, so that τ_t^π is a stable subordinator with index 1/2. Finaly, $B^1(\tau_t^\pi) = C^{-1}t$ (the stable subordinator with index 1 !) and $B^2(\tau_t^\pi)$ is a symmetric Cauchy process. The latter fact was first discovered by Spitzer [Sp1] and has been used since by many authors.

In conclusion, when $\alpha = \pi$, the different assertions of Theorem 5 are well-know facts. It turns out that all of them carry over to the general case $\alpha \in (\frac{\pi}{2},\pi]$. It is interesting to note that the last assertions of Theorem 5 give a probabilistic description of the random sets H_α and Δ^α.

<u>Proof of Theorem 5</u> : Let T be a stopping time such that $T \in H_\alpha$ a.s. Let $B^{(T)}$ denote the Brownian motion $B_t^{(T)} = B_{T+t} - B_T$ $(t \ge 0)$.

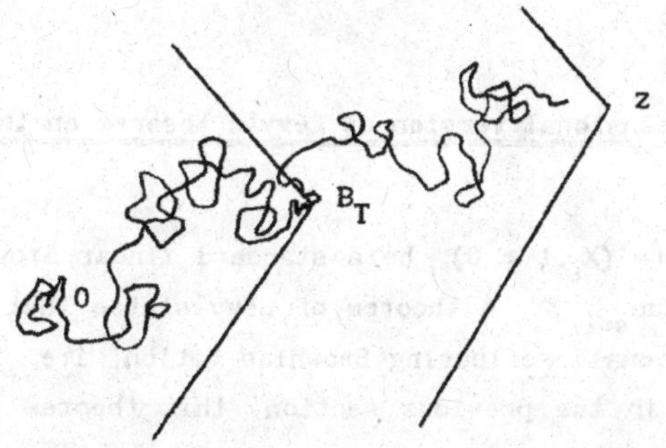

Fig. 3

Then $B^{(T)}$ is independent of $\mathcal{F}_T$. Moreover, a simple geometric argument shows that :

$$\{(t-T)_+ , t \in H_\alpha\} = H_\alpha^{(T)} ,$$

with an obvious notation. The fact that H_α is an $\mathcal{F}_t$-regenerative set follows at once.

Note that $\ell_t^\alpha = \mu_\alpha(B[0,t])$ increases only on H_α. Furthermore, the construction of μ_α easily gives :

$$\mu_\alpha(B[0,T+t]) = \mu_\alpha(B[0,T]) + \mu_\alpha(B[T,T+t]) = \mu_\alpha(B[0,T]) + \mu_\alpha^{(T)}(B^{(T)}[0,T]).$$

Therefore $(\ell_{T+t}^\alpha - \ell_T^\alpha)$ is independent of $\mathcal{F}_T$ and distributed as (ℓ_t^α). It follows that (ℓ_t^α) is a local time for H_α.

By the general theory of regenerative sets, (τ_t^α) is an $\mathcal{F}_{\tau_t^\alpha}$-subordinator (this also follows from the previous arguments) so that $(B(\tau_t^\alpha))$ is also an $\mathcal{F}_{\tau_t^\alpha}$-Lévy process. Next, fix $\lambda > 0$ and set

$$\tilde{B}_t = \lambda B_{t/\lambda^2} .$$

Then, for any $\varepsilon > 0$,

$$\tilde{\Delta}_\varepsilon^\alpha = \lambda \Delta_{\varepsilon/\lambda}^\alpha ,$$

and after some easy manipulations,

$$\tilde{\tau}_t^\alpha = \lambda^2 \tau^\alpha_{t/\lambda^{2-\pi/\alpha}} , \quad \tilde{B}(\tilde{\tau}_t^\alpha) = \lambda B(\tau^\alpha_{t/\lambda^{2-\pi/\alpha}}).$$

It follows that τ^α is stable with index $1 - \pi/2\alpha$ and $B \circ \tau^\alpha$ is stable with index $2 - \pi/\alpha$.

Geometric considerations entail that $B^1 \circ \tau^\alpha$ is a subordinator and $B^2 \circ \tau^\alpha$ is symmetric. Finally, the general theory of regenerative sets shows that H_α is the closure of the range of τ^α. □

3. A two-dimensional version of Lévy's theorem on the Brownian supremum process.

Let $X = (X_t, t \geq 0)$ be a standard linear Brownian motion started at 0 and $S_t = \sup_{s\leq t} X_s$. A theorem of Lévy states that the process $S - X$ is a (one-dimensional) reflecting Brownian motion, i.e. is distributed as $|X|$. As we noticed in the previous section, this theorem is closely related to the structure of H_π, which coincides with the zero set of $\sup_{s\leq t} B_s^1 - B_t^1$. We will now prove that for any $\alpha \in (\pi/2,\pi)$, H can also be interpreted as the zero set of a two-dimensional reflecting Brownian motion in the wedge W_α. This result is related to a two-dimensional version of Lévy's theorem.

We first recall a few basic facts about reflecting Brownian motion in a wedge.

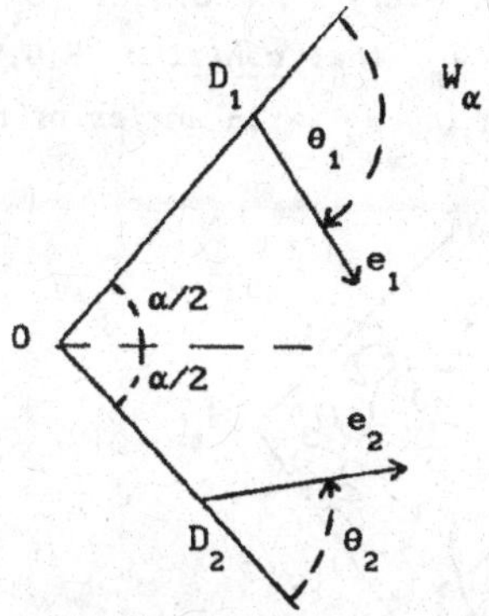

Fig. 4

We set $D_1 = \{r\, e^{i\alpha/2}\ ;\ r \geq 0\}$, $D_2 = \{r\, e^{-i\alpha/2}\ ;\ r \geq 0\}$. Let $\theta_1, \theta_2 \in (0,\pi)$, and

$$e_1 = e^{i(\alpha/2-\theta_1)}\ ,\ e_2 = e^{-i(\alpha/2-\theta_2)}\ .$$

A process $Z = (Z_t\ ;\ t \geq 0)$ with valued in W_α is called reflecting Brownian motion with angles of reflection θ_1, θ_2 if :

$$Z_t = Y_t + A^1_t\, e_1 + A^2_t\, e_2$$

where

- Y is a two-dimensional Brownian motion
- A^1, A^2 are two continuous non-decreasing processes adapted to the filtration of Y, and A^1 (resp. A^2) increases only when $Z_t \in D_1$ (resp. $Z_t \in D_2$).

This is not the most general presentation of reflecting Brownian motion in a wedge. It will however be sufficient to our purposes. Notice that it is far from obvious (and in fact not true) that such a process Z exists for all values of θ_1, θ_2. Assuming that Z exists, it can be proved that $\{t,\ Z_t = 0\}$ contains non-zero times iff $\theta_1 + \theta_2 > \pi$. To check the sufficiency of this condition, one introduces the function: For $r > 0$, and $|\theta| \leq \alpha/2$,

$$\psi(re^{i\theta}) = r^\xi \sin(\ \xi\theta + \frac{\theta_1-\theta_2}{2}\ )\ ,\quad \text{where}\ \ \xi = \frac{\theta_1+\theta_2-\pi}{\alpha} > 0\ .$$

An application of Itô's formula shows that $\psi(Z_t)$ is a local martingale on the time interval $[0,\tau)$, where $\tau = \inf\{\ s\ ;\ Z_s = 0\ \}$. The proof can then be completed by standard arguments (see [VW] for details).

If K is a compact subset of $\mathbb{C}$, the intersection of all cones of the type $z - W_\alpha$ that contain K is again a cone of the same type, which is the smallest one that contains K.

<u>Theorem 6</u> : *Let* $\alpha \in (0,\pi)$. *For every* $t \geq 0$, *let* S_t *be the vertex of the smallest cone of the type* $z - W_\alpha$ *that contains* $B[0,t]$. *The process* $S - B$ *is a reflecting Brownian motion in* W_α *with angles of reflection* $\theta_1 = \theta_2 = \alpha$.

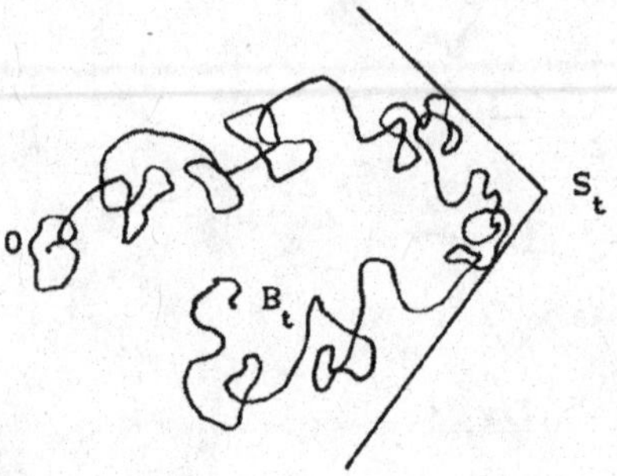

Fig. 5

<u>Corolllary 7</u> : $\Delta_\alpha \neq \{0\}$ *iff* $\alpha > \pi/2$.

<u>Proof of Corollary 7</u> : Note that H_α is exactly the zero set of S-B. Then we may apply the previous criterion observing that $\theta_1 + \theta_2 = 2\alpha > \pi$ iff $\alpha > \pi/2$. In fact, we do not need this criterion. The case $\alpha > \pi/2$ was treated in Corollary 4. Then it suffices to check that $\Delta_{\pi/2} = \{0\}$. However when $\alpha = \pi/2$, W_α is a quadrant and the directions of reflection are normal. It follows that $S_t - B_t = |\beta_t|\, e^{i\pi/4} + |\gamma_t|\, e^{-i\pi/4}$, where β, γ are two independent linear Brownian motions. By Corollary II-2, $\{t\ ;\ S_t - B_t = 0\} = \{0\}$. □

<u>Proof of Theorem 6</u> : Set $f_1 = e^{i\alpha/2}$, $f_2 = e^{-i\alpha/2}$. We have :

$$B_t = U_t f_1 + V_t f_2 ,$$

where U, V are two (correlated) linear Brownian motions. It is easy to check that

$$S_t = \hat{U}_t f_1 + \hat{V}_t f_2 ,$$

where $\hat{U}_t = \sup_{s \leq t} U_s$, $\hat{V}_t = \sup_{s \leq t} V_s$. Then,

$$S_t - B_t = -B_t + \hat{U}_t f_1 + \hat{V}_t f_2 .$$

Now notice that $\hat{U}_t$ increases only when $U_t = \hat{U}_t$ that is when $S_t - B_t \in D_2$, and similarly for $\hat{V}_t$. This gives the desired representation with $e_1 = f_2$, $e = f$, hence $\theta = \theta = \alpha$. □

If we combine Theorem 6 and Theorem 5 we get that, for a certain class of reflecting Brownian motions in a wedge, the zero set is exactly the closure of the range of a stable subordinator. This result in fact holds in great generality (see Williams [Wi3]).

As a by-product of the previous statements, we get the following result. Suppose that β, γ are two linear Brownian motions started at 0, correlated in the sense that $\langle\beta,\gamma\rangle_t = \rho\, t$, for some constant ρ. If $\rho > 0$, the set $\{ t \ge 0,\ \beta_t = \sup_{s\le t} \beta_s$ and $\gamma_t = \sup_{s\le t} \gamma_s \}$ is non-empty and is distributed as the range of a stable subordinator. If $\rho \le 0$, this set is empty.

Using the arguments of the proof of Theorem III.3 it is easy to deduce from Corollary 7 that there are no one-sided cone points with angle $\alpha < \pi/2$. The problem of the existence of one-sided cone points with angle $\pi/2$ remains open.

4. More about the first intersection of a line with the Brownian path.

At the end of Chapter III we obtained the following result. For $t \ge 0$ set

$$x_t = x_t(0) = \sup(B[0,t] \cap \mathbb{R}).$$

Then, for a fixed $t > 0$, with probability 1 for any $\beta > 0$,

$$(x_t + \mathring{W}_\beta) \cap B[0,t] \neq \emptyset \tag{2}$$

($\mathring{W}_\beta$ denotes the interior of W_β).

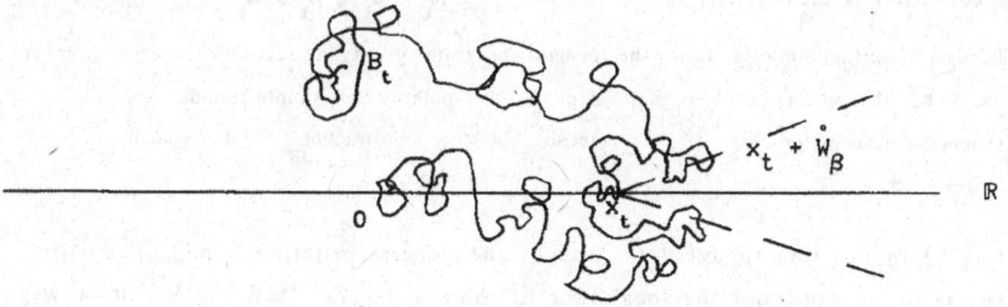

Fig. 6

We will now show that this property fails to hold at certain exceptional times t : these exceptional times will be such that $B_t \in \mathbb{R}$ and B_t is a one-sided cone point with angle $\alpha \in (\pi,2\pi)$.

Fix $\alpha \in (0,2\pi)$. If $\beta = 2\pi-\alpha$, property (2) is equivalent to the fact that $B[0,t]$ is not contained in $x_t - W_\alpha$.

For any $t \ge 0$ denote by $R_t \in \mathbb{R}$ the vertex of the smallest cone of the type $r - W$ $(r \in \mathbb{R})$ that contains $B[0,t]$.

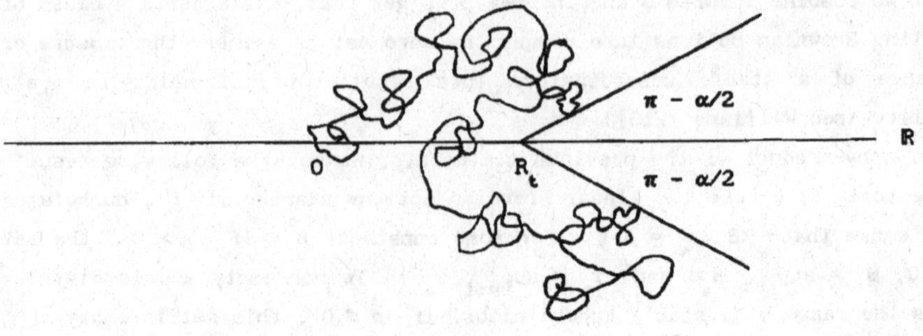

Fig. 7

Theorem 8 : The process $R_t - B_t$ is a reflecting **Brownian** motion in the wedge W_α with angles of reflection $\theta_1 = \theta_2 = \alpha/2$ (equivalently $e_1 = e_2 = 1$). In particular, the zero set of $R - B$ contains non-zero times iff $\alpha > \pi$.

Observe that if $R_t - B_t = 0$ for $t > 0$ then obviously $X_t = R_t$ and $B[0,t] \subset X_t - W_\alpha$, so that property **(2)** does not hold for $\beta \leq 2\pi - \alpha$.

Proof of Theorem 8 : The proof is similar to that of Theorem 6. It is easy to check that $R_t = \sup_{s\leq t} (B^1_s + |B^2_s| \cot g\, \alpha/2)$ where $\cot g\, x = \frac{\cos x}{\sin x}$. Next observe that R_t increases only when

$$B^1_t + |B^2_t| \cot g\, \alpha/2 = \sup_{s\leq t} (B'_s + |B^2_s| \cot g\, \alpha/2)$$

and this condition is clearly equivalent to $R_t - B_t \in D_1 \cup D_2$.

The last assertion follows from the general criterion given in Section 3 (when $\alpha = \pi$, the given result is equivalent to the polarity of simple points for the symmetric Cauchy process : recall Spitzer's construction of the Cauchy process...). $\square$

It is again possible to avoid the use of the general criterion. One possibility is to **contruct** the local time of the set $\{t, R_t = B_t\}$ in a way similar to what we did in Section 1. The analogues of the sets $\Delta^\alpha_\varepsilon$ are then subsets of $\mathbb{R}_+$ and the key technical ingredient is the fact that the function $|x|^{-\pi/\alpha}$ is **locally** integrable on $\mathbb{R}$ if $\alpha > \pi$.

Still another method would be to extend Corollary 4 to the case $\alpha > \pi$ (this can be done but is non-trivial). We get that $\dim \Delta^\alpha > 1$ if $\alpha > \pi$. Then some Hausdorff measure arguments show that $\Delta^\alpha - \{0\}$ must intersect any fixed horizontal line with positive probability. Finally the zero-one law entails that $\Delta^{\alpha} - \{0\}$ intersects $\mathbb{R}$ w.p. 1.

Bibliographical Notes . *One-sided cone points with angle less than* π *were discovered simultaneously by Burdzy* [B1] *and Shimura* [Sh2] *(see also* [Sh1] *for a related work). A very simple proof of their existence has been given by Adelman* [A2]. *The approach developed in this chapter follows closely* [L7], *with the important simplification that we deal only with the case* $\alpha < \pi$. *This aproach is certainly not the shortest one, but it leads to the local time of cone points, which plays an important role in many applications. In particular, the local time allows one to understand how the process behaves just before arriving at a cone point (see* [L7] *and also* [B3] *for certain related results). Sharp results about the Hausdorff measure of cone points are given in Evans* [Ev1]. *The construction of the symmetric Cauchy process recalled in Section 2 was given by Spitzer* [S1]. *The idea of Theorems 6 and 8 was discovered independently in El Bachir* [EB] *and in* [L7] . *However, it seems that this idea was incorrectly applied in* [EB], *where the oblique reflection property of the process* $S - B$ *was unnoticed. See also Burdzy* [B3] *for applications of this idea and for many results related to Theorem 8. Information about reflected Brownian motion in a wedge may be found in Varadhan and Williams* [VW] *and in Williams* [Wi1], [Wi2]. *The fact that the inverse local time at the vertex is a stable subordinator is proved in great generality in Williams* [Wi3].

CHAPTER V

Burdzy's theorem on twist points.

1. Twist points of the planar Brownian motion.

We consider a standard complex-valued Brownian motion $(B_t, \ t \geq 0)$ started at 0. We denote by F the unbounded connected component of $\mathbb{C} \setminus B[0,1]$ $(B[0,1] = \{B_s, \ 0 \leq s \leq 1\})$. Then ∂F consists of all points of B[0,1] that can be reached from the "exterior" of B[0,1] along a continuous curve. In other words, $z \in B[0,1]$ is in ∂F iff there exists a continuous function $\varphi : [0,1] \longrightarrow \mathbb{C}$ such that :

$$(1) \quad \varphi(s) \in F \quad , \quad \forall s \in [0,1) ,$$

$$(2) \quad \varphi(1) = z .$$

Let $z \in \partial F$. We say that z is a twist point of ∂F if there exists a continuous function φ satisfying (1) and (2) and such that:

$$\limsup_{s\to1, s<1} \arg(\varphi(s) - z) = +\infty ,$$

$$\liminf_{s\to1, s<1} \arg(\varphi(s) - z) = -\infty .$$

Here and in what follows $\arg(\varphi(s)-z)$ denotes a continuous determination of the argument of $\varphi(s) - z$. Fig. 1 gives a very crude idea of the shape of the boundary near a twist point.

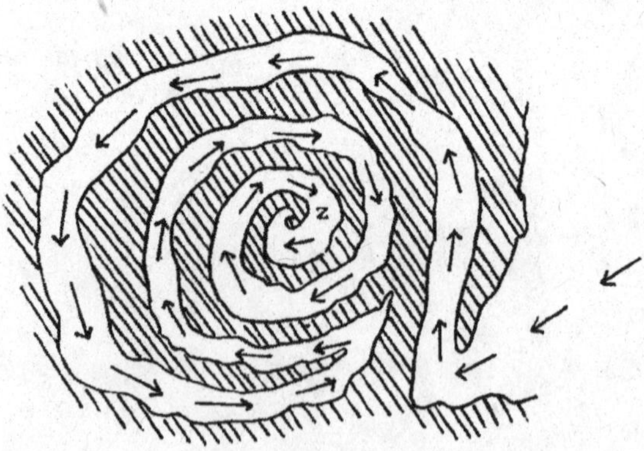

Fig. 1

It is a simple exercise to check that any point of ∂F that is also a two-sided cone point is not a twist point. Two-sided cone points form a dense subset of ∂F (indeed this is true for two-sided cone points with angle π). Nonetheless the next theorem shows that, in a sense, most of the points of ∂F are twist points.

Theorem 1 : *With probability 1, in the sense of harmonic measure almost all points of ∂F are twist points.*

We can rephrase Theorem 1 as follows. Let B' be another complex Brownian motion, independent of B and started at $z_1 \neq 0$. Let

$$T = \inf\{ t \geq 0,\ B'_t \in B[0,1] \}.$$

Then the point B'_T is w.p. 1 a twist point of ∂F.

The proof of Theorem 1 uses the following three ingredients.

(a) A theorem of McMillan in complex analysis.

(b) Certain estimates on harmonic measure.

(c) The bounds on the Hausdorff dimension of two-sided cone points derived in Chapter III.

2. Some results in complex analysis.

It will be convenient to work on the Riemann sphere $\hat{\mathbb{C}} = \mathbb{C} \cup \{ \infty \}$. Then, $\hat{F} := F \cup \{ \infty \}$ is a simply connected open subset of $\hat{\mathbb{C}}$. By the Riemann mapping theorem, we may find a one-to-one analytic mapping f from the open unit disk D onto $\hat{F}$. By Fatou's theorem, for $d\theta$ a.a. $\theta \in [0,2\pi]$, the radial limit

$$\lim_{\substack{r\to 1\\ r<1}} f(re^{i\theta})$$

exists. This limit is simply denoted by $f(e^{i\theta})$.

In our situation, $\mathbb{C} \setminus F$ is locally connected and it can be shown (see Pommerenke [Po, Chapter IX]) that the radial limit exists for every $\theta \in [0,2\pi)$, and that the extended mapping $f : \bar{D} \longrightarrow \hat{F} \cup \partial F$ is continuous and onto. Notice that this extended mapping needs not be one-to-one (in fact, in the present setting, f will not be one-to-one : it can be shown that $f_{|\partial D}$ is one-to-one iff ∂F has no cut points, and two-dimensional Brownian paths do have cut points, as was recently shown by Burdzy).

For any $\zeta \in \partial D$ and $r \in (0,1)$, we define the Stolz angle $S(\zeta,r)$ as the interior of the convex hull of $\{\zeta\} \cup D(0,r)$. We say that f has angular derivative ω at ζ if, for any $r \in (0,1)$,

$$\lim_{\substack{z \to \zeta \\ z \in S(\zeta,r)}} \frac{f(z) - f(\zeta)}{z - \zeta} = \omega$$

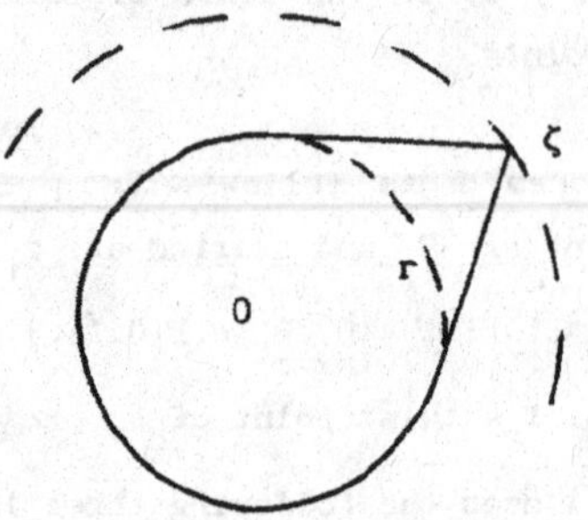

Fig. 2

Finally, we say that $\zeta \in \partial D$ is an f-twist point if $\arg(f(z) - f(\zeta))$ is unbounded above and below along every curve in D ending at ζ. Clearly, if f has a nonzero angular derivative at ζ, ζ cannot be an f-twist point.

The following theorem due to McMillan [MM] plays a basic role in the proof of Theorem 1.

Theorem. *For a.a. $\zeta \in \partial D$, either f has a non-zero angular derivative at ζ or ζ is an f-twist point.*

Let us turn to the proof of Theorem 1, using McMillan's theorem. We denote by T_f the set of all f-twist points and by A_f the set of all points $\zeta \in \partial D$ such that f has a non-zero angular derivative at ζ. We also denote by $T_{\partial F}$ the set of all twist points of ∂F. We observe that:

$$f^{-1}(\partial F \setminus T_{\partial F}) \subset (\partial D \setminus T_f) .$$

Indeed, let $\zeta \in \partial D$ be such that $f(\zeta)$ is not a twist point of ∂F. Then, for any curve $(\varphi(t), 0 \le t < 1)$ in D ending at ζ, $(f(\varphi(t)), 0 \le t < 1)$ is a curve in F ending at $f(\zeta)$, so that $\arg(f(\varphi(t)) - f(\zeta))$ must be bounded above or below.

Using the conformal invariance of harmonic measure (see Section II-2), it is then enough to check that $\partial D \setminus T_f$ has Lebesgue measure 0. However, McMillan's theorem states that $\partial D \setminus (T_f \cup A_f)$ has measure zero. To complete the proof, it suffices to prove that A_f has measure 0, or, by the conformal invariance of harmonic measure again, that $f(A_f)$ is contained in a set of harmonic measure zero. We need the following elementary lemma.

<u>Lemma 2</u> : *Suppose that* $z = f(\zeta)$ *for some* $\zeta \in A_f$. *Then, for any* $\alpha < \pi$, *there exist* $\varepsilon > 0$ *and an open wedge* W_α *with vertex* z *and angle* α *such that* $(W_\alpha \cap D(z,\varepsilon)) \subset F$.

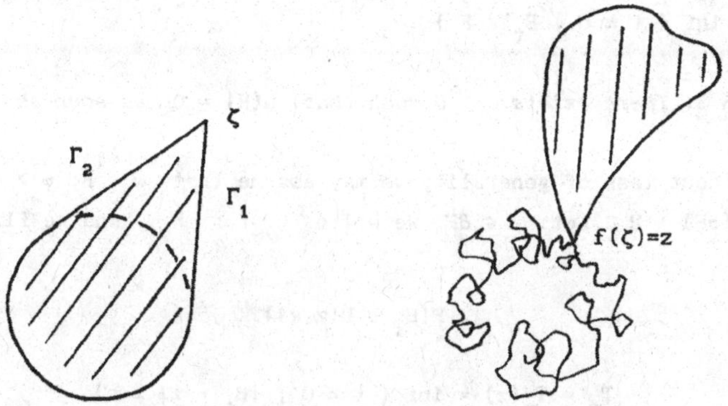

Fig. 3

<u>Proof</u>: Fix $r \in (0,1)$ and denote by $\Gamma_1 \cup \Gamma_2 \cup \Gamma_3$ the boundary of the Stolz angle $S(\zeta,r)$, as on Figure 3. Then $f(S(\zeta,r))$ is a simply connected subset of F with boundary $f(\Gamma_1) \cup f(\Gamma_2) \cup f(\Gamma_3)$. The fact that f has a non-zero angular derivative at ζ implies that $f(S(\zeta,r))$ contains $D(z,\varepsilon_r) \cap W_{\alpha_r}$ for some $\varepsilon_r > 0$ and some open wedge W_{α_r} with vertex z and angle α_r. Moreover by choosing r close to one, we can get α_r as close to π as desired. □

It follows from Lemma 2 that $f(A_f)$ is contained in the set of two-sided cone points with angle β of the Brownian path B, for any $\beta < \pi$. The results of Chapter III give

$$\dim f(A_f) = 0.$$

To complete the proof of Theorem 1, it suffices to prove that, for any subset H of ∂F such that $\dim H = 0$, the harmonic measure of H is zero. This follows from Makarov's theorem, which states (in particular) that the harmonic measure of H is 0 as soon as $\dim H < 1$. Clearly, we do not need the full strength of Makarov's theorem. In the next section, we will give a probabilistic proof of a much weaker statement, which nonetheless suffices to complete the proof of Theorem 1.

3. An estimate for harmonic measure

The results of this section apply to any simply connected open set $F \subset \hat{\mathbb{C}}$ such that $\hat{\mathbb{C}} \setminus F$ contains more than one point. We fix $z_0 \in F$ and we let $\mu = \mu_{z_0}$ be the associated harmonic measure on ∂F. In probabilistic terms

$$\mu(A) = P_{z_0}[B_T \in A]$$

where $T = \inf\{ t \geq 0 , B_t \notin F \}$.

Proposition 3: *There exists* $\alpha > 0$ *such that* $\mu(H) = 0$ *as soon as* $\dim H < \alpha$.

Proof : Without loss of generality we may assume that $\infty \in F$, $\infty > d(z_0, \partial F) > 1$ and $\operatorname{diam}(\partial F) > 2$. Let $z \in \partial F$. We write P for P_{z_0} and we first look for a bound on :

$$P[B_T \in D(z,\varepsilon)]$$

Set

$$T_\varepsilon = T_\varepsilon(z) = \inf\{ t \geq 0 ; |B_t - z| \leq \varepsilon\}$$

and

$$L_1 = \sup\{ t \leq T_\varepsilon ; |B_t - z| = 1\}$$

We claim that, on $\{ B_T \in D(z,\varepsilon)\}$, z belongs to the unbounded component of $\mathbb{C} \setminus B[L_1, T_\varepsilon]$. Indeed, if this were not the case, the component of z would be contained in $D(z,1)$, and so would be the connected set $\mathbb{C} \setminus F$ (which is contained in $\mathbb{C} \setminus B[L_1, T_\varepsilon]$ on $\{ B_T \in D(z,\varepsilon)\}$). This gives a contradiction since we have assumed $\operatorname{diam}(\partial F) > 2$.

For every integer $m \geq 1$, set

$$T_{(m)} = \inf\{ t \geq 0 ; |B_t - z| \leq 2^{-m} \}$$

$$L_{(m)} = \sup\{ t < T_{(m)} ; |B_t - z| = 2^{-m+1} \}$$

By the previous arguments,

$$P[B_T \in D(z,2^{-m})] \leq P[\bigcap_{k=1}^{m} A_k]$$

where

$$A_k = \{ z \text{ belongs to the unbounded component of } \mathbb{C} \setminus B[L_{(k)}, T_{(k)}]\}.$$

However, the strong Markov property implies that the events A_k, $k = 1,2,\ldots$ are independent, and a scaling argument shows that they have the same probability $c < 1$ (use the skew-product representation to check that $c < 1$). Therefore,

$$P[B_T \in D(z,2^{-m})] \leq c^m$$

and also for $\varepsilon \in (0,1/2)$,

$$P[B_T \in D(z,\varepsilon)] \le \varepsilon^a ,$$

for some constant $a > 0$.

It is then easy to check that Proposition 3 holds with $\alpha = a$. Indeed if $\dim H < a$ we may find a covering of H by disks $D(z_i,\varepsilon_i)$ with $z_i \in H$, $\varepsilon_i \in (0,1/2)$, in such a way that

$$\sum_i (\varepsilon_i)^a \le \delta$$

where δ is any fixed positive number. Then

$$P_{z_0}[B_T \in H] \le \sum_i P_{z_0}[B_T \in D(z_i,\varepsilon_i)] \le \sum_i (\varepsilon_i)^a \le \delta$$

and so $P_{z_0}[B_T \in H] = 0$, since δ was arbitrary. □

Bibliographical notes. Theorem 1 is due to Burdzy [B3] . *The idea of this result was already present, in a heuristic form, in Lévy* [Lé, p.239] *(see Chapter I). Our proof is somewhat different from Burdzy's one and perhaps simpler. The needed results of complex analysis, including the proof of McMillan's theorem, may be found in Pommerenke* [Po]. *Burdzy* [B4] *proves the existence of cut points on two-dimensional Brownian paths. Proposition 3 is a first step towards a probabilistic proof of Makarov's theorem* [Ma]. *K. Burdzy has pointed out that his recent work with G.F. Lawler* [BL1,BL2] *allows one to prove Proposition 3 with* $\alpha = 1/\pi^2$. *See also Bishop* [Bi] *for some recent related work. An interesting problem is to determine the Hausdorff dimension of* ∂F *(in the notation of Section 1). Mandelbrot has conjectured that the dimension of* ∂F *is* $4/3$. *See Burdzy and Lawler* [BL2] *for some recent progress on this problem.*

CHAPTER VI

Asymptotics for the Wiener sausage.

1. The definition of the Wiener sausage.

In this chapter, B is a Brownian motion in $\mathbb{R}^d$. As usual we make the convention that B starts from y under the probability P_y, and we write P for P_o.

Definition : *Let* K *be a compact subset of* $\mathbb{R}^d$ *and* $a, b \in \mathbb{R}_+$, $a \le b$. *The Wiener sausage* $S_K(a,b)$ *is defined by*

$$S_K(a,b) = \{y \in \mathbb{R}^d \; ; \; y - B_s \in K \text{ for some } s \in [a,b]\} = \bigcup_{a\le s\le b} (B_s + K)$$

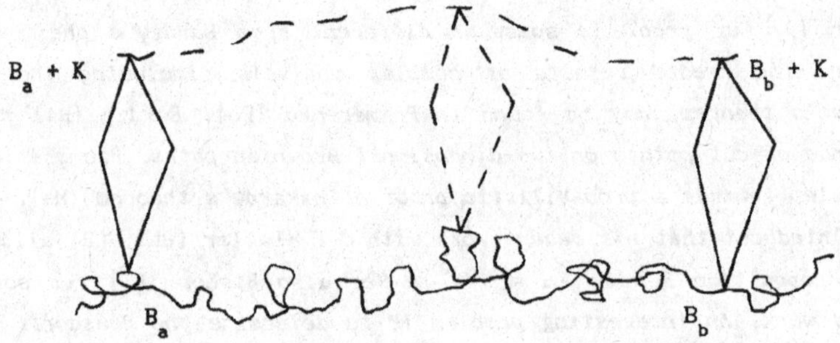

Fig. 1

When K is a closed ball centered at 0, $S_K(a,b)$ is a tubular neighborhood of $B[a,b]$.

We shall be interested in the following two problems:

(i) What is the asymptotic behavior of $m(S_K(0,t))$ as $t \to \infty$?

(ii) What is the asymptotic behavior of $m(S_{\varepsilon K}(0,1))$ as $\varepsilon \to 0$?

Notice that a scaling transformation gives

$$m(S_K(0,t)) \overset{(d)}{=} t^{d/2}\, m(S_{t^{-1/2}K}(0,1))$$

so that, up to some extent, questions (i) and (ii) are equivalent.

Let us briefly discuss question (i). The process $m(S_K(0,t))$ is subadditive, meaning that

$$m(S_K(0,t+s)) \le m(S_K(0,t)) + m(S_K(0,s)) \circ \theta_t$$

where θ_t is the usual shift on Brownian paths. This property is obvious since $m(S_K(0,s)) \circ \theta_t = m(S_K(t,t+s))$ and $S_K(0,t+s) = S_K(0,t) \cup S_K(t,t+s)$. Then Kingman's subadditive ergodic theorem gives :

$$\text{(1)} \qquad \frac{1}{t}\, m(S_K(0,t)) \xrightarrow{\text{a.s.},L^1} C_K$$

for some constant $C_K \ge 0$. If $d \ge 3$, C_K can be identified as the Newtonian capacity of K. However, if $d = 1$ or 2 ($d = 2$ is the only interesting case) $C_K = 0$ for any compact set K, so that (1) does not give much information on the limiting behavior of $m(S_K(0,t))$.

In this chapter we will put the emphasis on question (ii). Our approach is independent of Kingman's theorem and applies as well to any dimension $d \ge 2$. Furthermore, it may be extended to diffusion processes more general than Brownian motion.

For simplicity we write $S_{\varepsilon K} = S_{\varepsilon K}(0,1)$. Our approach consists of two steps of independent interest:

1. Estimation of the mean value $E[m(S_{\varepsilon K})]$.
2. Bounds on the fluctuations of $m(S_{\varepsilon K})$.

The proofs make use of certain results of probabilistic potential theory that are recalled in the next section.

2. Potential-theoretic preliminaries.

Let ζ denote an exponential time with parameter $\lambda > 0$, independent of B. It will be convenient to work with the process B killed at time ζ, which is a symmetric Markov process with Green function :

$$G_\lambda(x,y) = G_\lambda(y-x) = \int_0^\infty ds\, e^{-\lambda s}\, p_s(x,y)$$

where $p_s(x,y) = (2\pi s)^{-d/2} \exp - |y-x|^2/2s$. It is easily checked that :

- if $d \ge 3$,

$$\text{(2)} \qquad G_\lambda(x,y) \underset{|y-x|\to 0}{\sim} G_0(x,y) = C_d |y-x|^{2-d} \;;$$

- if $d = 2$,

$$\text{(3)} \qquad G_\lambda(x,y) \underset{|y-x|\to 0}{\sim} \frac{1}{\pi} \log \frac{1}{|y-x|}\,.$$

Let K be a compact subset of $\mathbb{R}^d$. Assume that K is non-polar and set

$$T_K = \inf\{t \,;\, B_t \in K\} \leq +\infty.$$

A basic formula of probabilistic potential theory gives the hitting probability of K for the process B killed at time ζ. For any $y \in \mathbb{R}^d \setminus K$,

$$P_y(T_K < \zeta) = \int_K G_\lambda(y,z)\, \mu_K^\lambda(dz) \tag{4}$$

where μ_K^λ is a finite measure supported on K, the λ-equilibrium measure of K. The total mass of μ_K^λ is denoted by $C_\lambda(K)$ and called the λ-capacity of K. The fact that K is non-polar is equivalent to $C_\lambda(K) > 0$ for some (or for any) $\lambda > 0$. Finally,

$$C_\lambda(K) = \left(\inf_{\mu\in\mathcal{P}(K)} \int \mu(dy)\,\mu(dz)\, G_\lambda(y,z)\right)^{-1} \tag{5}$$

where $\mathcal{P}(K)$ denotes the set of all probability measures supported on K.

The previous results also hold for $\lambda = 0$, i.e. $\zeta \equiv +\infty$, when $d \geq 3$. The quantity $C_0(K)$ is the Newtonian capacity of K.

We now observe that $y \in S_{\varepsilon K}$ iff $T_{y-\varepsilon K} \leq 1$. It will therefore be important to get information on the distribution function of $T_{y-\varepsilon K}$.

__Lemma 1__ : *Suppose that* K *is non-polar.*

(i) If $d \geq 3$,

$$C_\lambda(\varepsilon K) \underset{\varepsilon\to 0}{\sim} \varepsilon^{d-2}\, C_0(K)$$

and, for any $y \neq 0$,

$$\lim_{\varepsilon\to 0} \varepsilon^{2-d}\, P[T_{y-\varepsilon K} < \zeta] = C_0(K)\, G_\lambda(0,y)$$

(ii) If $d = 2$,

$$C_\lambda(\varepsilon K) \underset{\varepsilon\to 0}{\sim} \pi\,(\log 1/\varepsilon)^{-1}$$

and for any $y \neq 0$

$$\lim_{\varepsilon\to 0} (\log 1/\varepsilon)\, P[T_{y-\varepsilon K} < \zeta] = \pi\, G_\lambda(0,y)$$

(iii) There exists a constant $C_{\lambda,K,d}$ *such that, for any* $\varepsilon \in (0,1/2)$, $y \in \mathbb{R}^d$,

$$P[T_{y-\varepsilon K} < \zeta] \leq C_{\lambda,K,d}\, G_\lambda(0,y/2) \times \begin{cases} (\log 1/\varepsilon)^{-1} & \text{if } d = 2\,, \\ \varepsilon^{d-2} & \text{if } d \geq 3\,. \end{cases}$$

Proof : First notice that by (5)

$$C_\lambda(\varepsilon K) = \left(\inf_{\mu\in\mathcal{P}(K)} \int \mu(dy)\,\mu(dz)\,G_\lambda(\varepsilon y,\varepsilon z)\right)^{-1}.$$

If $d \geq 3$, the desired result follows from (2). If $d = 2$, (3) gives

$$C_\lambda(\varepsilon K) \underset{\varepsilon\to 0}{\sim} \left(\frac{1}{\pi}\log\frac{1}{\varepsilon} + \inf_{\mu\in\mathcal{P}(K)} \int \mu(dy)\mu(dz)\log\frac{1}{|y-z|}\right)^{-1} = \left(\frac{1}{\pi}\log\frac{1}{\varepsilon} + \text{const.}\right)^{-1}.$$

To get the other assertions of (i), (ii), simply write

$$P[T_{y-\varepsilon K} < \zeta] = \int_{y-\varepsilon K} G_\lambda(0,z)\,\mu^\lambda_{y-\varepsilon K}(dz) \underset{\varepsilon\to 0}{\sim} G_\lambda(0,y)\,\mu^\lambda_{y-\varepsilon K}(y-\varepsilon K),$$

and note that $\mu^\lambda_{y-\varepsilon K}(y-\varepsilon K) = C_\lambda(y-\varepsilon K) = C_\lambda(\varepsilon K)$.

Finally (iii) follows easily from (4) and (i), (ii) when $|y| > 2\varepsilon$, and is trivial if $|y| \leq 2\varepsilon$. □

3. Estimates for $E[m(S_{\varepsilon K})]$.

We have

$$E[m(S_{\varepsilon K})] = E\left[\int dy\, 1_{S_{\varepsilon K}}(y)\right] = \int dy\, P[T_{y-\varepsilon K} \leq 1].$$

Therefore we need estimates for $P[T_{y-\varepsilon K} \leq 1]$ as $\varepsilon \to 0$. However these estimates are easily derived from Lemma 1. In this section and the next ones, K is a non-polar subset of $\mathbb{R}^d$, $d \geq 2$ (when K is polar, it is immediate that $E[m(S_K)] = \int dy\, P[T_{y-\varepsilon K} \leq 1] = 0$, so that $m(S_K) = 0$ a.s.).

Lemma 2 : *Let* $t > 0$, $y \in \mathbb{R}^d \setminus \{0\}$.

(i) *If* $d \geq 3$,

$$\lim_{\varepsilon\to 0} \varepsilon^{2-d}\, P[T_{y-\varepsilon K} \leq t] = C(K) \int_0^t p_s(0,y)\,ds.$$

(ii) *If* $d = 2$,

$$\lim_{\varepsilon\to 0} \left(\log\frac{1}{\varepsilon}\right) P[T_{y-\varepsilon K} \leq t] = \pi \int_0^t p_s(0,y)\,ds.$$

Proof : Let us concentrate on the case $d = 2$ (the case $d \geq 3$ is similar). Denote by $\gamma_\varepsilon(ds)$ the law of $T_{y-\varepsilon K}$. Lemma 1 (ii) gives

$$\lim_{\varepsilon\to 0} \left(\log\frac{1}{\varepsilon}\right) \int_0^\infty e^{-\lambda s}\,\gamma_\varepsilon(ds) = \pi\, G_\lambda(0,y) = \pi \int_0^\infty e^{-\lambda s}\,p_s(0,y)ds.$$

Since this result holds for any $\lambda > 0$ it follows that the sequence of measures $|\log \varepsilon|\gamma_\varepsilon(ds)$ converges weakly towards the measure $\pi\, p_s(0,y)ds$. In particular,

$$\lim_{\varepsilon\to 0} (\log \frac{1}{\varepsilon})\, \gamma_\varepsilon([0,t]) = \pi \int_0^t p_s(0,y)ds. \quad \square$$

<u>**Theorem 3**</u> : *(i) If* $d \geq 3$,

$$\lim_{\varepsilon\to 0} \varepsilon^{2-d}\, E[m(S_{\varepsilon K})] = C(K)$$

(ii) If $d = 2$,

$$\lim_{\varepsilon\to 0} (\log \frac{1}{\varepsilon})\, E[m(S_{\varepsilon K})] = \pi.$$

<u>Proof</u> : Consider the case $d \geq 3$ (the case $d = 2$ is exactly similar). Then

$$\lim_{\varepsilon\to 0} \varepsilon^{2-d}\, E[m(S_{\varepsilon K})] = \lim_{\varepsilon\to 0} \varepsilon^{2-d} \int dy\, P[T_{y-\varepsilon K} \leq 1] = \int dy\, C(K) \int_0^1 p_s(0,y)ds = C(K).$$

Note that the use of dominated convergence is justified by Lemma 1 (iii), the bound

$$P[T_{y-\varepsilon K} \leq 1] \leq e^{\lambda}\, P[T_{y-\varepsilon K} < \zeta]$$

and the fact that the function $G_\lambda(0,y/2)$ is integrable over $\mathbb{R}^d$. $\square$

The previous arguments yield as well the following slightly stronger result. Take $d \geq 3$ for instance. Let f be a bounded Borel function on $\mathbb{R}^d$. Then

$$\lim_{\varepsilon\to 0} \varepsilon^{2-d}\, E[\int dy\, f(y)\, 1_{S_{\varepsilon K}}(y)] = C(K) \int dy\, f(y) \int_0^1 ds\, p_s(0,y) = C(K)\, E\Big[\int_0^1 ds\, f(B_s)\Big].$$

<u>Remark</u> : The previous proofs, as well as those of the next section, depend heavily on the tools of probabilitic potential theory that we have recalled in Section 2. When K is a ball, say when K is the unit ball of $\mathbb{R}^d$, it is possible to give elementary proofs of all the previous results. Note that in this case

$$T_{y-\varepsilon K} = T_\varepsilon(y) := \inf\{t\ ;\ |B_t - y| \leq \varepsilon\}.$$

The idea is then to compute the expected time spent in the ball of radius ε centered at y, in two different ways. Take $d = 2$ for definiteness. Then,

$$E\Big[\int_0^\zeta 1_{(|B_s - y|\leq\varepsilon)}ds\Big] = \int_0^\infty ds\, e^{-\lambda s} \int_{|z-y|\leq\varepsilon} dz\, p_s(0,z) \underset{\varepsilon\to 0}{\sim} \pi\, \varepsilon^2\, G_\lambda(0,y).$$

On the other hand, assuming that $|y| \geq \varepsilon$, we have by the Markov property at time $T_\varepsilon(y)$,

$$E\left[\int_0^\zeta 1_{(|B_s-y|\leq\varepsilon)} ds\right] = P[T_\varepsilon(y) < \zeta]\, E_{y_\varepsilon}\left[\int_0^\zeta 1_{(|B_s-y|\leq\varepsilon)} ds\right],$$

where y_ε is such that $|y_\varepsilon - y| = \varepsilon$. Easy calculations show that

$$E_{y_\varepsilon}\left[\int_0^\zeta 1_{(|B_s-y|\leq\varepsilon)} ds\right] = \int_{|z-y|\leq\varepsilon} dz\, G_\lambda(y_\varepsilon, z) \underset{\varepsilon\to 0}{\sim} \varepsilon^2 \log\frac{1}{\varepsilon}$$

and we recover Lemma 1 (ii) in this special case.

4. Bounds on $\mathrm{var}(m(S_{\varepsilon K}))$.

It turns out that, in order to get bounds on $\mathrm{var}(m(S_{\varepsilon K}))$, it is important to estimate the volume of the intersection of the Wiener sausages corresponding to two disjoint time intervals. We start with a lemma which gives bounds on the volume of the intersection of two independent Wiener sausages. We denote by B' another Brownian motion independent of B and also started at 0 under P. The associated Wiener sausage is denoted by $S'_{\varepsilon K}$.

Lemma 4 : *There exists a constant* $c = c_{d,K}$ *such that, for* $\varepsilon \in (0,1/2)$,

$$E[m(S_{\varepsilon K} \cap S'_{\varepsilon K})^2]^{1/2} \leq \begin{cases} c\,(\log 1/\varepsilon)^{-2} & \text{if } d = 2\,, \\ c\,\varepsilon^2 & \text{if } d = 3\,, \\ c\,\varepsilon^4 \log 1/\varepsilon & \text{if } d = 4\,, \\ c\,\varepsilon^d & \text{if } d \geq 5\,. \end{cases}$$

Proof : We have :

$$\begin{aligned} E[m(S_{\varepsilon K} \cap S'_{\varepsilon K})^2] &= \int dy\, dz\, P[y \in S_{\varepsilon K} \cap S'_{\varepsilon K},\ z \in S_{\varepsilon K} \cap S'_{\varepsilon K}] \\ &= \int dy\, dz\, P[y \in S_{\varepsilon K},\ z \in S_{\varepsilon K}]^2 \\ &= \int dy\, dz\, P[T_{y-\varepsilon K} \leq 1,\ T_{z-\varepsilon K} \leq 1]^2 . \end{aligned}$$

However

$$P[T_{y-\varepsilon K} \leq 1, T_{z-\varepsilon K} \leq 1] = P[T_{y-\varepsilon K} \leq T_{z-\varepsilon K} \leq 1] + P[T_{z-\varepsilon K} < T_{y-\varepsilon K} \leq 1].$$

The Markov property gives the bound

$$P[T_{y-\varepsilon K} \le T_{z-\varepsilon K} \le 1] \le E\Big[1_{\{T_{y-\varepsilon K}\le 1\}}\, E_{B(T_{y-\varepsilon K})}\,[T_{z-\varepsilon K} \le 1]\Big]$$

$$\le c\, s(\varepsilon)^2\, G_\lambda(0,\tfrac{y}{2})\, G_\lambda(0,\tfrac{z-y}{2})$$

where $s(\varepsilon) = (\log 1/\varepsilon)^{-1}$ if $d = 2$, ε^{d-2} if $d \ge 3$. The last bound follows from Lemma 1(iii) by dealing separately with the cases $|z-y| > 4\varepsilon$, $|z-y| \le 4\varepsilon$. Then,

$$E[m(S_{\varepsilon K} \cap S'_{\varepsilon K})^2] \le \int dydz\Big(\Big(c\, s(\varepsilon)^2\, (G_\lambda(0,\tfrac{y}{2})G_\lambda(0,\tfrac{z-y}{2}) + G_\lambda(0,\tfrac{z}{2})G_\lambda(0,\tfrac{y-z}{2}))\Big)^2 \wedge 1\Big)$$

and after some easy calculations we get the desired bounds (note that for $d \ge 4$, $G_\lambda(0,y/2)$ is not square-integrable). □

Theorem 5 : *There exists a constant* $c = c_{d,K}$ *such that, for* $\varepsilon \in (0,1/2)$

$$(\mathrm{var}\, m(S_{\varepsilon K}))^{1/2} \le \begin{cases} c(\log 1/\varepsilon)^{-2} & \text{if } d = 2\,, \\ c\,\varepsilon^2 \log 1/\varepsilon & \text{if } d = 3\,, \\ c\,\varepsilon^{d-1} & \text{if } d \ge 4\,. \end{cases}$$

Proof : Set $h(\varepsilon) = (\mathrm{var}\, m(S_{\varepsilon K}))^{1/2}$. Crude bounds show that h is bounded over $[0,1]$. The basic idea of the proof is to get a bound for $h(\varepsilon)$ in terms of $h(\varepsilon\sqrt{2})$. Our starting point is the trivial identity

$$m(S_{\varepsilon K}) = m(S_{\varepsilon K}(0,1/2)) + m(S_{\varepsilon K}(1/2,1)) - m(S_{\varepsilon K}(0,1/2) \cap S_{\varepsilon K}(1/2,1)).$$

Set $B'_t = B_{1/2-t} - B_{1/2}$, $B''_t = B_{1/2+t} - B_{1/2}$ for $0 \le t \le 1/2$. Then B', B'' are two independent Brownian motions started at 0, run on the time interval $[0,1/2]$. Furthermore, with an obvious notation,

$$m(S_{\varepsilon K}(0,1/2)) \cap S_{\varepsilon K}(1/2,1)) = m(S'_{\varepsilon K}(0,1/2) \cap S''_{\varepsilon K}(0,1/2))$$

and we can apply the bounds of Lemma 4 to the latter quantity.

On the other hand, the variables $m(S_{\varepsilon K}(0,1/2))$, $m(S_{\varepsilon K}(1/2,1))$ are independent and identically distributed, and a scaling argument gives :

$$m(S_{\varepsilon K}(0,1/2)) \overset{(d)}{=} 2^{-d/2}\, m(S_{\varepsilon\sqrt{2}K}).$$

Then, by the triangle inequality,

$$(\mathrm{var}\, m(S_{\varepsilon K}))^{1/2} \le (2\,\mathrm{var}\, m(S_{\varepsilon K}(0,1/2)))^{1/2} + (\mathrm{var}\, m(S_{\varepsilon K}(0,1/2) \cap S_{\varepsilon K}(1/2,1)))^{1/2}$$

so that:

$$h(\varepsilon) \le 2^{(1-d)/2}\, h(\varepsilon\sqrt{2}) + E[m(S'_{\varepsilon K}(0,1/2) \cap S''_{\varepsilon K}(0,1/2))^2]^{1/2}.$$

It remains to apply the bounds of Lemma 4 and to discuss according to the value of d.

If $d = 2$, we get :

$$h(\varepsilon) \le 2^{-1/2} h(\varepsilon\sqrt{2}) + c(\log 1/\varepsilon)^{-2}.$$

Set $k(\varepsilon) = (\log 1/\varepsilon)^2 h(\varepsilon)$. For any $\rho \in (2^{-1/2},1)$, for ε small, we have

$$k(\varepsilon) \le \rho\, k(\varepsilon\sqrt{2}) + c.$$

This implies that k is bounded over $(0,1/2)$.

If $d = 3$,

$$h(\varepsilon) \le \frac{1}{2} h(\varepsilon\sqrt{2}) + c\,\varepsilon^2.$$

Set $k(\varepsilon) = \varepsilon^{-2} h(\varepsilon)$. Then

$$k(\varepsilon) \le k(\varepsilon\sqrt{2}) + c,$$

which implies

$$k(\varepsilon) \le c' \log 1/\varepsilon.$$

The case $d \ge 4$ is similar. □

5. The main results.

Theorem 6 : *If* $d = 2$,

$$\lim_{\varepsilon\to 0} (\log 1/\varepsilon)\, m(S_{\varepsilon K}) = \pi.$$

If $d \ge 3$,

$$\lim_{\varepsilon\to 0} \varepsilon^{2-d} m(S_{\varepsilon K}) = C_o(K).$$

In both cases, the convergence holds in the L^2*-norm, and a.s. if* K *is star-shaped, that is if* $\varepsilon K \subset K$ *for* $\varepsilon \in (0,1)$.

Proof : The L^2-convergence is easy from Theorem 3 and Theorem 5. Simply observe that :

$$\lim_{\varepsilon\to 0} E\left[\left(\frac{m(S_{\varepsilon K})}{E[m(S_{\varepsilon K})]} - 1\right)^2\right] = 0.$$

When K is star-shaped, $m(S_{\varepsilon K})$ is a monotone increasing function of ε. We may therefore use a monotonicity argument to restrict our attention to a suitable sequence (ε_p). For instance, if $d = 2$, we take $\varepsilon_p = \exp - p^2$. Theorems 3 and 5 then imply that :

$$\sum_{p=1}^{\infty} E\left[\left(\frac{m(S_{\varepsilon_p K})}{E[m(S_{\varepsilon_p K})]} - 1\right)^2\right] < \infty$$

which gives

$$\lim_{p\to\infty} \frac{m(S_{\varepsilon_p K})}{E[m(S_{\varepsilon_p K})]} = 1, \quad \text{a.s.} \quad \square$$

The limiting behavior of $m(S_K(0,t))$ as $t \longrightarrow \infty$ can be deduced from Theorem 6 by the usual scaling transformation . The results are even better since $m(S_K(0,t))$ is always a monotone function of t.

<u>**Theorem 6'**</u> : *If* $d = 2$,

$$\lim_{t\to\infty} \frac{\log t}{t}\, m(S_K(0,t)) = 2\pi.$$

If $d \geq 3$,

$$\lim_{t\to\infty} \frac{1}{t}\, m(S_K(0,t)) = C_o(K).$$

In both cases the convergence holds a.s. and in the L^2*-norm.*

<u>Remarks</u> : It is interesting to observe that, when $d = 2$, the limiting behavior of $m(S_{\varepsilon K})$ as ε tends to 0 does not depend on K (provided K is non-polar). This fact is closely related to the recurrence properties of planar Brownian motion. It can be explained as follows. Let H, K be two compact subsets of $\mathbb{R}^2$ such that $H \subset K$ and H is non-polar. Then the conditional probability of the event $\{T_{y-\varepsilon H} \leq 1\}$ knowing that $\{T_{y-\varepsilon K} \leq 1\}$ tends to 1 as ε tends to 0. This can be checked by applying the Markov property at time $T_{y-\varepsilon K}$ and then using a suitable scaling argument and the recurrence of planar Brownian motion.

Theorem 6 is also related to the fact that the Hausdorff dimension of the Brownian curve is 2. In particular, for $d \geq 3$, the order of magnitude of the volume of a tubular neighborhood of the Brownian path is the same as would be that of a portion of plane. Note that for a C^1 curve the volume of a tubular neighborhood is of order ε^{d-1}.

6. <u>A heat conduction problem.</u>

The previous results are closely related to the following heat conduction problem. Assume that the compact set K is held at the temperature 1 from time $t = 0$ to $+\infty$, whereas the surrounding medium $\mathbb{R}^d \setminus K$ is at the temperature 0 at time $t = 0$. Clearly the temperature in the surrounding medium will increase, and one is interested in the total energy flow in time t from K to the surrounding medium. More precisely, the temperature at time t, at $x \in \mathbb{R}^d \setminus K$ solves the heat equation :

$$\frac{\partial u}{\partial t} = \frac{1}{2} \Delta u$$

with boundary conditions

$$u(0,x) = 0,$$

$$\lim_{x \to x_o} u(t,x) = 1,$$

for any $t > 0$, x_o regular point of ∂K.

Then $u(t,x)$ has the following probabilistic interpretation :

$$u(t,x) = P_x[T_K \leq t] .$$

The quantity of interest is

$$E_K(t) = \int_{\mathbb{R}^d \setminus K} u(t,x)dx .$$

Now observe that :

$$m(K) + E_K(t) = \int_{\mathbb{R}^d} P_x[T_K \leq t]dx = \int_{\mathbb{R}^d} P[T_{x-K} \leq t]dx = E[m(S_K(0,t))].$$

Therefore the limiting behavior of $E_K(t)$ is given by that of $m(S_K(0,t))$:

- if $d = 2$, $E_K(t) \underset{t \to \infty}{\sim} 2\pi \frac{t}{\log t}$;

- if $d \geq 3$, $E_K(t) \underset{t \to \infty}{\sim} C(K)\, t$.

__Bibliographical notes__. The strong law of large numbers for the Wiener sausage in $\mathbb{R}^d$ *,* $d \geq 3$ *, was first derived by Kesten, Spitzer and Whitman (cf [IMK, p.252-253] , [S3, p.40]). See Spitzer [S4] for a derivation using Kingman's subadditive ergodic theorem. Our approach is inspired from [L3], [L10] and Sznitman [Sz] . The relevant results of probabilistic potential theory may be found in the book of Port and Stone [PS] . Sharp estimates for the expected volume of the Wiener sausage were first derived by Spitzer [S2]. These estimates are refined in [L11] for* $d \geq 3$ *and in [L12] for* $d = 2$ *(see also Chapter XI of the present work). The bounds of Lemma 4 and Theorem 5 are* *sharp: see* [L3], [L5] *(and Chapter VIII) for additional information about intersections of independent Wiener sausages. The application developed in Section 6 is taken from Spitzer* [S2]. *Other applications may be found in Kac* [K]. *Certain large deviations results for the volume of the Wiener sausage, also motivated by physical applications, are proved in Donsker and Varadhan* [DV]. *The results of this chapter can be extended to processes more general than Brownian motion in* $\mathbb{R}^d$ *.See Chavel and Feldman* [CF1], [CF2] *for the case of Brownian motion on a Riemannian manifold. Sznitman* [Sz] *deals with elliptic diffusion processes in* $\mathbb{R}^d$ *: roughly speaking, the behavior of the sausage of small radius remains the same as for Brownian motion. However, if one considers hypoelliptic diffusion processes (that is diffusion processes whose generator satisfies the strong Hörmander condition), then the volume of a tubular neighborhood of the path may become much smaller: see Chaleyat-Maurel and Le Gall* [CML]. *Hawkes* [H] *considers the sausage associated with Lévy processes: Kingman's theorem can still be applied to the behavior in large time of the volume of the sausage. Weinryb* [W1] *extends Theorem 6 by considering* $\mu(m(S_{\varepsilon K}))$ *for certain measures* μ *such as the Lebesgue measure on a hyperplane. Finally, a discrete analogue of the volume of the Wiener sausage is the number of distinct sites (or the range) visited by a random walk. Discrete versions of Theorem 6' are proved in Dvoretzky and Erdös* [DE] *(see also Spitzer* [S3, p. 38-40] *and Jain and Pruitt* [JP]).

CHAPTER VII

Connected components of the complement of a planar Brownian path.

Let $B = (B_t, t \geq 0)$ be a complex-valued Brownian motion, and $B[0,1] = \{B_s; 0 \leq s \leq 1\}$. It seems very likely, and can be proved rigorously, that with probability 1 the open set $\mathbb{C} \setminus B[0,1]$ has an infinite number of connected components. The following question was raised by Mandelbrot. Let N_ε denote the number of connected components whose area is greater than $\varepsilon > 0$. What is the limiting behavior of N_ε as ε goes to 0 ? Mandelbrot conjectured that $N_\varepsilon \sim \varepsilon^{-1} L(\varepsilon)$ for some slowly varying function L such that $\int_0^1 u^{-1} L(u)du < \infty$. The goal of this chapter is to prove that Mandelbrot's conjecture holds with $L(\varepsilon) = 2\pi(\log \varepsilon)^{-2}$.

The problem of determining the asymptotics of N_ε is closely related to the study of the planar Wiener sausage. To explain this, denote by W_ε the union of all connected components whose area is smaller than $\pi\varepsilon^2$. It is obvious that W_ε is contained in $S_\varepsilon(0,1)$ (the Wiener sausage of radius ε associated with the unit disk). The converse inclusion is also "almost true". Precisely, if $y \in \mathbb{C}$ belongs to $S_\varepsilon(0,1)$, then, with a probability close to 1, y will also belong to W_ε. This fact is explained by the recurrence properties of planar Brownian motion: if B comes within a distance ε of y before time 1 then B will come much closer with great probability, and it will be very likely that the connected component of y is contained in $D(y,\varepsilon)$ (these heuristic arguments can easily be made rigorous). We conclude that

$$m(W_\varepsilon) \sim m(S_\varepsilon(0,1)) \sim \frac{\pi}{|\log \varepsilon|},$$

by Theorem VI-6.

It turns out, but is now non-trivial, that much more is true. For any fixed $\lambda \in (0,1)$,

$$m(W_\varepsilon) - m(W_{\lambda\varepsilon}) \underset{\varepsilon \to 0}{\sim} \frac{\pi}{|\log \varepsilon|} - \frac{\pi}{|\log \lambda\varepsilon|} \sim \frac{\pi|\log \lambda|}{|\log \varepsilon|^2}. \tag{1}$$

Notice that $m(W_\varepsilon) - m(W_{\lambda\varepsilon})$ is closely related to the number of connected components with area between $\pi(\lambda\varepsilon)^2$ and $\pi\varepsilon^2$. Most of this chapter is devoted to a rigorous formulation and proof of (1) . The asymptotics of N_ε then follow rather easily.

1. Estimates for the probability distribution of the area of the connected component of a given point.

Throughout this section, we assume that the Brownian motion B starts at 1 and, for any $R > 0$, we set : $T_R = \inf\{t, |B_t| = R\}$. We denote by $\mathcal{C}_R$ the connected component of $\mathbb{C} \setminus B[0,T_R]$ that contains 0.

Lemma 1 : *There exists a positive constant* α *such that, for any* $R \geq 2$,

$$P[\mathcal{C}_R \text{ is unbounded}] \leq R^{-\alpha}.$$

Proof : Fix $\rho > 1$ and denote by $\mathcal{C}_{(n)}$ the connected component of $\mathbb{C} \setminus B[T_{\rho^n}, T_{\rho^{n+1}}]$ that contains 0. For any $n \geq 1$,

$$P[\mathcal{C}_{\rho^n} \text{ is unbounded}] \leq P\left[\bigcap_{k=0}^{n-1} \{\mathcal{C}_{(k)} \text{ is unbounded}\}\right].$$

The strong Markov property shows that the events $\{\mathcal{C}_{(k)}$ is unbounded$\}$ are independent. By scaling they also have the same probability $c < 1$. Hence,

$$P[\mathcal{C}_{\rho^n} \text{ is unbounded}] \leq c^n.$$

Lemma 1 now follows easily. □

Remark. If $\mathcal{C}_R$ is bounded then obviously it is contained in $D(0,R)$. Using a scaling argument we obtain the following result. Suppose now that $|B_0| \leq \varepsilon$ and let $\delta \geq 2\varepsilon$, then, with a probability greater than $1 - (\varepsilon/\delta)^\alpha$, the connected component of $\mathbb{C} \setminus B[0,T_\delta]$ that contains 0 is contained in $D(0,\delta)$. This form of Lemma 1 will be used on several occasions.

For any $r \in (0,1)$, set

$$P(R,r) = P[m(\mathcal{C}_R) \leq \pi r^2].$$

Our first goal is to obtain good estimates for $P(R,r)$ as $R \longrightarrow \infty$ and $r \longrightarrow 0$. We follow the ideas described in the introduction. Firstly,

$$P(R,r) \leq P[T_r < T_R] = \frac{\log R}{\log R - \log r}.$$

On the other hand, we may get a lower bound on $P(R,r)$ by conditioning on $\{T_{r/2^n} < T_R\}$, applying the Markov property at time $T_{r/2^n}$ and using the remark after Lemma 1 to bound the probability that $\mathcal{C}_R$ is not contained in $D(0,r)$. For every $n \geq 1$,

$$P(R,r) \geq P[T_{r/2^n} < T_R]\,(1-2^{-n\alpha}) = \frac{\log R}{\log R - \log r + n \log 2}\,(1 - 2^{-n\alpha}).$$

If we choose $n = [K \log\log R]$ with K large enough we get :

$$P(R,r) \geq \frac{\log R}{\log R - \log r + O(\log\log R)} (1 + O((\log R)^{-M}))$$

for any $M > 0$.

In what follows, we shall be interested in estimates as $R \longrightarrow \infty$, holding uniformly in r. We will always assume that

$$(\log R)^{\gamma} \leq |\log r| \leq (\log R)^{1/2},$$

for some $\gamma > 0$. The previous bounds give

$$P(R,r) = 1 + \frac{\log r}{\log R} + O(\frac{\log\log R}{\log R}). \tag{2}$$

We now fix $\lambda \in (0,1)$ and set

$$Q(R,r) = P(R,r) - P(R,\lambda r).$$

<u>Lemma 2</u> : *As* $R \longrightarrow +\infty$, $Q(R,r) = \frac{|\log \lambda|}{\log R} + O(\frac{\log\log R}{(\log R)^{5/4}})$, *uniformly for* $(\log R)^{\gamma} \leq |\log r| \leq (\log R)^{1/2}$.

<u>Proof</u> : Notice that a brutal application of (2) gives nothing. The idea of the proof is to compare $Q(R,r)$ and $Q(R,\lambda r)$ using a scaling transformation, and only then to apply (2).

The event $A_r = \{\pi(\lambda r)^2 \leq m(\mathscr{C}_R) \leq \pi r^2\}$ is contained in $\{T_r < T_R\}$. By the remark following Lemma 1, and the previous arguments,

$$P(A_r) = P(A_r \cap \{\mathscr{C}_R \subset D(0,1)\}) + O((\log R)^{-M}).$$

(use the bound $|\log r| \geq (\log R)^{\gamma}$). We now want to estimate $Q(R,\lambda r) = P(A_{\lambda r})$. Note that $A_{\lambda r}$ is trivially contained in $\{T_\lambda < T_r\}$. We define $\mathscr{C}'_R$ as the connected component of $\mathbb{C} \setminus B[0,T_{\lambda R}]$ that contains 0, and we set :

$$A'_r = \{\pi(\lambda^2 r)^2 \leq m(\mathscr{C}'_R) \leq \pi(\lambda r)^2\}.$$

As previously,

$$P(A'_r) = P(A'_r \cap \{\mathscr{C}'_R \subset D(0,\lambda)\}) + O((\log R)^{-M}).$$

However, by the Markov property and scaling,

$$P(A'_r \cap \{\mathscr{C}'_R \subset D(0,\lambda)\}) = P(T_\lambda < T_R)\, P(A_r \cap \{\mathscr{C}_R \subset D(0,1)\}).$$

The point is that, on the set $\{\mathscr{C}'_R \subset D(0,\lambda)\}$, $\mathscr{C}'_R$ is also the connected component of $\mathbb{C} \setminus B[T_\lambda, T_{\lambda R}]$ that contains 0 .

We now want to compare the sets A'_r and $A_{\lambda r}$. The problem is that $\mathscr{C}_R$ may be smaller than $\mathscr{C}'_R$ because of the portion of the path between times $T_{\lambda R}$

and T_R. However, by Lemma 1 we may choose $K > 0$ large enough so that $\mathcal{C}'_R$ is contained in $D(0,(\log R)^K)$ except on a set of probability $O((\log R)^{-10})$. It follows that,

$$P[A'_r \setminus A_{\lambda r}] \le O((\log R)^{-10}) + P(A'_r)\, P\big(\inf_{[T_{\lambda R},T_R]} |B_u| < (\log R)^K\big)$$

$$\le O((\log R)^{-10}) + (1-P(\lambda R,\lambda^2 r))\,\frac{|\log \lambda|}{\log R - K\log\log R} = O((\log R)^{-3/2})$$

by (2) and our assumption $|\log r| \le (\log R)^{1/2}$. A similar reasoning gives

$$P(A_{\lambda r} \setminus A'_r) = O((\log R)^{-3/2}),$$

and we get:

$$P[A_{\lambda r}] = P[A'_r] + O((\log R)^{-3/2})$$

From the previous considerations, we obtain

$$P[A_{\lambda r}] = P[T_\lambda < T_R]\, P[A_r] + O((\log R)^{-3/2})$$

or equivalently

$$Q(R,\lambda r) = Q(R,r)\Big(1 + \frac{\log \lambda}{\log R}\Big) + O((\log R)^{-3/2}). \tag{3}$$

Now let $N \ge 1$ be an integer such that $N \le (\log R)^{1/2}$. By (3),

$$P(R,r) - P(R,\lambda^N r) = \sum_{k=0}^{N-1} Q(R,\lambda^k r) = \frac{\log R}{|\log \lambda|}\Big(1 - \Big(1 + \frac{\log \lambda}{\log R}\Big)^N\Big) + O(N^2(\log R)^{-3/2})$$

uniformly for $(\log R)^\gamma \le |\log r| \le (\log R)^{1/2}$. Furthermore,

$$1 - \Big(1 + \frac{\log \lambda}{\log R}\Big)^N = N\,\frac{|\log \lambda|}{\log R} + O\Big(\frac{N^2}{(\log R)^2}\Big),$$

which gives

$$P(R,r) - P(R,\lambda^N r) = N\Big(1 + O\Big(\frac{N}{\log R}\Big)\Big)\, Q(R,r) + O(N^2(\log R)^{-3/2}).$$

However, by (2),

$$P(R,r) - P(R,\lambda^N r) = \frac{N|\log \lambda|}{\log R} + O\Big(\frac{\log\log R}{\log R}\Big),$$

so that we obtain :

$$\Big(1 + O\Big(\frac{N}{\log R}\Big)\Big)\, Q(R,r) = \frac{|\log \lambda|}{\log R} + O\Big(\frac{\log\log R}{N \log R} + N(\log R)^{-3/2}\Big).$$

We now take $N = [(\log R)^{1/4}]$ to complete the proof. □

2. Asymptotics for N_ε.

We now take $B_0 = 0$. We will apply the previous estimates to the asymptotics of N_ε. Most of this section is devoted to a rigorous proof of (1). For simplicity, we set

$$U_\varepsilon = W_\varepsilon \setminus W_{\lambda\varepsilon}$$

so that U_ε is the union of all components whose area is between $\pi(\lambda\varepsilon)^2$ and $\pi\varepsilon^2$. We will obtain the limiting behavior of $m(U_\varepsilon)$ by a method similar to the one we used for the area of the Wiener sausage in Chapter VI.

Proposition 3 : *As* $\varepsilon \longrightarrow 0$,

$$E[m(U_\varepsilon)] = \frac{\pi|\log\lambda|}{(\log\varepsilon)^2} + o\Big(\frac{1}{(\log\varepsilon)^2}\Big).$$

Proof : For $\varepsilon > 0$ small enough we define $\delta = \delta(\varepsilon) > \varepsilon$ by the condition

$$\frac{\delta}{\exp(|\log\delta|^{1/4})} = \varepsilon.$$

Note that $|\log\delta| \sim |\log\varepsilon|$ as $\varepsilon \longrightarrow 0$. Let $y \in \mathbb{C} \setminus D(0,\delta)$. Set

$$T_\delta(y) = \inf\{s \ge 0 \ ; \ |B_s - y| < \delta\}$$

and

$$R_\delta(y) = \inf\{s \ge T_\delta(y) \ ; \ |B_s - y| > (\log\delta)^{-4}\} \ .$$

Notice that $\{y \in U_\varepsilon\} \subset \{T_\delta(y) \le 1\}$. We denote by $\mathcal{C}(y)$, resp. $\mathcal{C}_\delta(y)$, the connected component of $\mathbb{C} \setminus B[0,1]$, resp. $\mathbb{C} \setminus B[T_\delta(y),R_\delta(y)]$, that contains y. Then,

$$(4) \qquad |P[y \in U_\varepsilon] - P[T_\delta(y) \le 1 \ ; \ \pi(\lambda\varepsilon)^2 \le m(\mathcal{C}_\delta(y)) \le \pi\varepsilon^2]|$$

$$\le P[T_\delta(y) \le 1 \ ; \ \pi(\lambda\varepsilon)^2 \le m(\mathcal{C}_\delta(y)) \le \pi\varepsilon^2 \ ; \ \mathcal{C}(y) \ne \mathcal{C}_\delta(y)]$$

$$+ P[T_\delta(y) \le 1 \ ; \ \pi(\lambda\varepsilon)^2 \le m(\mathcal{C}(y)) \le \pi\varepsilon^2 \ ; \ \mathcal{C}(y) \ne \mathcal{C}_\delta(y)].$$

We proceed to bound the right side of (4). We have

$$(5) \quad P[T_\delta(y) \le 1 \ ; \ \pi(\lambda\varepsilon)^2 \le m(\mathcal{C}_\delta(y)) \le \pi\varepsilon^2 \ ; \ \mathcal{C}(y)\ne\mathcal{C}_\delta(y)] \le P[T_\delta(y) \le 1 \le R_\delta(y)]$$

$$+ P[T_\delta(y) \le R_\delta(y) \le 1 ; (B[0,T_\delta(y)] \cup B[R_\delta(y),1]) \cap \mathcal{C}_\delta(y) \ne \emptyset \ ; \pi(\lambda\varepsilon)^2 \le m(\mathcal{C}_\delta(y)) \le \pi\varepsilon^2].$$

It is very easy to check that :

$$P[T_\delta(y) \le 1 \le R_\delta(y)] \le P[|B_1 - y| \le (\log\delta)^{-4}] \le (\log\delta)^{-3}\,\psi_1(y),$$

for some <u>integrable</u> function $\psi_1 : \mathbb{C} \longrightarrow \mathbb{R}_+$. Next, Lemma 1 gives

$$P[m(\mathscr{C}_\delta(y)) \le \pi\varepsilon^2 \ , \ \mathscr{C}_\delta(y) \cap (\mathbb{C} \setminus D(y,\delta)) \ne \varnothing \mid \mathscr{F}_{T_\delta(y)}] = O(|\log \delta|^{-M})$$

uniformly in $y \in \mathbb{C}$. It follows that the second term of the right side of (5) is bounded by :

$$P[T_\delta(y)\le 1]O(|\log \delta|^{-M})+ P[T_\delta(y)\le 1 \ ; \inf_{[R_\delta(y),R_\delta(y)+1]} |B_u|<\delta \ ; \pi(\lambda\varepsilon)^2\le m(\mathscr{C}_\delta(y))\le\pi\varepsilon^2]$$

$$= P[T_\delta(y)\le 1] \left(O(|\log \delta|^{-M}) + Q(\frac{(\log \delta)^{-4}}{\delta}, \frac{\varepsilon}{\delta}) \ P\left[\inf_{[R_\delta(y);R_\delta(y)+1]} |B_u| < \delta\right]\right)$$

using the Markov property at $T_\delta(y)$ and at $R_\delta(y)$. It follows from Lemma VI-1 (iii) that

$$P\left[\inf_{[R_\delta(y),R_\delta(y)+1]} |B_u| < \delta\right] = O(\frac{\log|\log \delta|}{|\log \delta|}).$$

Then using Lemma 2 and Lemma VI-1 (iii) again we conclude that the right side of (5) is bounded by

$$\frac{\log|\log \delta|}{|\log \delta|^3}\,\psi_2(y)$$

for some integrable function $\psi_2 : \mathbb{C} \longrightarrow \mathbb{R}_+$.

Similar arguments show that the second term of the right side of (4) is bounded by :

$$\frac{\log|\log \delta|}{|\log \delta|^3}\,\psi_3(y)$$

for some integrable function $\psi_3 : \mathbb{C} \longrightarrow \mathbb{R}_+$. It follows that :

$$\left|\int dy\ P[y \in U_\varepsilon] - \int dy\ P[T_\delta(y) \le 1 \ ; \ \pi(\lambda\varepsilon)^2 \le m(\mathscr{C}_\delta(y)) \le \pi\varepsilon^2]\right| = O(\frac{\log|\log \delta|}{|\log \delta|^3}).$$

However, by the Markov property at time $T_\delta(y)$, if $|y| > \delta$,

$$P[T_\delta(y) \le 1 \ ; \ \pi(\lambda\varepsilon)^2 \le m(\mathscr{C}_\delta(y)) \le \pi\varepsilon^2] = P[T_\delta(y) \le 1]\ Q(\frac{(\log \delta)^{-4}}{\delta}, \frac{\varepsilon}{\delta})$$

so that

$$E[m(U_\varepsilon)] = \int dy\ P[y \in U_\varepsilon] = E[m(S_\delta(0,1))]\ Q(\frac{(\log \delta)^{-4}}{\delta}, \frac{\varepsilon}{\delta}) + o((\log \delta)^{-2}).$$

Proposition 3 now follows from Lemma 2 and Theorem VI-3 (ii). □

Proposition 4 : *There exists a constant* K *such that, for any* $\varepsilon \in (0,1/2)$,

$$\mathrm{var}(m(U_\varepsilon)) \le K\,|\log \varepsilon|^{-11/2} .$$

Proof : The main idea is the same as in the proof of Theorem VI-5. We let B^1, B^2 denote two independent complex-valued Brownian motions started at 0. For every $y \in \mathbb{C} \setminus (B^1[0,1/2] \cup B^2[0,1/2])$ we denote by $\mathscr{C}'(y)$, resp. $\mathscr{C}^1(y)$, $\mathscr{C}^2(y)$, the connected component of $\mathbb{C} \setminus (B^1[0,1/2] \cup B^2[0,1/2])$, resp. $\mathbb{C} \setminus B^1[0,1/2]$, $\mathbb{C} \setminus B^2[0,1/2]$, that contains y. We set :

$$U'_\varepsilon = \{y \,;\, \pi(\lambda\varepsilon)^2 \le m(\mathscr{C}'(y)) \le \pi\varepsilon^2\},$$

$$U^i_\varepsilon = \{y \,;\, \pi(\lambda\varepsilon)^2 \le m(\mathscr{C}^i(y)) \le \pi\varepsilon^2\} \qquad (i = 1,2).$$

Obviously $m(U'_\varepsilon)$ and $m(U_\varepsilon)$ are identically distributed, so that $\mathrm{var}\, m(U_\varepsilon) = \mathrm{var}\, m(U'_\varepsilon)$. The key step of the proof is to show that $m(U'_\varepsilon)$ is not too different from $m(U^1_\varepsilon) + m(U^2_\varepsilon)$.

Lemma 5 : *There exists a constant* K' *such that, for* $\varepsilon \in (0,1/2)$,

$$E[(m(U'_\varepsilon) - m(U^1_\varepsilon) - m(U^2_\varepsilon))^2] \le K'\,|\log \varepsilon|^{-11/2} .$$

Proof : We first observe that :

$$(6)\quad |m(U'_\varepsilon)-m(U^1_\varepsilon)-m(U^2_\varepsilon)| \le m(U'_\varepsilon\setminus(U^1_\varepsilon \cup U^2_\varepsilon)) + m(U^1_\varepsilon\setminus U'_\varepsilon) + m(U^2_\varepsilon\setminus U'_\varepsilon) + m(U^1_\varepsilon \cap U^2_\varepsilon).$$

Let us bound

$$E[m(U^1_\varepsilon \setminus U'_\varepsilon)^2] = E\Big[\int dy\, dz\, 1_{U^1_\varepsilon\setminus U'_\varepsilon}(y)\, 1_{U^1_\varepsilon\setminus U'_\varepsilon}(z)\Big] .$$

We take $\delta = \delta(\varepsilon)$ as in the proof of Proposition 3. It follows from Lemma 1 (and the remark after this lemma) that:

$$E[m(U^1_\varepsilon \setminus U'_\varepsilon)^2] = E\Big[\int dy\, dz\, 1_{\{\mathscr{C}_\delta(y)\subset D(y,\delta);\,\mathscr{C}_\delta(z)\subset D(z,\delta)\}}\, 1_{U^1_\varepsilon\setminus U'_\varepsilon}(y)\, 1_{U^1_\varepsilon\setminus U'_\varepsilon}(z)\Big]$$
$$+\ O(|\log \varepsilon|^{-M}) .$$

Now notice that, if $y \in U^1_\varepsilon \setminus U'_\varepsilon$ and $\mathscr{C}_\delta(y) \subset D(y,\delta)$, then $B^2[0,1/2]$ must intersect $D(y,\delta)$. It follows that

$$E[m(U^1_\varepsilon \setminus U'_\varepsilon)^2] \le E\Big[\int dy\, dz\, 1_{U^1_\varepsilon}(y)\, 1_{U^1_\varepsilon}(z)\, 1_{S^2_\delta}(y)\, 1_{S^2_\delta}(z)\Big] + O(|\log \varepsilon|^{-M})$$

$$= \int dy\, dz\, P[y \in U^1_\varepsilon,\ z \in U^1_\varepsilon]\, P[y \in S^2_\delta,\ z \in S^2_\delta] + O(|\log \varepsilon|^{-M}),$$

where $S^2_\delta = S^2_\delta(0,1/2)$ denotes the Wiener sausage associated with B^2. Recall from Chapter VI (see the proof of Lemma VI-4) the bound

(7) $P[y \in S^2_\delta,\ z \in S^2_\delta] \le C(\log\delta)^{-2}\,(G_1(0,y/2) + G_1(0,z/2))\,G_1(0,(z-y)/2).$

The problem is then to get a suitable bound on $P[y \in U^1_\varepsilon,\ z \in U^1_\varepsilon]$. Let $T^1_\delta(y)$, $R^1_\delta(y)$ be as previously, with B replaced by B^1. We suppose that $|z-y| \ge |\log\delta|^{-3}$ and we restrict our attention to the case $T^1_\delta(y) \le T^1_\delta(z)$. Lemma 2 and the Markov property at time $T^1_\delta(z)$ give :

$$P[y \in U^1_\varepsilon, z \in U^1_\varepsilon\ ;\ T^1_\delta(y) \le T^1_\delta(z)]$$

$$\le P[T^1_\delta(y) \le T^1_\delta(z) \le \tfrac{1}{2}\ ;\ m(\mathcal{C}^1(y)) \ge \pi(\lambda\varepsilon)^2\ ;\ m(\mathcal{C}^1(z)) \ge \pi(\lambda\varepsilon)^2\ ]$$

$$\le P[T^1_\delta(y) \le T^1_\delta(z) \le \tfrac{1}{2};\ m(\tilde{\mathcal{C}}^1(y)) \ge \pi(\lambda\varepsilon)^2]\ (1 - P(\frac{(\log\delta)^{-4}}{\delta}, \frac{\lambda\varepsilon}{\delta})) + P[T^1_\delta(z) \le \tfrac{1}{2} \le R^1_\delta(z)]$$

$$\le C\ |\log\delta|^{-3/4}\ P[T^1_\delta(y) \le T^1_\delta(z) \le \tfrac{1}{2}\ ;\ m(\tilde{\mathcal{C}}^1(y)) \ge \pi(\lambda\varepsilon)^2] + P[T^1_\delta(z) \le \tfrac{1}{2} \le R^1_\delta(z)],$$

using (2). Here $\tilde{\mathcal{C}}^1(y)$ denotes the connected component of $\mathbb{C} \setminus B[0,R^1_\delta(y)]$ that contains y. Clearly the term $P[T^1_\delta(z) \le \frac{1}{2} \le R^1_\delta(z)]$ is bounded by $C|\log\delta|^{-8}$. Therefore it suffices to bound

$$P[T^1_\delta(y) \le T^1_\delta(z) \le \tfrac{1}{2}\ ;\ m(\tilde{\mathcal{C}}^1(y)) \ge \pi(\lambda\varepsilon)^2]$$

$$\le C\,\frac{G_1(0,(z-y)/2)}{|\log\delta|}\ P[T^1_\delta(y) \le \tfrac{1}{2}\ ;\ m(\tilde{\mathcal{C}}^1(y)) \ge \pi(\lambda\varepsilon)^2]$$

$$\le C'\,|\log\delta|^{-11/4}\ G_1(0,y/2)\,G_1(0,(z-y)/2).$$

This first bound uses the Markov property at time $R^1_\delta(y)$ and Lemma VI-1 (iii). The second one follows from the same lemma, the Markov property at $T^1_\delta(y)$ and (2).

We conclude that, for any y,z such that $|z-y| \ge |\log\delta|^{-3}$,

(8) $$P[y \in U^1_\varepsilon, z \in U^1_\varepsilon\ ;\ T^1_\delta(y) \le T^1_\delta(z)]$$

$$\le C\ |\log\varepsilon|^{-7/2}\ G_1(0,y/2)\,G_1(0,(z-y)/2) + O(|\log\varepsilon|^{-8}).$$

Combining (7) and (8) gives the bound :

$$E[m(U^1_\varepsilon \setminus U'_\varepsilon)^2] \le C\ |\log\varepsilon|^{-11/2}.$$

Obviously the same bound holds for $E[m(U_\varepsilon^2 \setminus U'_\varepsilon)^2]$. Similar arguments give even better bounds for the other two terms of the right side of (6). □

We now complete the proof of Proposition 4. We set $h(\varepsilon) = (\mathrm{var}\ m(U_\varepsilon))^{1/2}$. Notice that $m(U_\varepsilon^1)$ and $m(U_\varepsilon^2)$ are independent and that

$$m(U_\varepsilon^1) \overset{(d)}{=} m(U_\varepsilon^2) \overset{(d)}{=} \frac{1}{2}\, m(U_{\varepsilon\sqrt{2}})$$

by a scaling argument. Then Lemma 5 implies

$$(\mathrm{var}\ m(U_\varepsilon))^{1/2} \le (\mathrm{var}(m(U_\varepsilon^1) + m(U_\varepsilon^2)))^{1/2} + O(|\log \varepsilon|^{-11/4})$$

$$= 2^{1/2}\,(\mathrm{var}\ m(U_{\varepsilon\sqrt{2}}))^{1/2} + O(|\log \varepsilon|^{-11/4}).$$

Therefore,

$$h(\varepsilon) \le 2^{-1/2}\, h(\varepsilon\sqrt{2}) + O(|\log \varepsilon|^{-11/4})$$

and Proposition 4 follows using arguments similar to those of the proof of Theorem VI-5. □

We may now state the main result of this Chapter. For $u < v$, $N_{[u,v)}$ denotes the number of connected components of $\mathbb{C} \setminus B[0,1]$ whose area belongs to the interval $[u,v)$ (in particular $N_\varepsilon = N_{[\varepsilon,\infty)}$).

Theorem 6 : *With probability* 1, *for any* $\delta > 0$,

$$\lim_{u\to 0} \left(\sup_{v \ge (1+\delta)u} \left| \frac{(\log u)^2\, N_{[u,v)}}{u^{-1} - v^{-1}} - 2\pi \right| \right) = 0.$$

In particular,

$$\lim_{\varepsilon\to 0} \varepsilon\, (\log \varepsilon)^2\, N_\varepsilon = 2\pi, \quad a.s.$$

Proof : Propositions 3 and 4 give

$$\lim_{n\to\infty} (\log \lambda^n)^2\, m(U_{\lambda^n}) = \pi |\log \lambda|, \quad a.s.$$

Let $\bar N_{[u,v)}$ be the number of connected components with area in $[\pi u^2, \pi v^2)$. Note that :

$$(\pi\lambda^{2n})^{-1}\, m(U_{\lambda^n}) \le \bar N_{[\lambda^{n+1},\lambda^n)} \le (\pi\lambda^{2n+2})^{-1}\, m(U_{\lambda^n}).$$

Therefore, w.p. 1,

$$(9)\qquad |\log\lambda| \le \liminf_{n\to\infty} \lambda^{2n}(\log\lambda^n)^2\,\bar N_{[\lambda^{n+1},\lambda^n)}$$

$$\le \limsup_{n\to\infty} \lambda^{2n}(\log\lambda^n)^2\,\bar N_{[\lambda^{n+1},\lambda^n)} \le \frac{|\log\lambda|}{\lambda^2}.$$

Fix an integer $p \ge 1$ and set $\lambda' = \lambda^{1/p}$. Since

$$\bar N_{[\lambda^{n+1},\lambda^n)} = \sum_{i=0}^{p-1} \bar N_{[\lambda'^{np+i+1},\lambda'^{np+i})},$$

it follows from (9) that :

$$\frac{|\log\lambda|}{p}\sum_{i=0}^{p-1}\lambda^{-2i/p} \le \liminf_{n\to\infty} \lambda^{2n}(\log\lambda^n)^2\,\bar N_{[\lambda^{n+1},\lambda^n)}$$

$$\le \limsup_{n\to\infty} \lambda^{2n}(\log\lambda^n)^2\,\bar N_{[\lambda^{n+1},\lambda^n)} \le \frac{|\log\lambda|}{p}\sum_{i=0}^{p-1}\lambda^{-(2i+2)/p}.$$

Choosing p large we conclude that

$$\lim_{n\to\infty} \lambda^{2n}(\log\lambda^n)^2\,\bar N_{[\lambda^{n+1},\lambda^n)} = |\log\lambda|\int_0^1 \lambda^{-2s}\,ds = \frac{1}{2}(\lambda^{-2}-1), \quad \text{a.s.}$$

A simple monotonicity argument allows us to improve this convergence to

$$\lim_{x\to 0} x^2(\log x)^2\,\bar N_{[x,\alpha x)} = \frac{1}{2}(1-\alpha^{-2}) \quad \text{a.s.},$$

for any $\alpha > 1$. Equivalently,

$$\lim_{u\to 0} u\,(\log u)^2\,N_{[u,\alpha u)} = 2\pi\Big(1-\frac{1}{\alpha}\Big), \quad \text{a.s.}$$

Theorem 5 follows easily from this last result. □

Bibliographical notes. Mandelbrot ([Ma], Chapter 25) raises some interesting questions about the connected components of the complement of a planar Brownian path. Motivated by these questions, Mountford [Mo] has obtained a weak form of Theorem 6. The main ideas and techniques of this chapter are taken from [Mo], although the form of Theorem 6 given above is from [L13]. We refer to the latter paper for additional details in the proofs (the estimates of [L13] are somewhat sharper than those presented here).

CHAPTER VIII

Intersection local times and first applications.

1. The intersection local time of p independent Brownian paths.

Let $p \geq 2$ be an integer, and let $B^1,\dots,B^p$ denote p independent Brownian motions in $\mathbb{R}^2$, started at $x^1,\dots,x^p$ respectively. The intersection local time of $B^1,\dots,B^p$ is a random measure $\alpha(ds_1,\dots ds_p)$ on $(\mathbb{R}_+)^p$, supported on

$$\left\{ (t_1,\dots,t_p) \in (\mathbb{R}_+)^p \ ; \ B^1_{t_1} = \dots = B^p_{t_p} \right\}.$$

The fact that the latter set is non-empty with probability 1 is more or less equivalent to the existence of p-multiple points for the planar Brownian path. In our approach, the non-emptiness of this set will follow from the fact that it supports a non-trivial measure.

The measure $\alpha(ds_1\dots ds_p)$ is formally defined by :

$$\alpha(ds_1\dots ds_p) = \delta_{(o)}\left(B^1_{s_1} - B^2_{s_2}\right)\dots\delta_{(o)}\left(B^{p-1}_{s_{p-1}} - B^p_{s_p}\right) ds_1\dots ds_p$$

where $\delta_{(0)}$ denotes the Dirac measure at 0. Equivalently,

$$\alpha(ds_1\dots ds_p) = \left(\int_{\mathbb{R}^2} dy\ \delta_{(y)}\left(B^1_{s_1}\right)\dots\delta_{(y)}\left(B^p_{s_p}\right)\right) ds_1\dots ds_p.$$

We will use the latter formal expression as a starting point for our construction. The idea is to replace the Dirac measure at y by a suitable approximation. We set :

$$\delta^{\varepsilon}_{(y)}(z) = (\pi\varepsilon^2)^{-1}\, 1_{D(y,\varepsilon)}(z)\ ,$$

and

$$\alpha_{\varepsilon}(ds_1\dots ds_p) = \varphi_{\varepsilon}\left(B^1_{s_1},\dots,B^p_{s_p}\right) ds_1\dots ds_p$$

where

$$\varphi_{\varepsilon}(z_1,\dots,z_p) = \int_{\mathbb{R}^2} \prod_{j=1}^{p} \delta^{\varepsilon}_{(y)}(z_j)\ dy.$$

Notice that $\varphi_{\varepsilon}(z_1,\dots,z_p) = \varphi_{\varepsilon}(z_1+x,\dots,z_p+x)$ for every $x \in \mathbb{R}^2$.

Theorem 1 : *There exists w.p. 1 a (random) measure* $\alpha(ds_1 \ldots ds_p)$ *on* $(\mathbb{R}_+)^p$ *such that, for any* $A^1, \ldots, A^p$ *bounded Borel subsets of* $\mathbb{R}_+$,

$$\lim_{\varepsilon \to 0} \alpha_\varepsilon(A^1 \times \ldots \times A^p) = \alpha(A^1 \times \ldots \times A^p)$$

in the L^n*-norm, for any* $n < \infty$.

The measure $\alpha(\cdot)$ *is a.s. supported on*

$$\left\{(s_1, \ldots, s_p) \; ; \; B^1_{s_1} = \ldots = B^p_{s_p}\right\} .$$

With probability 1 , *for any* $j \in \{1, \ldots, p\}$ *and any* $t \geq 0$,

$$\alpha(\{s_j = t\}) = 0.$$

Finally,

$$(1) \qquad E\left[\alpha(A^1 \times \ldots \times A^p)^n\right] = \int_{(\mathbb{R}^2)^n} dy_1 \ldots dy_n$$

$$\times \prod_{j=1}^{p} \left(\int_{(A^j)^n_<} ds_1 \ldots ds_n \sum_{\sigma \in \Sigma_n} \left(p_{s_1}(x^j, y_{\sigma(1)}) \prod_{k=2}^{n} p_{s_k - s_{k-1}}(y_{\sigma(k-1)}, y_{\sigma(k)})\right)\right)$$

where Σ_n *is the set of all permutations of* $\{1, \ldots, n\}$ *and*

$$(A^j)^n_< = \left\{(s_1, \ldots, s_n) \in (A^j)^n \; ; \; 0 \leq s_1 < \ldots < s_n\right\}.$$

Proof : First step. We first check the L^2-convergence of $\alpha_\varepsilon(A^1 \times \ldots \times A^p)$. It suffices to prove that

$$\lim_{\varepsilon, \varepsilon' \to 0} E[\alpha_\varepsilon(A^1 \times \ldots \times A^p) \, \alpha_{\varepsilon'}(A^1 \times \ldots \times A^p)]$$

exists and is finite. By Fubini's theorem,

$$(2) \qquad E[\alpha_\varepsilon(A^1 \times \ldots \times A^p) \, \alpha_{\varepsilon'}(A^1 \times \ldots \times A^p)]$$

$$= \int dy dy' \prod_{j=1}^{p} \left(\int_{(A^j)^2} ds ds' \; E\left[\delta^\varepsilon_{(y)}(B^j_s) \, \delta^{\varepsilon'}_{(y')}(B^j_{s'})\right]\right)$$

$$= \int dy dy' \prod_{j=1}^{p} \left(\int_{(A^j)^2_<} ds ds' \; E\left[\delta^\varepsilon_{(y)}(B^j_s) \delta^{\varepsilon'}_{(y')}(B^j_{s'}) + \delta^{\varepsilon'}_{(y')}(B^j_s) \delta^\varepsilon_{(y)}(B^j_{s'})\right]\right).$$

It is obvious that, for $(s, s') \in (A^j)^2_<$,

$$\lim_{\varepsilon, \varepsilon' \to 0} E\left[\delta^\varepsilon_{(y)}(B^j_s) \, \delta^{\varepsilon'}_{(y')}(B^j_{s'})\right] = p_s(x^j, y) \, p_{s'-s}(y, y').$$

The only problem is thus to justify the use of dominated convergence. We will find a function $\varphi(y, y', s, s')$ such that, for every $M > 0$,

(3) $$\int dy dy' \left(\iint_{[0,M]^2_<} ds\, ds'\ \varphi(y,y',s,s') \right)^p < \infty$$

and for any $y, y' \in \mathbb{R}^2$, $(s,s') \in (0,\infty)^2_<$, $\varepsilon, \varepsilon' \in (0,1)$,

(4) $$E\left[\delta^{\varepsilon}_{(y)}(B^j_s)\ \delta^{\varepsilon'}_{(y')}(B^j_{s'})\right] \le \varphi(y-x^j, y', s, s').$$

The existence of such a function justifies the passage to the limit under the integral sign in the right side of (2).

Clearly we may assume that $x^j = 0$, and we drop the superscript j in what follows. We first consider $E[\delta^{\varepsilon}_{(y)}(B_s)]$ for $s > 0$. If $|y| \ge 2\varepsilon$, then obviously

$$E[\delta^{\varepsilon}_{(y)}(B_s)] \le p_s(0, y/2).$$

If $|y| < 2\varepsilon$, then

$$E[\delta^{\varepsilon}_{(y)}(B_s)] \le (\pi\varepsilon^2)^{-1} \wedge (2\pi s)^{-1} \le 4(|y|^{-2} \wedge s^{-1}).$$

Therefore,

$$E[\delta^{\varepsilon}_{(y)}(B_s)] \le \psi(y,s)$$

where :

$$\psi(y,s) = 4\, 1_{(|y| \le 2)}\ |y|^{-2} \wedge s^{-1} + p_s(0, y/2).$$

Notice that :

(5) $$\int_0^M \psi(y,s)\, ds \le C_M\, G_1(0,y)$$

where $G_1(x,y) = \int_0^\infty e^{-s}\, p_s(x,y)\, ds$.

We now bound $E[\delta^{\varepsilon}_{(y)}(B_s)\ \delta^{\varepsilon'}_{(y')}(B_{s'})]$. The easy case is when $|y'-y| \ge 2(\varepsilon+\varepsilon')$. Then the Markov property at time s gives :

$$E[\delta^{\varepsilon}_{(y)}(B_s)\ \delta^{\varepsilon'}_{(y')}(B_{s'})] \le E[\delta^{\varepsilon}_{(y)}(B_s)]\ p_{s'-s}(0, \tfrac{y'-y}{2}) \le \psi(y,s)\ p_{s'-s}(0, \tfrac{y'-y}{2})$$

Suppose now that $|y'-y| < 2(\varepsilon+\varepsilon')(\le 4)$. If $\varepsilon \le \varepsilon'$, the Markov property gives

$$E[\delta^{\varepsilon}_{(y)}(B_s)\ \delta^{\varepsilon'}_{(y')}(B_{s'})] \le E[\delta^{\varepsilon}_{(y)}(B_s)] \left((\pi\varepsilon'^2)^{-1} \wedge (2\pi(s'-s))^{-1} \right)$$

$$\le 16\ \psi(y,s)\ (|y'-y|^{-2} \wedge (s'-s)^{-1}).$$

If $\varepsilon' < \varepsilon$, then we discuss separately each of the cases $s'-s > |y'-y|^2$, $s'-s \le |y'-y|^2$. If $s'-s > |y'-y|^2$, then obviously

$$E[\delta^{\varepsilon}_{(y)}(B_s)\ \delta^{\varepsilon'}_{(y')}(B_{s'})] \le E[\delta^{\varepsilon}_{(y)}(B_s)]\ (2\pi(s'-s))^{-1} \le \psi(y,s)\ (|y'-y|^{-2} \wedge (s'-s)^{-1}).$$

Finally, if $s'-s \le |y'-y|^2$, $\varepsilon' < \varepsilon$,

$$E[\delta^{\varepsilon}_{(y)}(B_s)\,\delta^{\varepsilon'}_{(y')}(B_{s'})] \le (\pi\varepsilon^2)^{-1}E[\delta^{\varepsilon'}_{(y')}(B_{s'})] \le 16\ |y'-y|^{-2}\ \psi(y',s')$$
$$\le 16\ (|y'-y|^{-2}\wedge(s'-s)^{-1})\ \psi(y',s').$$

The previous estimates show that (4) holds with

$$\varphi(y,y',s,s') = (\psi(y,s) + \psi(y',s'))\Big(p_{s'-s}(0,(y'-y)/2) + 16(|y'-y|^{-2}\wedge(s'-s)^{-1})\Big).$$

Note that

$$\int_{[0,M]^2_<} ds\,ds'\ \varphi(y,y',s,s') \le C'_M\ \Big(G_1(0,\tfrac{y}{2}) + G_1(0,\tfrac{y'}{2})\Big)\ G_1\Big(0,\ \frac{y'-y}{2}\Big)$$

so that (3) is clearly satisfied (the key ingredient is the fact that $G_1(0,y)$ is in L^p for any $p < \infty$).

<u>Second step</u> : The first step allows us to set :

$$\tilde{\alpha}(A^1 \times \ldots \times A^p) = L^2 - \lim_{\varepsilon\to 0}\ \alpha_\varepsilon(A^1 \times \ldots \times A^p).$$

We now check that the convergence holds in L^n, and that the nth-moment of $\tilde{\alpha}(A^1 \times \ldots \times A^p)$ is the right side of (1). To this end, it is enough to obtain the convergence of

(6) $\quad E[\alpha_\varepsilon(A^1 \times \ldots \times A^p)^n]$

$$= \int_{(\mathbb{R}^2)^n} dy_1\ldots dy_n \prod_{j=1}^{p} \Big(\int_{(A^j)^n} ds_1\ldots ds_n\ E\Big[\prod_{k=1}^{n} \delta^{\varepsilon}_{(y_k)}(B^j_{s_k})\Big]\Big)$$

$$= \int_{(\mathbb{R}^2)^n} dy_1\ldots dy_n \prod_{j=1}^{p} \Big(\sum_{\sigma\in\Sigma_n} \int_{(A^j)^n_<} ds_1\ldots ds_n\ E\Big[\prod_{k=1}^{n} \delta^{\varepsilon}_{(y_{\sigma(k)})}(B^j_{s_k})\Big]\Big).$$

Clearly, for $(s_1,\ldots,s_n) \in (A^j)^n_<$,

$$\lim_{\varepsilon\to 0} E\Big[\prod_{k=1}^{n} \delta^{\varepsilon}_{(y_{\sigma(k)})}\ (B^j_{s_k})\Big] = p_{s_1}(x^j,y_{\sigma(1)}) \prod_{k=2}^{n} p_{s_k-s_{k-1}}(y_{\sigma(k-1)},y_{\sigma(k)}).$$

Again we have to justify dominated convergence. This is similar (in fact easier) to what we did in the first step. Indeed, the Markov property at times $s_{n-1}, s_{n-2}, \ldots, s_1$ leads to :

$$E\Big[\prod_{k=1}^{n} \delta^{\varepsilon}_{(y_{\sigma(k)})}(B^j_{s_k})\Big] \le 4\ \psi\Big(y_{\sigma(1)}-x^j,s_1\Big)\ \times \ldots \times 4\ \psi\Big(y_{\sigma(n)}-y_{\sigma(n-1)},s_n-s_{n-1}\Big)$$

with the same function ψ as above (consider separately the cases $|y_{\sigma(k)}-y_{\sigma(k-1)}| \ge 4\varepsilon$, and $|y_{\sigma(k)}-y_{\sigma(k-1)}| < 4\varepsilon$). Then the bound (5) justifies the passage to the limit in the right side of (6).

Third step : We will now construct a random measure $\alpha(\cdot)$ such that for any $A^1,\dots, A^p$, $\alpha(A^1 \times \dots \times A^p) = \tilde{\alpha}(A^1 \times \dots \times A^p)$ a.s. We first consider the case $A^j = [a_j,b_j]$, $a_j \le b_j \le M$. Then by applying the (generalized) Hölder inequality to the right side of (1) we get :

$$E\Big[\tilde{\alpha}(A_1 \times \dots \times A_p)^n\Big]$$

$$\le (n!)^p \prod_{j=1}^{p} \left(\int_{(\mathbb{R}^2)^n} dy_1\dots dy_n \left(\int_{(A^j)^n_<} ds_1\dots ds_n\, p_{s_1}(x^j,y_1)\prod_{k=2}^{n} p_{s_k-s_{k-1}}(y_{k-1},y_k)\right)^p\right)^{1/p}$$

$$\le (n!)^p \prod_{j=1}^{p} \left(\int_{(\mathbb{R}^2)^n} dy_1\dots dy_n\, G^{a_j,b_j}(x^j,y_1)^p \prod_{k=2}^{n} G^{0,b_j-a_j}(y_{k-1},y_k)^p\right)^{1/p}$$

where :

$$G^{u,v}(x,y) = \int_u^v ds\, p_s(x,y).$$

It is easy to check that :

$$\int dy\, G^{u,v}(x,y)^p \le C_p(v-u)$$

for some constant C_p (use scaling when $u = 0$). We conclude that :

(7) $$E\Big[\tilde{\alpha}(A^1 \times \dots \times A^p)^n\Big] \le (C_p)^n\,(n!)^p \prod_{j=1}^{p} (b_j-a_j)^{n/p}.$$

It follows from (7) and the multidimensional version of Kolmogorov's lemma that the mapping

$$(a_1,b_1,a_2,b_2,\dots,a_p,b_p) \longrightarrow \tilde{\alpha}\Big([a_1,b_1] \times\dots\times [a_p,b_p]\Big)$$

has a continuous version, denoted by $\alpha([a_1,b_1] \times\dots\times [a_p,b_p])$. Notice that w.p 1, $\alpha([a_1,b_1] \times\dots\times [a_p,b_p])$ is a nondecreasing finitely additive function of $[a_1,b_1] \times\dots\times [a_p,b_p]$ (consider first the case of rational a_j, b_j). Standard measure-theoretic arguments allow us to extend $\alpha(\cdot)$ to a Radon measure on $(\mathbb{R}_+)^p$.

If $A^{(n)} = A_1^{(n)} \times A_2 \times \dots \times A_p$ increases (resp. decreases) towards $A = A_1 \times A_2 \times \dots \times A_p$ then $\tilde{\alpha}(A^{(n)})$ converges in L^2 towards $\tilde{\alpha}(A)$, (by (1)), whereas $\alpha(A^{(n)})$ converges a.s towards $\alpha(A)$. This observation and the monotone class theorem easily give :

$$\alpha(A_1 \times\dots\times A_p) = \tilde{\alpha}(A_1 \times\dots\times A_p)\ ,\ \text{a.s}$$

for any $A_1,\dots,A_p$ bounded Borel subsets of $\mathbb{R}_+$.

<u>Fourth step</u> : It remains to check that α has the desired properties. The fact that $\alpha(\{s_j = t\}) = 0$ for every $t \geq 0$, a.s, is obvious from the continuity of

$$(a_1, b_1, \ldots, a_p, b_p) \longrightarrow \alpha([a_1, b_1] \times \ldots \times [a_p, b_p]).$$

Finally, suppose that $A = [a_1, b_1] \times \ldots \times [a_p, b_p]$ is a closed rectangle with rational coordinates. Then on the set

$$\mathcal{A} = \left\{ \omega \; ; \; A \cap \{(s_1, \ldots, s_p) \; ; \; B^1_{s_2} = B^2_{s_2} = \ldots = B^p_{s_p}\} = \emptyset \right\}$$

we have $\alpha_\varepsilon(A) = 0$ for ε small, by the definition of α_ε. Therefore $\alpha(A) = 0$ a.s on $\mathcal{A}$. Since this is true w.p. 1 for any rectangle with rational coordinates, the support property of α follows at once. □

<u>Remark</u>. It was convenient in the previous proof to assume that the starting point of each Brownian motion was deterministic. However it is immediate that Theorem 1 still holds in the more general situation where the starting points may be random. The right side of (1) should then be integrated with respect to $\mu^1(dx^1)\ldots\mu^p(dx^p)$ where $\mu^j(dx^j)$ stands for the initial distribution of B^j.

Proposition 2 : *Suppose that* $B^1_0 = \ldots = B^p_0$.

(i) *For any* $t \geq 0$, $\lambda > 0$, $\alpha([0,\lambda t]^p) \overset{(d)}{=} \lambda\, \alpha([0,t]^p)$.

(ii) *With probability one, for every* $t > 0$, $\alpha([0,t]^p) > 0$.

<u>Proof</u> : Without loss of generality we may take $B^1_0 = \ldots = B^p_0 = 0$. Property (i) follows from a simple scaling argument. Set

$$\tilde{B}^j_t = \lambda^{-1/2} B^j_{\lambda t} \qquad (j = 1, \ldots ,p).$$

Then $\tilde{\alpha}_\varepsilon([0,t]^p) = \lambda^{-1} \alpha_{\lambda^{1/2}\varepsilon}([0,\lambda t]^p)$, which implies

$$\tilde{\alpha}\left([0,t]^p\right) = \lambda^{-1} \alpha\left([0,\lambda t]^p\right) \quad \text{a.s.}$$

To prove (ii), notice that the events $\{\alpha([0,t]^p) > 0\}$ decrease as t decreases. It follows from (i) that

$$P[\alpha([0,1]^p) > 0) = P(\alpha([0,t]^p) > 0) = P[\bigcap_{s>0} \{\alpha([0,s]^p) > 0\}] \; .$$

However $P[\alpha([0,1]^p) > 0] > 0$ since $E[\alpha([0,1]^p)] > 0$. The zero-one law yields the desired result. □

Proposition 2 (ii) and Theorem 1 imply that, provided $B^1_0 = \ldots = B^p_0$, for any $\varepsilon > 0$, there exist $t_1, \ldots, t_p \in (0,\varepsilon)$ such that

$$B^1_{t_1} = \dots = B^p_{t_p} .$$

In the case of arbitrary starting points, one can use a scaling argument to check that these equalities hold for some $t_1, \dots, t_p \in (0,\infty)$, w.p 1. Therefore the paths of $B^1,\dots,B^p$ have a common point (different from their starting point).

2. Intersections of independent Wiener sausages.

Our goal in this section is to provide an approximation of the intersection local time in terms of Wiener sausages. This approximation is similar to the well-known approximations of the usual Brownian local time.

We fix a non-polar compact subset K of R^2 and for $j \in \{1,\dots,p\}$ we denote by $S^j_{\varepsilon K}(0,t)$ the Wiener sausage associated with the Brownian motion B^j and the compact set εK, on the time interval $[0,t]$:

$$S_{\varepsilon K}(0,t) = \bigcup_{0\le s\le t} (B_s + \varepsilon K) .$$

As we have seen in chapter VI,

$$\lim_{\varepsilon\to 0} (\log 1/\varepsilon)\, m(S^j_{\varepsilon K}(0,t)) = \pi t ,$$

in the L^2-norm.

Theorem 3 : *We have :*

$$\lim_{\varepsilon\to 0} (\log 1/\varepsilon)^p\, m(S^1_{\varepsilon K}(0,t) \cap \dots \cap S^p_{\varepsilon K}(0,t)) = \pi^p\, \alpha([0,t]^p).$$

in the L^2*-norm.*

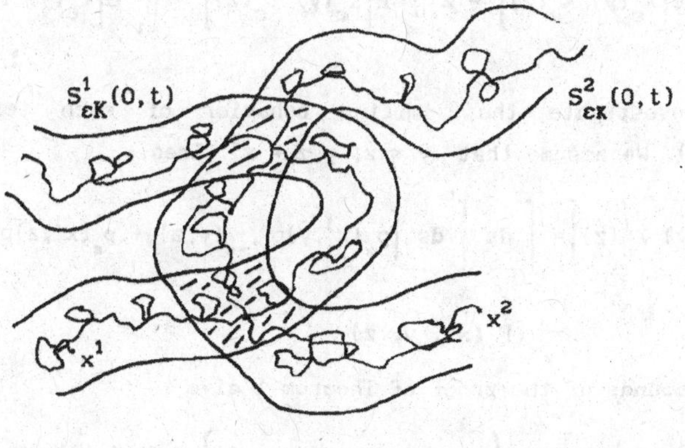

Fig. 1

Remark. The convergence of Theorem 3 holds in the L^n-norm for any $n < \infty$. However we shall restrict our attention to the L^2-convergence.

Proof : To simplify notation, we will assume that the starting points $B_0^1 = x^1$, ..., $B_0^p = x^p$ are deterministic. We fix $t > 0$. Then

$$\alpha_\varepsilon([0,t]^p) = \int_{\mathbb{R}^2} ds\, X_\varepsilon(y)$$

where

$$X_\varepsilon(y) = \prod_{j=1}^{p} \int_0^t ds\, \delta^\varepsilon_{(y)}(B_s^j).$$

Similarly,

$$\pi^{-p}(\log 1/\varepsilon)^p\, m\left(S^1_{\varepsilon K}(0,t) \cap \ldots \cap S^p_{\varepsilon K}(0,t)\right) = \int_{\mathbb{R}^2} dy\, Y_\varepsilon(y)$$

where

$$Y_\varepsilon(y) = \prod_{j=1}^{p} \left(\pi^{-1}(\log 1/\varepsilon)\, I(y \in S^j_{\varepsilon K}(0,t))\right).$$

Therefore Theorem 3 is equivalent to:

$$\text{(8)} \qquad \lim_{\varepsilon\to 0} E\left[\left(\int_{\mathbb{R}^2} dy\left(X_\varepsilon(y) - Y_\varepsilon(y)\right)\right)^2\right] = 0.$$

Write

$$X_\varepsilon(y) = \prod_{j=1}^{p} X^j_\varepsilon(y)\ , \quad Y_\varepsilon(y) = \prod_{j=1}^{p} Y^j_\varepsilon(y)$$

with an obvious notation. Then

$$\text{(9)} \quad E\left[\left(\int_{\mathbb{R}^2} dy\, (X_\varepsilon(y) - Y_\varepsilon(y))\right)^2\right]$$

$$= \int_{(\mathbb{R}^2)^2} dydz\left(\prod_{j=1}^{p} E\left[X^j_\varepsilon(y)\, X^j_\varepsilon(z)\right] - 2\prod_{j=1}^{p} E\left[X^j_\varepsilon(y)\, Y^j_\varepsilon(z)\right] + \prod_{j=1}^{p} E\left[Y^j_\varepsilon(y)\, Y^j_\varepsilon(z)\right]\right).$$

We will investigate the limiting behavior of each term of the right side of (9). We assume that $y \neq z$, $y, z \neq x^j$. Then,

$$\text{(10)} \quad \lim_{\varepsilon\to 0} E\left[X^j_\varepsilon(y)\, X^j_\varepsilon(z)\right] = \int_0^t ds \int_s^t ds' \left(p_s(x^j,y)p_{s'-s}(y,z) + p_s(x^j,z)p_{s'-s}(z,y)\right)$$

$$=: F_t(x^j,\, y,\, z).$$

Furthermore the bounds of the proof of Theorem 1 give :

$$E\left[X^j_\varepsilon(y)\, X^j_\varepsilon(z)\right] \le C\left(G_1(0,y/2) + G_1(0,z/2)\right) G_1(0,(z-y)/2).$$

Next, we have :

$$E\left[X_\varepsilon^j(y)\, Y_\varepsilon^j(z)\right] = \pi^{-2}\, \varepsilon^{-2}(\log 1/\varepsilon)\, E\left[\left(\int_0^t ds\, 1_{D(y,\varepsilon)}(B_s^j)\right) I\left(z \in S_{\varepsilon K}(0,t)\right)\right].$$

Set $T_{\varepsilon K}^j(z) = \inf\{t \geq 0\ ;\ B_t^j \in z-\varepsilon K\}$. Then,

$$E\left[I(z \in S_{\varepsilon K}^j(0,t)) \int_0^t ds\, 1_{D(y,\varepsilon)}(B_s^j)\right]$$

$$= E\left[I(T_{\varepsilon K}^j(z) \leq t) \int_{T_{\varepsilon K}^j(z)}^t ds\, 1_{D(y,\varepsilon)}(B_s^j)\right] + E\left[\int_0^t ds\, 1_{D(y,\varepsilon)}(B_s^j)\, I\left(s < T_{\varepsilon K}^j(z) \leq t\right)\right].$$

Lemma VI.2 and the Markov property at time $T_{\varepsilon K}^j(z)$ give

$$\lim_{\varepsilon\to 0} \pi^{-2}\varepsilon^{-2}(\log 1/\varepsilon)\, E\left[I\left(T_{\varepsilon K}^j(z) \leq t\right) \int_{T_{\varepsilon K}^j(z)}^t ds\, 1_{D(y,\varepsilon)}(B_s^j)\right]$$

$$= \int_0^t ds'\, p_{s'}(x^j,z) \int_{s'}^t ds\, p_{s-s'}(z,y).$$

Next,

$$E\left[\int_0^t ds\, 1_{D(y,\varepsilon)}(B_s^j)\, I\left(s < T_{\varepsilon K}^j(z) \leq t\right)\right] = E\left[\int_0^t ds\, 1_{D(y,\varepsilon)}(B_s^j)\, I\left(z \in S_{\varepsilon K}^j(s,t)\right)\right]$$

$$- E\left[\int_0^t ds\, 1_{D(y,\varepsilon)}(B_s^j)\, I\left(z \in S_{\varepsilon K}^j(0,s) \cap S_{\varepsilon K}^j(s,t)\right)\right].$$

On one hand, Lemma VI.2 (ii) implies

$$\lim_{\varepsilon\to 0} \pi^{-2}\varepsilon^{-2}(\log 1/\varepsilon)\, E\left[\int_0^t ds\, 1_{D(y,\varepsilon)}(B_s^j)\, I\left(z \in S_{\varepsilon K}^j(s,t)\right)\right]$$

$$= \int_0^t ds\, p_s(x^j,y) \int_s^t ds'\, p_{s'-s}(y,z).$$

In fact we need a little more than the convergence of Lemma VI.2 (ii): A simple compactness argument shows that this convergence holds uniformly when y varies over a compact subset of $\mathbb{R}^2 \setminus \{0\}$. On the other hand, the Markov property at $T_{\varepsilon K}(z)$ and the bounds of Lemma VI.1 give :

$$\varepsilon^{-2}(\log 1/\varepsilon)\, E\left[\int_0^t ds\, 1_{D(y,\varepsilon)}(B_s^j)\, I\left(z \in S_{\varepsilon K}^j(0,s) \cap S_{\varepsilon K}^j(s,t)\right)\right] = O\left((\log 1/\varepsilon)^{-1}\right)$$

as ε tends to 0. We conclude that

(11) $$\lim_{\varepsilon\to 0} E\left[X^j_\varepsilon(y)\ Y^j_\varepsilon(z)\right] = F_t(x^j,y,z).$$

Moreover, Lemma VI.1 and the previous arguments show that $E\left[X^j_\varepsilon(y)\ Y^j_\varepsilon(z)\right]$ satisfies the same bound as $E\left[X^j_\varepsilon(y)\ X^j_\varepsilon(z)\right]$.

Finally we consider

$$E\left[Y^j_\varepsilon(y)\ Y^j_\varepsilon(z)\right] = \pi^{-2}(\log 1/\varepsilon)^2\ P\left[T^j_{\varepsilon\kappa}(y) \le t,\ T^j_{\varepsilon\kappa}(z) \le t\right].$$

We have already noticed in the proof of Lemma VI.4 that this quantity satisfies the same bound as $E\left[X^j_\varepsilon(y)\ X^j_\varepsilon(z)\right]$. Since

$$P\left[T^j_{\varepsilon\kappa}(y) \le t,\ T^j_{\varepsilon\kappa}(z) \le t\right]$$

$$\le P\left[T^j_{\varepsilon\kappa}(y) \le t,\ z \in S^j_{\varepsilon\kappa}(T^j_{\varepsilon\kappa}(y),t)\right] + P\left[T^j_{\varepsilon\kappa}(z) \le t,\ y \in S^j_{\varepsilon\kappa}(T^j_{\varepsilon\kappa}(z),t)\right],$$

Lemma VI.2 and the Markov property give

(12) $$\limsup_{\varepsilon\to 0} E\left[Y^j_\varepsilon(y)\ Y^j_\varepsilon(z)\right] \le F_t(x^j,y,z).$$

We now pass to the limit in the right side of (9), using (10), (11), (12). Observe that the use of dominated convergence is justified by our bounds and the fact that $G_1(0,y)$ is in L^n for any $n < \infty$. It follows that :

$$\limsup_{\varepsilon\to 0} E\left[\left(\int_{\mathbb{R}^2} dy\ (X_\varepsilon(y) - Y_\varepsilon(z))\right)^2\right] \le 0.$$

This completes the proof of (8), and that of Theorem 3. □

3. Self-intersection local times.

We now consider only one Brownian motion B started at 0. We are interested in p-multiple points of the process B. The (p-multiple) self-intersection local time of B is the Radon measure on

$$\mathcal{T}_p := \{(s_1,\dots,s_p) \in (\mathbb{R}_+)^p\ ;\ 0 \le s_1 < s_2 <\dots< s_p\},$$

formally defined by :

$$\beta(ds_1 \dots ds_p) = \delta_{(0)}(B_{s_1} - B_{s_2})\dots\ \delta_{(0)}(B_{s_{p-1}} - B_{s_p})\ ds_1\dots ds_p.$$

To construct β rigorously we proceed as in Section 1. For $\varepsilon > 0$, we set

$$\beta_\varepsilon(ds_1\dots ds_p) = 1_{\mathcal{T}_p}(s_1,\dots,s_p)\ \ \varphi_\varepsilon(B_{s_1},\dots,B_{s_p})\ ds_1\dots ds_p$$

where

$$\varphi_\varepsilon(z_1,\dots,z_p) = \int dy\ \prod_{j=1}^p \delta^\varepsilon_{(y)}(z_j).$$

We then have the following analogue of Theorem 1.

Theorem 4 : *There exists w.p 1 a Radon measure* $\beta(ds_1\dots ds_p)$ *on* $\mathcal{T}_p$ *such that, for any compact subset of* $\mathcal{T}_p$ *of the form* $A_1 \times A_2 \dots \times A_p$,

$$\beta(A_1 \times A_2 \times \dots \times A_p) = \lim_{\varepsilon\to 0} \beta_\varepsilon(A_1 \times A_2 \times \dots \times A_p) ,$$

in the L^n*-norm for any* $n < \infty$.

The measure $\beta(\cdot)$ *is w.p. 1 supported on :*

$$\{(s_1,\dots,s_p) \in \mathcal{T}_p \ ; \ B_{s_1} = \dots = B_{s_p}\} .$$

Moreover, $\beta(\{s_j=t\}) = 0$ *for any* $j \in \{1,\dots,p\}$ *and any* $t \ge 0$, *a.s.*

Proof : We may find a countable collection of compact rectangles

$$I^m = [a_1^m,b_1^m] \times\dots\times [a_p^m,b_p^m] \quad (0 \le a_1^m \le b_1^m < a_2^m \le b_2^m < \dots < a_p^m \le b_p^m)$$

such that

(i) $\mathcal{T}_p = \bigcup_{m=1}^{\infty} I^m$;

(ii) if $m \ne m'$, $I^m \cap I^{m'}$ is contained in a finite union of "hyperplanes" $\{s_j = t\}$;

(iii) any compact subset of $\mathcal{T}_p$ intersects only a finite number of the rectangles I_m .

Fix one of these rectangles $I = [a_1,b_1] \times\dots\times [a_p,b_p]$. For every $j \in \{1,\dots,p\}$, define a process $(\Gamma_t^j,\ 0 \le t \le b_j - a_j)$ by :

$$\Gamma_t^j = B_{a_j+t} \ , \text{ for } t \in [0,\ b_j-a_j] .$$

Of course the processes $\Gamma^1,\dots,\Gamma^p$ are not independent. However, the distribution of $(\Gamma^1,\dots,\Gamma^p)$ is absolutely continuous with respect to that of p independent Brownian motions. More precisely, define p probability measures $\mu_1,\dots,\mu_p$ on $\mathbb{R}^2$ by

$$\mu_1(dy) = \begin{cases} \delta_{(0)}(dy) & \text{if } a_1 = 0 \\ (2\pi)^{-1} \exp -|y| \ dy & \text{if } a_1 > 0 \end{cases}$$

and for $j \ge 2$,

$$\mu_j(dy) = (2\pi)^{-1} \exp -|y| \ dy .$$

Denote by $W(dw_1\dots dw_p)$ the joint distribution of $(B^1,\dots,B^p)$, where $B^1,\dots,B^p$ are independent and each B^j is a planar Brownian motion defined on the time interval $[0,\ b_j-a_j]$, with initial distribution μ_j. The distribution of $(\Gamma^1, \ldots, \Gamma^p)$ is then absolutely continuous w.r.t. W . The associated Radon-Nikodym density can be written explicitly, and straightforward estimates show that it belongs to $L^2(W)$.

It follows from this observation and Theorem 1 (use also the remark after Theorem 1) that there exists a (random) measure $\beta^I(ds_1 \ldots ds_p)$ supported on I such that, for any compact subset of I of the form $A_1 \times \ldots \times A_p$,

$$\beta^I(A_1 \times \ldots \times A_p) = \lim_{\varepsilon \to 0} \beta_\varepsilon(A_1 \times \ldots \times A_p)$$

in the L^n-norm for any $n < \infty$. Furthermore, β^I does not charge the hyperplanes. and is supported on $\{(s_1, \ldots, s_p) \in I \,;\, B_{s_1} = \ldots = B_{s_p}\}$.

To complete the proof of Theorem 4, we simply set :

$$\beta = \sum_{m=1}^{\infty} \beta^{I_m}$$

It is easy to check that β has the desired properties. In particular, property (iii) ensures that β is a Radon measure.

<u>Proposition</u> 5. ***With*** probability 1, for any $0 \leq a < b$,

$$\beta\left(\mathcal{T}_p \cap [a,b]^p\right) = +\infty .$$

<u>Proof</u> : We take $a = 0$, $b = 1$ (the extension is trivial). Set

$$I = I_0^0 = [0, \tfrac{1}{2p}] \times [\tfrac{2}{2p}, \tfrac{3}{2p}] \times \ldots \times [\tfrac{2(p-1)}{2p}, \tfrac{2p-1}{2p}]$$

and more generally for any $k \geq 0$, $\ell \in \{0, 1, \ldots, 2^k - 1\}$,

$$I_\ell^k = [\ell 2^{-k}, \ell 2^{-k} + \frac{2^{-k}}{2p}] \times \ldots \times [\ell 2^{-k} + \frac{2(p-1)}{2p} 2^{-k}, \ell 2^{-k} + \frac{2p-1}{2p} 2^{-k}].$$

It is obvious that for any fixed k , the random variables $(\beta(I_\ell^k), \ell = 0, 1, \ldots, 2^k - 1)$ are independent and identically distributed. Moreover, the scaling argument of the proof of Proposition 2 gives

$$\beta(I_\ell^k) \overset{(d)}{=} 2^{-k} \beta(I).$$

It follows that :

$$E\left(\sum_{\ell=1}^{2^k - 1} \beta(I_\ell^k)\right) = E(\beta(I)) = C > 0$$

$$\mathrm{var}\left(\sum_{\ell=1}^{2^k - 1} \beta(I_\ell^k)\right) = 2^{-k} \mathrm{var}(\beta(I)) = 2^{-k} C' .$$

Therefore,

$$\beta\left(\mathcal{T}_p \cap [0,1]^p\right) \geq \sum_{k=0}^{\infty} \left(\sum_{\ell=1}^{2^k - 1} \beta(I_\ell^k)\right) = +\infty . \quad \square$$

<u>Remark.</u> As a consequence of Theorem 4 and Proposition 5, we get the existence

of p-tuples $(s_1,\ldots,s_p) \in \mathcal{T}_p$ such that $B_{s_1}=\ldots=B_{s_p}$, that is, the existence of p-multiple self-intersections. Our derivation of this result is certainly not the shortest one. The construction of the self-intersection local time however yields much useful information about multiple points (see in particular Chapter IX).

Proposition 5 leads us to the so-called renormalization problems. For certain physical questions (especially in polymer models) it is desirable to define a random variable "measuring the number" of p-multiple self-intersections of the Brownian path, say on the time interval $[0,1]$. The natural candidate would be $\beta(\mathcal{T}_p \cap [0,1]^p)$ if this variable were finite. This raises the question of whether it is possible to define a "renormalized self-intersection local time" whose value on the set $\mathcal{T}_p \cap [0,1]^p$ would be finite. The answer is yes. The case $p = 2$ is easy with the tools developed up to now, and will be treated in the next section. The general case is much harder and will be considered in Chapter X.

4. Varadhan's renormalization and an application to the Wiener sausage.

In this section, we take $p = 2$ and we set

$$\mathcal{T} = \mathcal{T}_2 \cap [0,1]^2 .$$

For any $k \geq 0$ and $\ell \in \{0,\ldots,2^k-1\}$ we set

$$A^k_\ell = [\frac{2\ell}{2^{k+1}}, \frac{2\ell+1}{2^{k+1}}) \times (\frac{2\ell+1}{2^{k+1}}, \frac{2\ell+2}{2^{k+1}}]$$

Notice that the sets A^k_ℓ form a partition of $\mathcal{T}$ (see fig. 2).

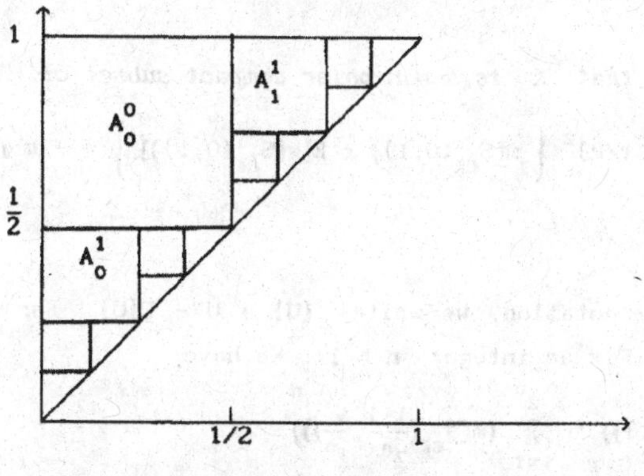

Fig. 2

Proposition 6. *For any Borel subset* A *of* $\mathcal{T}$, *the series*

$$\sum_{k=0}^{\infty}\left(\sum_{\ell=0}^{2^k-1}(\beta(A\cap A_\ell^k)-E[\beta(A\cap A_\ell^k)])\right)$$

converges a.s. and in L^2. *The sum of this series is denoted by* $\gamma(A)$, *and the mapping* $A \longrightarrow \gamma(A)$ *is called the renormalized self-intersection local time of* B.

Proof : Let $\alpha(\cdot)$ denote the intersection local time of two independent planar Brownian motions started at 0. We first observe that $\beta(A_\ell^k) \overset{(d)}{=} \alpha([0,2^{-k-1}]^2)$. Indeed, take $k = \ell = 0$. The processes $B_t^1 = B_{1/2-t} - B_{1/2}$, $B_t^2 = B_{1/2+t} - B_{1/2}$ are two independent Brownian motions (defined on the time interval [0,1/2]) and, from our construction, it is obvious that $\beta(A_0^0)$ coincides with the intersection local time of B^1 and B^2 , on the square $[0,1/2]^2$.

Then, for any fixed k , the random variables $\beta(A\cap A_\ell^k)$, $\ell \in \{0,..,2^k-1\}$ are independent. This is clear since $\beta(A \cap A_\ell^k)$ only depends on the increments of B between times $2\ell\, 2^{-k-1}$ and $(2\ell+2)2^{-k-1}$.

To complete the proof we bound

$$\mathrm{var}\left(\sum_{\ell=0}^{2^k-1}\beta(A\cap A_\ell^k)\right)=\sum_{\ell=0}^{2^k-1}\mathrm{var}\left(\beta(A\cap A_\ell^k)\right)\le\sum_{\ell=0}^{2^k-1}E\left[\beta(A_\ell^k)^2\right]=2^kE\left[\alpha([0,2^{-k-1}]^2)\right]=C2^{-k},$$

by Proposition 2. □

We will now apply Proposition 5 to a theorem concerning the fluctuations of the area of the two-dimensional Wiener sausage. By Theorem VI.6, this area is of order $\pi/(\log 1/\varepsilon)$ for ε small. The next theorem shows that the fluctuations of this area around its expected value are related to the (double) self-intersections of the process.

Theorem 7 : *Suppose that* K *is a non-polar compact subset of* $\mathbb{R}^2$. *Then,*

$$\lim_{\varepsilon\to 0}(\log 1/\varepsilon)^2\left(m(S_{\varepsilon K}(0,1))-E[m(S_{\varepsilon K}(0,1))]\right)=-\pi^2\gamma(\mathcal{T})$$

in the L^2*-norm.*

Proof : To simplify notation, we write $\{U\} = U - E[U]$ for any integrable random variable U. Fix an integer $n \geq 1$. We have:

$$(13)\qquad \{m(S_{\varepsilon K}(0,1))\}=\sum_{i=1}^{2^n}\{m(S_{\varepsilon K}(\tfrac{i-1}{2^n},\tfrac{i}{2^n}))\}$$

$$-\sum_{k=0}^{n-1}\sum_{\ell=0}^{2^k-1}\{m(S_{\varepsilon K}(\tfrac{2\ell}{2^{k+1}},\tfrac{2\ell+1}{2^{k+1}})\cap S_{\varepsilon K}(\tfrac{2\ell+1}{2^{k+1}},\tfrac{2\ell+2}{2^{k+1}}))\}.$$

Note that the variables $m(S_{\varepsilon K}(\frac{i-1}{2^n},\frac{i}{2^n}))$, $i \in \{1,\dots,2^n\}$ are independent. Then, by scaling and Theorem VI.5,

$$(14)\quad E[(\sum_{i=1}^{2^n} \{m(S_{\varepsilon K}(\frac{i-1}{2^n}, \frac{i}{2^n}))\})^2]^{1/2} = 2^{n/2}\ E[\{m(S_{\varepsilon K}(0, \frac{1}{2^n}))\}^2]^{1/2}$$

$$= 2^{-n/2}\ E[\{m(S_{\varepsilon 2^{-n/2}K}(0,1))\}^2]^{1/2}$$

$$\leq C\, 2^{-n/2}\ (\log 1/\varepsilon)^{-2}\ ,$$

for ε small (depending on n). On the other hand, by Theorem 3 and the arguments of the proof of Proposition 6,

$$(15)\quad L^2\text{-}\lim_{\varepsilon\to 0} (\log 1/\varepsilon)^2 \sum_{k=0}^{n-1} \sum_{\ell=0}^{2^k-1} \{m(S_{\varepsilon K}(\frac{2\ell}{2^{k+1}}, \frac{2\ell+1}{2^{k+1}}) \cap S_{\varepsilon K}(\frac{2\ell+1}{2^{k+1}}, \frac{2\ell+2}{2^{k+1}}))\}$$

$$= \pi^2 \sum_{k=0}^{n-1} \sum_{\ell=0}^{2^k-1} \{\beta(A_\ell^k)\}\ ,$$

and the latter sum is close to $\gamma(\mathcal{T})$ when n is large, by the definition of $\gamma(\mathcal{T})$.

To complete the proof, fix $\delta > 0$. We can choose n so that the right side of (14) is smaller than $(\delta/3)\ (\log 1/\varepsilon)^{-2}$ for ε small, and the L^2-norm of

$$\pi^2(\gamma(\mathcal{T}) - \sum_{k=0}^{n-1} \sum_{\ell=0}^{2^k-1} \{\beta(A_\ell^k)\})$$

is less than $\delta/3$. Then by (13) and (15) the L^2-norm of

$$(\log 1/\varepsilon)^2\ \{m(S_{\varepsilon K}(0,1))\} - \pi^2\ \gamma(\mathcal{T})$$

will be smaller than δ , for ε small. □

Remark. The minus sign in $-\ \pi^2\ \gamma(\mathcal{T})$ corresponds to the intuitive idea that if there are many self-intersections then the area of the sausage will be smaller.

Spitzer [Sp2] obtains the following expansion for the expected area of the two-dimensional Wiener sausage:

$$E[m(S_{\varepsilon K}(0,1))] = \frac{\pi}{\log 1/\varepsilon} + \frac{\pi}{(\log 1/\varepsilon)^2}\left(\frac{1+\kappa-\log 2}{2} + R(K)\right) + o\left(\frac{1}{(\log 1/\varepsilon)^2}\right),$$

where κ denotes Euler's constant, and $R(K)$ is the logarithm of the logarithmic capacity of K (see Chapter XI for a precise definition). We can combine this expansion with Theorem 7 to get:

$$m(S_{\varepsilon K}(0,1)) = \frac{\pi}{\log 1/\varepsilon} + \frac{\pi}{(\log 1/\varepsilon)^2}\left(\frac{1+\kappa-\log 2}{2} + R(K) - \pi\,\gamma(\mathcal{T})\right) + \mathcal{R}(\varepsilon,K) ,$$

where

$$\lim_{\varepsilon\to 0} (\log 1/\varepsilon)^2\, \mathcal{R}(\varepsilon,K) = 0 ,$$

in the L^2-norm. This result will be extended in Chapter XI , where we will obtain a full asymptotic expansion of $m(S_{\varepsilon K}(0,1))$. The kth term of this expansion is of order $|\log \varepsilon|^{-k}$ and involves a random variable related to the k-multiple self-intersections of B .

Bibliographical notes. The notion of intersection local time was motivated by physical problems: see in particular Edwards [E] and Symanzik [Sy]. In appendix to Symanzik's paper [Sy], Varadhan gave a construction of the renormalized variable $\gamma(\mathcal{T})$ (in the more difficult case of the planar Brownian bridge), without introducing the intersection local time. The first work on intersection local times is probably due to Wolpert [Wo]. Dynkin [Dy1] gave a general construction of additive functionals of several independent Markov processes, which includes intersection local times as a particular case. See also [Dy4] for results in the special case of Brownian motion. Using a different approach, depending on the Gaussian character of Brownian motion, Geman, Horowitz and Rosen [GHR] derived precise information about the intersection local time of independent Brownian motions. The self-intersection local time of Brownian motion has been studied extensively by Rosen ([R1], [R2]) and Yor ([Y3], [Y4]). Rosen [R5] has extended some of his results to diffusion processes more general than Brownian motion. See also [L8] for the intersection local time of Lévy processes. In the case of Brownian motion, Rosen [R3] and Yor [Y1] prove Tanaka-like formulas for the intersection local time, and apply these formulas to the Varadhan renormalization. See also Yor [Y2] for a weak analogue of the Varadhan renormalization in three dimensions. The results of Section 2 are from [L3] (at least in the case K = D), where they were applied to estimates concerning the Hausdorff measure of multiple points (see also Weinryb [W2] for some extensions). The methods of Sections 3 and 4 are taken from [L2]. Theorem 7 was proved in [L2] in the special case K = D , and then extended in [L10] . The latter paper also contains fluctuation theorems for the Wiener sausage in higher dimensions. See also Chavel, Feldman and Rosen [CFR] for an extension of Theorem 7 to Brownian motion on Riemannian

CHAPTER IX

Points of infinite multiplicity of the planar Brownian motion.

As a simple consequence of Theorem VIII-4 and Proposition VIII-5 we get that a planar Brownian path has p-multiple points for any integer p, w.p.1. Intersection local times certainly do not provide the shortest way of arriving at this result. Nonetheless, they can be used to get much useful information about multiple points. In this chapter we will use self-intersection local times to prove the existence of points of infinite multiplicity. The proof involves no technical estimate, mainly because the hard work has already been done in the previous chapter. The first section develops certain tools which are of independent interest.

1. The behavior of Brownian motion between the successive hitting times of a given multiple point.

Throughout this chapter, $B = (B_t, t \geq 0)$ is a planar Brownian motion started at 0. For every integer $p \geq 2$, we denote by β_p the random measure that was constructed in Theorem VIII-4.

Consider a double point $z = B_s = B_t$ for some $s < t$. One may expect that the path of B between times s and t looks like a Brownian loop with initial point z and length $t-s$. Recall that a Brownian loop with length T and initial point z is by definition a Brownian motion started at z and conditioned to be at z at time T. A simple example will show that some care is needed in order to make the previous affirmation rigorous. The easiest way of constructing a double point is to set

$$T = \inf\{t \geq 1 \; ; \; B_t \in B[0,1/2]\} ,$$

and

$$S = \sup\{s \leq 1/2 \; ; \; B_s = B_T\} .$$

Notice that $S < 1/2 < 1 < T$ a.s., and that S is certainly not a stopping time. The process $(B_{S+u}, 0 \leq u \leq T-S)$ turns out to be very different from a Brownian loop. Indeed, this process cannot perform small closed loops around its starting point as a Brownian loop would do, because this would contradict the definition of S.

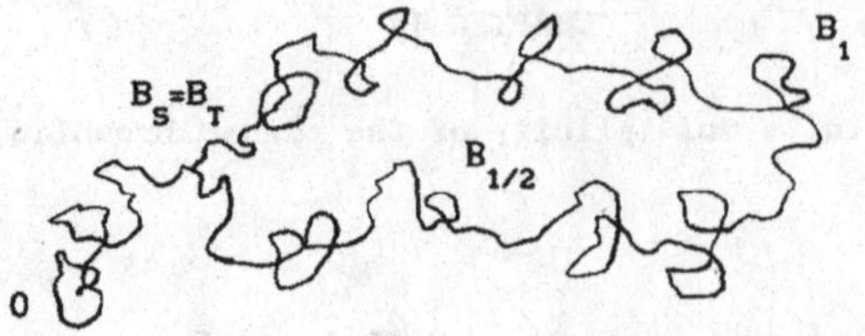

Fig. 1

This example does not mean that our previous heuristic affirmation is incorrect. It can be explained by the fact that the double point $B_S = B_T$ is in some sense exceptional. To avoid these exceptional double points we will have to average over all double points. Averaging will simply mean integrating with respect to the self-intersection local time.

We need some notation. For $0 \le u \le v$ we set :

$$ {}_uB_v(t) = B_{(u+t)\wedge v} - B_u \,, $$

$$ {}_vB_u(t) = B_{(v-t)\vee u} - B_v \,, $$

so that ${}_uB_v, {}_vB_u$ define (random) elements of the space $C(\mathbb{R}^2)$ of all continuous functions from $\mathbb{R}_+$ into $\mathbb{R}^2$.

If $L = (L(t), 0 \le t \le r)$ is a Brownian loop with initial point z and length r, we set $L(t) = z$ for $t > r$, by convention.

Finally, for any process Γ, we denote by Γ^t the process Γ stopped at time t $(\Gamma^t(s) = \Gamma(s \wedge t))$.

Theorem 1 : *Let* $p \ge 2$. *Then, for any Borel subset* A *of* $\mathcal{T}_p$ *and any non-negative measurable function* F *on* $C(\mathbb{R}^2)^{p+1}$,

$$ E\Big[\int_A \beta_p(ds_1 \ldots ds_p)\, F({}_0B_{s_1}, {}_{s_1}B_{s_2}, \ldots, {}_{s_p}B_\infty)\Big] $$

$$ = \int_A \frac{ds_1 \ldots ds_p}{(2\pi)^{p-1}(s_2-s_1)\ldots(s_p-s_{p-1})}\, E[F(\Gamma^{s_1}, L_{1,s_2-s_1}, \ldots, L_{p-1,s_p-s_{p-1}}, \Gamma')] $$

where the processes $\Gamma, \Gamma', L_{j,s_j-s_{j-1}}$ $(j \in \{2,\ldots,p\})$ *are independent,* Γ, Γ' *are two planar Brownian motions started at* 0 , *and, for* $j \in \{2,\ldots,p\}$, $L_{j,s_j-s_{j-1}}$ *is a Brownian loop with initial point* 0 *and length* $s_j - s_{j-1}$.

In particular, let H *be a Borel subset of* $C(\mathbb{R}^2)^{p+1}$ *such that, for* $ds_1 \ldots ds_p$ *a.a.* $(s_1,\ldots,s_p) \in \mathcal{T}_p$,

$$ P[(\Gamma^{s_1}, L_{1,s_2-s_1}, \ldots, L_{p-1,s_p-s_{p-1}}, \Gamma') \in H] = 1. $$

Then, w.p. 1, for $\beta_p(ds_1\ldots ds_p)$ *a.a.* $(s_1,\ldots,s_p)\in\mathcal{T}_p$,

$$({}_0B_{s_1},{}_{s_1}B_{s_2},\ldots,{}_{s_p}B_\infty)\in H.$$

Proof : The second assertion follows from the first one by taking for F the indicator function of the complement of H. We prove the first assertion in the case $p = 2$ (the general case is similar). We may assume that F is bounded and continuous and that A is a compact rectangle. Then it easily follows from Theorem VIII-4 that

$$E\left[\int_A \beta_2(ds_1ds_2)\, F({}_0B_{s_1},{}_{s_1}B_{s_2},{}_{s_2}B_\infty)\right]=\lim_{\varepsilon\to 0} E\left[\int_A \beta_2^\varepsilon(ds_1ds_2)\, F({}_0B_{s_1},{}_{s_1}B_{s_2},{}_{s_2}B_\infty)\right]$$

(we write β_p^ε instead of β_ε in Chapter VIII). However,

$$E\left[\int_A \beta_2^\varepsilon(ds_1ds_2)\, F({}_0B_{s_1},{}_{s_1}B_{s_2},{}_{s_2}B_\infty)\right]$$

$$= \int_A ds_1ds_2\ (\pi\varepsilon^2)^{-2}\ E[m(D(B_{s_1},\varepsilon)\cap D(B_{s_2},\varepsilon))\, F({}_0B_{s_1},{}_{s_1}B_{s_2},{}_{s_2}B_\infty)].$$

Then use the trivial observation

$$m(D(B_{s_1},\varepsilon)\cap D(B_{s_2},\varepsilon)) = m(D(0,\varepsilon)\cap D(B_{s_2}-B_{s_1},\varepsilon))$$

and condition with respect to $B_{s_2}-B_{s_1}$. It follows that :

$$E\left[\int_A \beta_2^\varepsilon(ds_1ds_2)\, F({}_0B_{s_1},{}_{s_1}B_{s_2},{}_{s_2}B_\infty)\right] = \int_A ds_1ds_2\int_{|y|\le 2\varepsilon} dy\ p_{s_2-s_1}(0,y)$$

$$\times\ (\pi\varepsilon^2)^{-2}\ m(D(0,\varepsilon)\cap D(y,\varepsilon))\ E[F({}_0B_{s_1},{}_{s_1}B_{s_2},{}_{s_2}B_\infty)\mid B_{s_2}-B_{s_1}=y].$$

To complete the proof notice that

$$\lim_{y\to 0} E[F({}_0B_{s_1},{}_{s_1}B_{s_2},{}_{s_2}B_\infty)\mid B_{s_2}-B_{s_1}=y] = E[F(\Gamma^{s_1},L_{1,s_2-s_1},\Gamma'))]$$

and

$$\int dy\ (\pi\varepsilon^2)^{-2}\ m(D(y,\varepsilon)\cap D(0,\varepsilon)) = 1.\ \square$$

Theorem 1 is certainly not a deep result. If we replace $\beta_p(ds_1\ldots ds_p)$ by its formal definition

$$\beta_p(ds_1\ldots ds_p) = \delta_{(0)}(B_{s_2}-B_{s_1})\ldots\delta_{(0)}(B_{s_p}-B_{s_{p-1}})\ ds_1\ldots ds_p$$

then the first assertion of Theorem 1 becomes almost obvious. The second assertion of Theorem 1 will however be useful as it provides a (very weak) form of the Markov property, at times which are typically not stopping times. Indeed, it shows that, for a typical multiple point, the behavior of the process before or after the successive hitting times of this multiple point is similar to that of a Brownian motion or a Brownian loop. Notice that the notion of intersection local time is needed to say what a "typical multiple point" is.

As a first application of Theorem 1, we state a result which shows that the points of multiplicity $p+1$ are very rare among the p-multiple points.

Proposition 2 : *With probability* 1, *for* β_p*-a.a.* $(s_1,\dots,s_p)$, *the point* $B_{s_1}=\dots=B_{s_p}$ *is not a* $(p+1)$*-multiple point.*

Remark : Proposition 2 is also valid for $p = 1$, in which case β_1 should be interpreted as the Lebesgue measure on $\mathbb{R}_+$. If $\ell_p(dz)$ denotes the image measure of β_p under the mapping $(s_1,\dots,s_p) \longrightarrow B_{s_1}$, then ℓ_p is in some sense the canonical measure on the set of p-multiple points, and Proposition 2 shows that the measures ℓ_p $(p = 1,2,\dots)$ are singular w.r.t. each other.

Proof : For $\varphi \in C(\mathbb{R}^2)$ set

$$\zeta(\varphi) = \inf\{t > 0 \ ; \ \varphi \text{ is constant on } [t,\infty)\}$$

$(\inf \emptyset = +\infty)$ and

$$H = \{(\varphi_0,\varphi_1,\dots,\varphi_p) \in C(\mathbb{R}^2)^{p+1} \ ; \ \forall t \in [0,\zeta(\varphi_0)), \varphi_0(t) \neq \varphi_0(\zeta(\varphi_0)) \text{ and for } j = 1,\dots,p, \ \forall t \in (0,\zeta(\varphi_j)), \varphi_j(t) \neq \varphi_j(0)\}.$$

The polarity of single points for planar Brownian motion implies that H satisfies the assumption of Theorem 1. The desired result follows from Theorem 1. □

2. Points of infinite multiplicity.

We say that two compact subsets K,K' of $\mathbb{R}$ have the same order type if there exists an increasing homeomorphism φ of $\mathbb{R}$ such that $\varphi(K) = K'$.

Theorem 3 : *Let* K *be a totally disconnected compact subset of* $\mathbb{R}$. *Then with probability* 1 *there exists a point* z *of the plane such that* $\{t \geq 0, B_t = z\}$ *has the same order type as* K .

Note that when K is a finite set, Theorem 3 just says that there exist points of multiplicity (exactly) p for any p. As a consequence of Theorem 3 we get the existence of points of (exactly) countable multiplicity, as well as the existence of points of uncountable multiplicity.

The proof of Theorem 3 relies on a key lemma, which itself is an easy consequence of Theorem 1. Let us first explain the need for this lemma. We start with a double point $z_1 = B_r = B_s$ (with $r < s$). Choose $\varepsilon > 0$ small (at least smaller than $(s-r)/2$) and consider the 4 paths ${}_rB_{r-\varepsilon}$, ${}_rB_{r+\varepsilon}$, ${}_sB_{s-\varepsilon}$, ${}_sB_{s+\varepsilon}$. We would like to say that these 4 paths are "not too different" from those of 4 independent Brownian motions started at z_1. If this is the case, the results of Chapter VIII allow us to find a common point other than z to these 4 paths. That is, we may find $t \in (r-\varepsilon, r)$, $u \in (r, r+\varepsilon)$, $v \in (s-\varepsilon, s)$, $w \in (s, s+\varepsilon)$ such that $B_t = B_u = B_v = B_w = z_2$. We may even choose z_2 as close to z_1 as we wish. We can then by similar arguments construct a point z_3 of multiplicity 8 close to z_2. At the nth step we get a point z_n of multiplicity 2^n. It should then be clear that, if the construction is performed with enough care, the point $z = \lim z_n$ will be a point of infinite multiplicity (in fact $\{t \,;\, B_t = z\}$ will contain a Cantor set).

$B_r = B_s$
First step.

$B_t = B_u = B_v = B_w$
Second step.

Fig. 2

The only trouble in the previous arguments comes from the assertion "the 4 paths ${}_rB_{r-\varepsilon}, \ldots$ are not too different from 4 independent Brownian paths". The next lemma will demonstrate that, for most of the double points $B_r = B_s$, these 4 paths behave like 4 independent Brownian paths, at least for the properties that are of interest here.

<u>Lemma 4</u> : *With probability one, for* β_p*-a.a.* $(s_1, \ldots, s_p)$ *and for any* $\delta > 0$,

$$\beta_{2p}((s_1-\delta, s_1) \times (s_1, s_1+\delta) \times \ldots \times (s_p-\delta, s_p) \times (s_p, s_p+\delta)) > 0.$$

<u>Proof</u> : For any compact rectangle R in $\mathcal{T}_p$ we may find a sequence (ε_k) decreasing to 0 such that

(1) $$\beta_p(R) = \lim_{k\to\infty} \beta_p^{\varepsilon_k}(R), \quad \text{a.s.}$$

We may assume that the same sequence (ε_k) works for any p and for any rectangle with rational coordinates. Then a monotonicity argument shows that (1) holds simultaneously for all (compact or non-compact) rectangles of $\mathcal{T}_p$. Finally, let α_p be the intersection local time of p independent Brownian motions started at 0 (see section VIII-1). We may assume that, with the same sequence (ε_k),

(2) $$\alpha_p(R) = \lim_{k\to\infty} \alpha_p^{\varepsilon_k}(R),$$

for all compact rectangles in $(\mathbb{R}_+)^p$, a.s.

Let $f_0, f_1, \ldots, f_p \in C(\mathbb{R}^2)$ and let $\zeta_i = \zeta(f_i)$ be as in the proof of Proposition 2. If $\zeta(f_i) < \infty$ for every $i \in \{0,\ldots,p-1\}$, set

$$\ell_\delta(f_0,f_1,\ldots,f_p) = \liminf_{k\to\infty} \int_{[0,\delta]^{2p}} dt_1 \ldots dt_{2p}$$
$$\varphi^{2p}_{\varepsilon_k}(f_0(\zeta_0-t_1),f_1(t_2),f_1(\zeta_1-t_3),\ldots,f_{p-1}(t_{2p-2}),f_{p-1}(\zeta_{p-1}-t_{2p-1}),f_p(t_p))$$

where

$$\varphi^{2p}_\varepsilon(z_1,\ldots,z_{2p}) = \int_{\mathbb{R}} dy \prod_{j=1}^{2p} \delta^\varepsilon_{(y)}(z_j)$$

as in Chapter VIII. Otherwise, set $\ell_\delta(f_0,\ldots,f_p) = 0$.

By looking at the finite-dimensional marginal distributions, it is very easy to check that, if L is a Brownian loop with length a, for any $\delta < a/2$, the joint distribution of $(L(t),L(a-t)\ ;\ 0 \le t \le \delta)$ is absolutely continuous with respect to that of two independent Brownian paths. It follows from this observation, (2), and Proposition VIII-2 (ii) that the set

$$H = \{(f_0,f_1,\ldots,f_p)\ ;\ \ell_\delta(f_0,f_1,\ldots,f_p) > 0,\ \forall\delta > 0\}$$

satisfies the assumption of Theorem 1. Therefore w.p. 1 for β_p-a.a. $(s_1,\ldots,s_p)$,

$$({}_0B_{s_1}, {}_{s_1}B_{s_2}, \ldots, {}_{s_{p-1}}B_{s_p}, {}_{s_p}B_\infty) \in H.$$

Lemma 4 follows using (1) and the definition of β^ε_{2p}. □

<u>Proof of Theorem 3</u> : We will show in detail how to construct a point z such that $\{t\ ;\ B_t = z\}$ contains a Cantor set. We set $t^0_1 = 1/2$, $z_0 = B_{1/2}$, $\delta_0 = 1/4$. We observe that for any $\delta > 0$

$$\beta_2((\tfrac{1}{2} - \delta, \tfrac{1}{2}) \times (\tfrac{1}{2}, \tfrac{1}{2} + \delta)) > 0, \quad \text{a.s.},$$

by the arguments of the proof of Proposition VIII-6. By Lemma 4 applied with $p = 2$, we may find a pair $(t_1^1, t_2^1) \in (1/4, 1/2) \times (1/2, 3/4)$ such that :

$$B_{t_1^1} = B_{t_2^1} =: z_1$$

and, for any $\delta > 0$,

$$\beta_4((t_1^1 - \delta, t_1^1) \times (t_1^1, t_1^1 + \delta) \times (t_2^1 - \delta, t_2^1) \times (t_2^1, t_2^1 + \delta)) > 0.$$

We proceed by induction on n. At the nth step we have constructed

$$(t_1^n, \ldots, t_{2^n}^n) \in (t_1^{n-1} - \delta_{n-1}, t_1^{n-1}) \times \ldots \times (t_{2^{n-1}}^{n-1}, t_{2^{n-1}}^{n-1} + \delta_{n-1})$$

in such a way that

$$B_{t_1^n} = \ldots = B_{t_{2^n}^n} =: z_n \;,$$

and for any $\delta > 0$,

$$\beta_{2^{n+1}}((t_1^n - \delta, t_1^n) \times (t_1^n, t_1^n + \delta) \times \ldots \times (t_{2^n}^n - \delta, t_{2^n}^n) \times (t_{2^n}^n, t_{2^n}^n + \delta)) > 0.$$

We set $\delta_n = \frac{1}{4} (\delta_{n-1} \wedge \min(t_i^n - t_{i-1}^n \; ; \; i = 2, \ldots, 2^n))$. By the induction hypothesis and Lemma 4 we find

$$(t_1^{n+1}, \ldots, t_{2^{n+1}}^{n+1}) \in (t_1^n - \delta_n, t_1^n) \times \ldots \times (t_{2^n}^n, t_{2^n}^n + \delta_n)$$

such that

$$B_{t_1^{n+1}} = \ldots = B_{t_{2^{n+1}}^{n+1}} =: z_{n+1}$$

and for any $\delta > 0$

$$\beta_{2(n+2)} ((t_1^{n+1} - \delta, t_1^{n+1}) \times \ldots \times (t_{2^{n+1}}^{n+1}, t_{2^{n+1}}^{n+1} + \delta)) > 0.$$

Finally the continuity of paths implies that the sequence (z_n) converges towards some $z \in \mathbb{R}^2$. Furthermore $\{t \geq 0 \; ; \; B_t = z\}$ contains the closed set

$$K = \bigcap_{n=1}^{\infty} \overline{\left(\bigcup_{m=n}^{\infty} \{t_J^m \; ; \; J \in \{1, \ldots, 2^n\}\} \right)}.$$

Our construction (in particular the choice of the constants δ_n) ensures that K is a Cantor set.

By being a little more careful in the construction we can even get

$$K = \{t \geq 0, B_t = z\},$$

which gives Theorem 3 in the case of a Cantor set.

The general case requires some technical adjustments but no new idea. If for instance K is the union of a Cantor set and an isolated point located on the right of K, we proceed as follows. We construct t_1^1, t_2^1 as previously but in the second step we "forget" about the path during $(t_2^1-\delta_1, t_2^1)$ and we choose

$$(t_1^2, t_2^2, t_3^2) \in (t_1^1-\delta_1, t_1^1) \times (t_1^1, t_1^1+\delta_1) \times (t_2^1, t_2^1+\delta_1)$$

so that for any $\delta > 0$

$$\beta_5((t_1^2-\delta, t_1^2) \times (t_1^2, t_1^2+\delta) \times (t_2^2-\delta, t_2^2) \times (t_2^2, t_2^2+\delta) \times (t_3^2, t_3^2+\delta)) > 0$$

(this requires a new version of Lemma 4). At the $(n+1)^{th}$ step we construct

$$(t_1^{n+1}, \dots, t_{2^n+1}^{n+1}) \in (t_1^n-\delta_n, t_1^n) \times \dots \times (t_{2^{n-1}}^n, t_{2^{n-1}}^n+\delta_n) \times (t_{2^{n-1}+1}^n, t_{2^{n-1}+1}^n+\delta_n)$$

so that

$$B_{t_1^{n+1}} = \dots = B_{t_{2^n+1}^{n+1}} =: z_{n+1}$$

and, for any $\delta > 0$,

$$\beta_{2^{n+1}+1}((t_1^{n+1}-\delta, t_1^{n+1}) \times (t_1^{n+1}+\delta) \times \dots \times (t_{2^n}^{n+1}, t_{2^n}^{n+1}+\delta) \times (t_{2^n+1}^{n+1}, t_{2^n+1}^{n+1}+\delta)) > 0 .$$

The point $z = \lim z_n$ will satisfy the desired condition, again provided the construction is done with enough care.

Bibliographical notes. *The problem of the existence of points of finite multiplicity for a d-dimensional Brownian path was completely solved by Dvoretzky, Erdös and Kakutani* [DK1], [DK2] *and* [DKT] *in collaboration with Taylor. See Kahane* [Kh] *for an elegant modern approach. The existence of points of infinite multiplicity for a planar Brownian path was proved in* [DK3]. *However the given proof is not totally satisfactory: it seems that the authors apply the strong Markov property at certain random times that are typically not stopping times. The material of this Chapter is taken from* [L6], *to which we refer for a more detailed proof of Theorem 3. Proposition 2 is a rigorous form of Lévy's intuitive statement quoted in the introduction. See also Adelman and Dvoretzky* [AD] *for a weak form of this result. Another way of comparing the size of the sets of points of multiplicity* p *and* $p+1$, *that was suggested by Lévy* [Lé4, p. 325-329], *is to use Hausdorff measures. The exact Hausdorff measure function for the set of* p*-multiple points is* $\varphi_p(x) = x^2 (\log 1/x \text{ logloglog } 1/x)^p$ *(see* [L9], *for* $p = 1$, *this result is due to Taylor* [T1]*). A weaker form of this result had been conjectured by Taylor* [T2] *and proved in* [L3].

CHAPTER X

Renormalization for the powers of the occupation field of a planar Brownian motion

1. The main theorem.

Throughout this chapter $B = (B_t, t \geq 0)$ denotes a planar Brownian motion, which starts at z under the probability P_z. Let $p \geq 2$ be an integer. In chapter VIII, we introduced the (p-multiple) self-intersection local time of B as a Radon measure on

$$\mathcal{T}_p = \{(s_1,\ldots,s_p)\ ;\ 0 \leq s_1 <\ldots< s_p\},$$

supported on $\{(s_1,\ldots,s_p)\ ;\ B_{s_1} =\ldots= B_{s_p}\}$. This measure, denoted by β_p, is such that, for any compact rectangle $A \subset \mathcal{T}_p$,

(1) $$\beta_p(A) = \lim_{\varepsilon\to 0} \int dy \int_A \varphi^y_\varepsilon(B_{s_1})\ldots\varphi^y_\varepsilon(B_{s_p})\, ds_1\ldots ds_p$$

in the L^2-norm. Here,

$$\varphi^y_\varepsilon(z) = (\pi\varepsilon^2)^{-1}\, 1_{D(y,\varepsilon)}(z).$$

We know that, for every $M > 0$, $\beta_p(\mathcal{T}_p \cap [0,M]^p) = \infty$ a.s. Our goal in this chapter is to define a renormalized version of $\beta_p(\mathcal{T}_p \cap [0,M]^p)$.

By (1) we have the formal expression

$$\beta_p(\mathcal{T}_p \cap [0,M]^p) = \int dy \int_{\mathcal{T}_p\cap[0,M]^p} ds_1\ldots ds_p\ \delta_{(y)}(B_{s_1})\ldots\delta_{(y)}(B_{s_p})$$

$$\text{"="}\ \frac{1}{p!}\int dy \left(\int_0^M ds\ \delta_{(y)}(B_s)\right)^p.$$

More generally, we shall introduce renormalized versions of the quantities

$$\int dy\, f(y) \left(\int_0^M ds\ \delta_{(y)}(B_s)\right)^p,$$

for $f : \mathbb{C} \longrightarrow \mathbb{R}$ bounded measurable. In this way we define what may be called the p-th power of the occupation field of B. Recall that the occupation field, or occupation measure, of B on $[0,M]$ is the measure

$$f \longrightarrow \int_0^M ds\, f(B_s),$$

whose formal density is

$$\int_0^M ds\, \delta_{(y)}(B_s).$$

As a matter of fact, the need for a renormalization of β_p is closely related to the singularity of the occupation measure with respect to Lebesgue measure.

We need some notation before stating our main result. First notice that in (1) φ_ε^y could be replaced by many other suitable approximations of the Dirac measure at y. In what follows, the most convenient approximation will be the uniform probability measure on the circle of radius ε centered at y, denoted by $C(y,\varepsilon)$. This leads us to the local time of B on $C(y,\varepsilon)$. This local time can be defined rigorously in several ways. The most elementary approach is to show that

$$\text{(2)} \qquad \lim_{\delta\to 0} \frac{1}{4\pi\varepsilon\delta} \int_0^t 1_{\{\varepsilon-\delta<|B_s-y|<\varepsilon+\delta\}}\, ds =: \ell_\varepsilon^y(t)$$

exists in the L^2-norm, for any $t \geq 0$, $\varepsilon \in (0,1)$ and $y \in \mathbb{C}$. Alternatively, $\ell_\varepsilon^y(t)$ may be defined as $(2\pi\varepsilon)^{-1}$ times the usual (semi-martingale) local time of $|B_s-y|$, at level ε and at time t. Kolmogorov's lemma yields the existence of a continuous version of $(\varepsilon,y,t) \longrightarrow \ell_\varepsilon^y(t)$. From now on we shall only deal with this version.

The methods of Chapter VIII can be adapted to give :

$$\text{(3)} \qquad \beta_p(A) = \lim_{\varepsilon\to 0} \int dy \int_A \ell_\varepsilon^y(ds_1)\ldots\ell_\varepsilon^y(ds_p)$$

for any compact rectangle A (here $\ell_\varepsilon^y(ds)$ denotes the measure on $\mathbb{R}_+$ associated with the continuous nondecreasing function $t \longrightarrow \ell_\varepsilon^y(t)$). We shall not use (3) , except to motivate the next results, and we leave the proof as an exercise for the reader.

For technical reasons that will appear later, it turns out to be very convenient to work with Brownian motion killed at an independent exponential time. Therefore we fix $\lambda > 0$ and we let ζ denote an exponential time with parameter λ, independent of B. For any $\varepsilon > 0$ we set :

$$h_\varepsilon = -E_\varepsilon[\ell_\varepsilon^0(\zeta)].$$

Notice that $E_z[\ell_\varepsilon^y(\zeta)] = -h_\varepsilon$ whenever $|z-y| = \varepsilon$, by the rotational invariance of planar Brownian motion. It easily follows from (2) that

$$-h_\varepsilon = \int_0^\infty ds\, e^{-\lambda s} \int_{C(0,\varepsilon)} \pi_\varepsilon(0,dy)\, p_s(\varepsilon,y) = \int_{C(0,\varepsilon)} \pi_\varepsilon(0,dy)\, G_\lambda(\varepsilon,y)$$

where $\pi_\varepsilon(0,dy)$ is the uniform probability measure on $C(0,\varepsilon)$. Recall that

$$\text{(4)} \qquad G_\lambda(y,z) = \frac{1}{\pi} K_0(\sqrt{2\lambda}|z-y|),$$

where K_0 is the usual modified Bessel function. It follows that

$$\text{(4')} \qquad G_\lambda(y,z) = \frac{1}{\pi} \log \frac{1}{|z-y|} + \frac{1}{\pi}\left(\frac{\log(2/\lambda)}{2} - \kappa\right) + O\left(|z-y|^2 \log \frac{1}{|z-y|}\right)$$

where κ denote Euler's constant. Hence,

$$\text{(5)} \qquad h_\varepsilon = -\frac{1}{\pi} \log \frac{1}{\varepsilon} - \frac{1}{\pi}\left(\frac{\log(2/\lambda)}{2} - \kappa\right) + O\left(\varepsilon^2 \log \frac{1}{\varepsilon}\right)$$

using the harmonicity of $y \longrightarrow \log|y|$. We set

$$\Delta_p = \mathcal{T}_p \cap [0,\zeta)^p.$$

It follows from (3) that :

$$\lim_{\varepsilon\to 0} \int dy \frac{1}{p!} \ell^y_\varepsilon(\zeta)^p = \lim_{\varepsilon\to 0} \int dy \int_{\Delta_p} \ell^y_\varepsilon(ds_1)\dots\ell^y_\varepsilon(ds_p) = \beta_p(\Delta_p) = \infty,$$

in probability (in fact this limit also holds a.s.). We get a renormalized version of $\beta_p(\Delta_p)$ by the following procedure. For every $\varepsilon > 0$, we replace $\ell^y_\varepsilon(\zeta)^p/p!$ by another polynomial of $\ell^y_\varepsilon(\zeta)$, with the same leading term, and coefficients of lower degree depending on ε. A suitable choice of these coefficients allows us to get an L^2-convergence as ε goes to 0.

Theorem 1 : *For every* $\varepsilon \in (0,1)$, $p \geq 1$ *set*

$$Q^p_\varepsilon(u) = \sum_{k=1}^{p} \binom{p-1}{k-1} (h_\varepsilon)^{p-k} \frac{u^k}{k!}\,.$$

For any bounded Borel function $f : \mathbb{C} \longrightarrow \mathbb{R}$, *set*

$$T^\varepsilon_p f = \int dy\, f(y)\, Q^p_\varepsilon(\ell^y_\varepsilon(\zeta)).$$

Then,

$$\lim_{\varepsilon\to 0} T^\varepsilon_p f =: T_p f$$

exists in the L^2*-norm.*

Remark : For $p = 1$, it can easily be checked that

$$T_1 f = \int_0^{\zeta} ds\, f(B_s)$$

(simply compute $E[(T_1^{\varepsilon}\varphi - T_1\varphi)^2]$, etc...).

Most of the remainder of this chapter is devoted to the proof of Theorem 1. Let us briefly discuss the contents of this result. In some sense, the random variables $T_p 1$ $(p = 2,3,\dots)$ provide the renormalized versions of $\beta_p(\Delta_p)$ that we aimed to define. In the next chapter, we will prove that these quantities appear in the different terms of a full asymptotic expansion for the area of the planar Wiener sausage. This result will also allow us to relate $T_2 1$ to the renormalized self-intersection local time (for double points) discussed in Chapter VIII. The proof of Theorem 1 for a general function f is not more difficult than in the special case $f = 1$.

The simple form of the polynomials Q_{ε}^p will be explained in the proof below. Notice that we could use other approximations for the Dirac measure at y : a result analogous to Theorem 1 would then hold, with (essentially) the same limiting variables $T_p f$, but the renormalization polynomials would usually be much more complicated. For instance, the approximation could be given by the function φ_{ε}^y, so that $\ell_{\varepsilon}^y(\zeta)$ should be replaced by

$$(\pi\varepsilon^{-2})^1 \int_0^{\zeta} 1_{D(y,\varepsilon)}(B_s)ds.$$

However, already in this simple case, the renormalization polynomials cannot be written explicitly (see [Dy3,Dy5]).

2. Preliminary estimates.

The proof of Theorem 1 depends on certain precise estimates that will be derived in this section. We start with a lemma which explains the form of the polynomials Q_{ε}^p.

Lemma 2 : *Set*

$$\Lambda_p = \{(s_1,\dots,s_p) \,;\, 0 \le s_1 \le s_2 \le \dots \le s_p < \zeta\}.$$

Then, for every $\varepsilon > 0$, $y \in \mathbb{C}$,

$$Q_{\varepsilon}^p(\ell_{\varepsilon}^y(\zeta)) = \int_{\Lambda_p} \ell_{\varepsilon}^y(ds_1) \prod_{i=2}^{p} (\ell_{\varepsilon}^y(ds_i) + h_{\varepsilon}\delta_{(s_{i-1})}(ds_i)).$$

Proof : First notice that the expression

$$\ell^y_\varepsilon(ds_1) \prod_{i=2}^{p} (\ell^y_\varepsilon(ds_i) + h_\varepsilon\, \delta_{(s_{i-1})}(ds_i))$$

gives a well-defined signed measure on the set Λ_p. Furthermore we may expand the product and get terms of the form

$$(h_\varepsilon)^k\, \ell^y_\varepsilon(ds_1)\ldots\ell^y_\varepsilon(ds_{j_1-1})\, \delta_{(s_{j_1-1})}(ds_{j_1})\ell^y_\varepsilon(ds_{j_1+1})\ldots\ell^y_\varepsilon(ds_{j_2-1})$$

$$\delta_{(s_{j_2-1})}(ds_{j_2})\, \ell^y_\varepsilon(ds_{j_2+1}) \ldots\ldots \delta_{(s_{j_k-1})}(ds_{j_k})\, \ell^y_\varepsilon(ds_{j_k+1})\ldots\ell^y_\varepsilon(ds_p)$$

where $k \in \{0,1,\ldots, p-1\}$ and $1 < j_1 < j_2 <\ldots< j_k \le p$. Next, if we integrate such a measure over Λ_p, the effect of the Dirac masses is to force $s_i = s_{i+1}$ for $i \in \{j_1,\ldots,j_k\}$, and we are left with the integral :

$$(h_\varepsilon)^k \int_{\Lambda_{p-k}} \ell^y_\varepsilon(dt_1)\, \ell^y_\varepsilon(dt_2)\ldots\ell^y_\varepsilon(dt_{p-k}) = (h_\varepsilon)^k \frac{\ell^y_\varepsilon(\zeta)^{p-k}}{(p-k)!}\,.$$

Finally, for every k, we have $\binom{p-1}{k-1}$ possible choices of $j_1,\ldots,j_k$. □

To simplify notation we write $G(z-y) = G_\lambda(y,z)$. Notice that $G(z)$ is a nonincreasing function of $|z|$. By (4) and well-known properties of the function K_0, we may find two positive constants C_*, η such that for any $y,z \neq 0$, with $|y|/2 < |z| < 2|y|$,

$$|G(z) - G(y)| \le C_*|z-y| \left(\frac{1}{|y|} \exp - \eta|y|\right)$$

(simply notice that the derivative of the function $r \longrightarrow K_0(r)$ is bounded by $C\, r^{-1} \exp - \alpha r$). We set $g(r) = 2\, C_*\, r^{-1} \exp(-\eta r/2)$. Note that g is nonincreasing and that, under the previous assumptions on z, y,

$$(6) \qquad |G(z) - G(y)| \le |z-y|\; g(2|y|)$$

Lemma 3 : *For every integer* $n \ge 1$, *for* $x,y,z \in \mathbb{C}$, $\varepsilon,\varepsilon' \in (0,1/2)$,, *set*

$$H^n_{\varepsilon,\varepsilon'}(x,y,z) = \begin{cases} E_x\left[\int_{\Lambda_n} \ell^y_\varepsilon(ds_1)\ell^z_{\varepsilon'}(ds_2)\ell^y_\varepsilon(ds_3)\ldots\ell^z_{\varepsilon'}(ds_n)\right] & \textit{if } n \textit{ is even,} \\ E_x\left[\int_{\Lambda_n} \ell^y_\varepsilon(ds_1)\ell^z_{\varepsilon'}(ds_2)\ell^y_\varepsilon(ds_3)\ldots\ell^y_\varepsilon(ds_n)\right] & \textit{if } n \textit{ is odd.} \end{cases}$$

Then, for any $x,y,z \in \mathbb{C}$, $\varepsilon, \varepsilon' \in (0,1/2)$ *such that* $|x-y| > 4(\varepsilon \vee \varepsilon')$, $|z-y| > 8(\varepsilon \vee \varepsilon')$ *and* $\frac{\varepsilon}{2} < \varepsilon' < 2\varepsilon$,

$$H^n_{\varepsilon,\varepsilon'}(x,y,z) \le G(\tfrac{y-x}{2})\, G(\tfrac{z-y}{4})^{n-1}$$

and

$$|H^n_{\varepsilon,\varepsilon'}(x,y,z) - G(y-x)G(z-y)^{n-1}|$$

$$\le \varepsilon\, (g(2|y-x|)G(z-y)^{n-1} + 2(n-1)\, G(\tfrac{y-x}{2})G(\tfrac{z-y}{4})^{n-2}g(|z-y|)).$$

Remark. When $n = 1$, $H^n_{\varepsilon,\varepsilon'}(x,y,z) = H^1_\varepsilon(x,y)$ depends only on ε, x, y and the bounds of Lemma 3 give:

$$H^1_\varepsilon(x,y) \le G(\tfrac{y-x}{2}),$$

$$|H^1_\varepsilon(x,y) - G(y-x)| \le \varepsilon\, g(2|y-x|).$$

Proof : We use induction on n. For $n = 1$,

$$H^1_{\varepsilon,\varepsilon'}(x,y,z) = H^1_\varepsilon(x,y) = E_x[\ell^y_\varepsilon(\zeta)] = \int \pi_\varepsilon(y,dw)G(w-x).$$

However, if $w \in C(y,\varepsilon)$,

$$G(w-x) \le G(\tfrac{y-x}{2}),$$

$$|G(w-x) - G(y-x)| \le \varepsilon\, g(2|y-x|)$$

by (6) and our assumptions on x, y, ε. The desired bounds follow.

Now let $n \ge 2$. Assume that Lemma 3 holds at the order $n-1$. Using the Markov property of Brownian motion killed at time ζ, we get

$$H^n_{\varepsilon',\varepsilon}(x,y,z) = E_x\left[\int_0^\zeta \ell^y_\varepsilon(ds_1)\, H^{n-1}_{\varepsilon',\varepsilon}(B_{s_1},z,y)\right].$$

Notice that $\ell^y_\varepsilon(ds_1)$ a.e., $B_{s_1} \in C(y,\varepsilon)$. The induction hypothesis gives for any $w \in C(y,\varepsilon)$,

$$H^{n-1}_{\varepsilon',\varepsilon}(w,z,y) \le G(\tfrac{z-w}{2})\, G(\tfrac{z-y}{4})^{n-2} \le G(\tfrac{z-y}{4})^{n-1}.$$

The first bound of the lemma follows by using the case $n = 1$.

Next, by the induction hypothesis again, we have for every $w \in C(y,\varepsilon)$,

$$|H^{n-1}_{\varepsilon',\varepsilon}(w,z,y) - G(z-w)\,G(z-y)^{n-2}|$$

$$\le \varepsilon\Big(g(2|z-w|)G(z-y)^{n-2} + 2(n-2)\, g(|z-y|)G(\tfrac{z-w}{2})G(\tfrac{z-y}{4})^{n-3} \Big)$$

$$\le \varepsilon\ (2(n-2)+1)\ g(|z-y|)G(\tfrac{z-y}{4})^{n-2} .$$

Furthermore,

$$|G(z-w) - G(z-y)| \le \varepsilon\ g(2|z-y|) .$$

It follows that

$$|H^{n-1}_{\varepsilon',\varepsilon}(w,z,y) - G(z-y)^{n-1}| \le \varepsilon\ (2(n-2)+2)\ g(|z-y|)G(\tfrac{z-y}{4})^{n-2} .$$

and, by the first bound of the lemma with $n = 1$,

$$|H^{n}_{\varepsilon,\varepsilon'}(x,y,z) - E_x[\ell^y_\varepsilon(\zeta)]\ G(z-y)^{n-1}| \le 2(n-1)\varepsilon\ G(\tfrac{y-x}{2})\ g(|z-y|)G(\tfrac{z-y}{4})^{n-2}.$$

The proof is now completed by using the second bound of the lemma with $n = 1$:

$$|E_x[\ell^y_\varepsilon(\zeta)] - G(y-x)| \le \varepsilon\ g(2|y-x|) . \ \square$$

3. Proof of Theorem 1.

The main step of the proof of Theorem 1 is the following key lemma.

Lemma 4 : *There exists a positive constant* C_p *such that, for any* $x,y,z \in \mathbb{C}$, $\varepsilon,\varepsilon' \in (0,1/2)$, *with* $|y-x| > 4(\varepsilon \vee \varepsilon')$, $|z-x| > 4(\varepsilon \vee \varepsilon')$, $|z-y| > 8(\varepsilon \vee \varepsilon')$ *and* $\varepsilon/2 \le \varepsilon' \le 2\varepsilon$,

$$|E_x[Q^p_\varepsilon(\ell^y_\varepsilon(\zeta))\ Q^p_{\varepsilon'}(\ell^z_{\varepsilon'}(\zeta))] - (G(y-x) + G(z-x))\ G(z-y)^{2p-1}|$$

$$\le C_p\ \varepsilon|\log\varepsilon|^{2p-2}((g(|y-x|) + g(|z-x|))\ G(z-y) + (G(\tfrac{y-x}{2}) + G(\tfrac{z-x}{2}))\ g(|z-y|)).$$

Proof : We use Lemma 2 to write :

$$|E_x[Q^p_\varepsilon(\ell^y_\varepsilon(\zeta))\ \ Q^p_{\varepsilon'}(\ell^z_\varepsilon(\zeta))] = E_x\Big[\int_{\Lambda_p} \ell^y_\varepsilon(ds_1) \prod_{j=2}^{p} (\ell^y_\varepsilon(ds_j) + h_\varepsilon\delta_{(s_{j-1})}(ds_j))$$

$$\times \int_{\Lambda_p} \ell^z_{\varepsilon'}(dt_1) \prod_{k=2}^{p} (\ell^z_{\varepsilon'}(dt_k) + h_{\varepsilon'}\delta_{(t_{k-1})}(dt_k))\Big]$$

$$= E_x\Big[\int_{\Lambda_p\times\Lambda_p} \mu^{y,z}_{\varepsilon,\varepsilon'}(ds_1\ldots ds_p\ dt_1\ldots dt_p)\Big].$$

Here $\mu^{y,z}_{\varepsilon,\varepsilon'}$ is a signed measure on the product $\Lambda_p \times \Lambda_p$. We now need to order $s_1,\dots,s_p$, $t_1,\dots,t_p$. Each possible order is associated with a nondecreasing function $\varphi : \{1,\dots,p\} \longrightarrow \{0,1,\dots,p\}$ in the following way. For any such function φ, let

$$\Gamma_\varphi = \{(s_1,\dots,s_p,\ t_1,\dots,t_p) \in \Lambda_p \times \Lambda_p\ ;\ \forall i \in \{1,\dots,p\},\ t_{\varphi(i)} < s_i < t_{\varphi(i)+1}\}$$

where by convention $t_0 = 0$, $t_{p+1} = \zeta$. If $\varphi \neq \varphi'$ the corresponding sets Γ_φ, $\Gamma_{\varphi'}$ are disjoint. Moreover $(\Lambda_p \times \Lambda_p) - \bigcup_\varphi \Gamma_\varphi$ is contained in

$$\bigcup_{i,j\in\{1,\dots,p\}} \{s_i = t_j\}$$

and the $\mu^{y,z}_{\varepsilon,\varepsilon'}$-measure of this set is zero because of our assumptions on $y,z,\varepsilon,\varepsilon'$ (observe that $\mu^{y,z}_{\varepsilon,\varepsilon'}$ is supported on :

$$\{(s_1,\dots,s_p,\ t_1,\dots,t_p)\ ;\ \forall i,\ |B_{s_i}-y| = \varepsilon,\ |B_{t_i}-z| = \varepsilon'\})$$

In view of the previous observations we may write

$$\text{(7)} \qquad E_x[Q^p_\varepsilon(\ell^y_\varepsilon(\zeta))\, Q^p_{\varepsilon'}(\ell^z_{\varepsilon'}(\zeta))] = \sum_\varphi E_x[\mu^{y,z}_{\varepsilon,\varepsilon'}(\Gamma_\varphi)]$$

Remark that we could as well have introduced

$$\bar{\Gamma}_\psi = \{(s_1,\dots,s_p,\ t_1,\dots,t_p) \in \Lambda_p \times \Lambda_p\ ;\ \forall i \in \{1,\dots,p\},\ s_{\psi(i)} < t_i < s_{\psi(i)+1}\}$$

and that $\bar{\Gamma}_\psi = \Gamma_\varphi$ if and only if $\psi = \bar{\varphi}$, where :

$$\bar{\varphi}(j) = \sup\{i,\ \varphi(i) < j\ \} \quad (\sup \emptyset = 0).$$

We first consider the simple situation where both φ and $\bar{\varphi}$ are strictly monotone (in other words s_i and s_{i+1} are always separated by at least one t_j, and conversely). This can only occur in the following two cases

$$\varphi(i) = \varphi_1(i) := i - 1,$$

$$\varphi(i) = \varphi_2(i) := i.$$

We have first

$$E_x[\mu^{y,z}_{\varepsilon,\varepsilon'}(\Gamma_{\varphi_1})] = E_x\Big[\int_0^\zeta \ell^y_\varepsilon(ds_1)\int_{s_1}^\zeta \ell^z_{\varepsilon'}(dt_1)\int_{t_1}^\zeta \cdots \int_{t_{p-1}}^\zeta \ell^y_\varepsilon(ds_p)\int_{s_p}^\zeta \ell^z_{\varepsilon'}(dt_p)\Big].$$

(notice that the Dirac masses give no contribution, because of the choice of φ and because of the support property of $\mu^{y,z}_{\varepsilon,\varepsilon'}$). By Lemma 3,

$$|E_x[\mu^{y,z}_{\varepsilon,\varepsilon'}(\Gamma_{\varphi_1})] - G(y-x)\,G(z-y)^{2p-1}|$$

$$\le \varepsilon\,(g(2|y-x|)\,G(z-y)^{2p-1} + 2(2p-1)G(\tfrac{y-x}{2})g(|z-y|)G(\tfrac{z-y}{4})^{2p-2}).$$

Similarly, Lemma 3 gives :

$$|E_x[\mu^{y,z}_{\varepsilon,\varepsilon'}(\Gamma_{\varphi_2})] - G(z-x)\,G(z-y)^{2p-1}|$$

$$\le \varepsilon\,(g(2|z-x|)\,G(z-y)^{2p-1} + 2(2p-1)G(\tfrac{z-x}{2})g(|z-y|)G(\tfrac{z-y}{4})^{2p-2}).$$

Our assumptions on y, z allow us to bound

$$G(z-y) \le G(\tfrac{z-y}{4}) \le G(\varepsilon) \le C\,|\log \varepsilon|\ .$$

The proof of Lemma 4 will be complete if we can check that the other terms of the right side of (7) give a negligible contribution. To understand why this is so, let us consider the easy case where φ is such that $\varphi(p) = \varphi(p-1) = p$, which implies that Γ_φ is contained in

$$\{(s_1,\dots,s_p,\ t_1,\dots,t_p)\ ;\ t_p < s_{p-1} < s_p\}.$$

Then,

$$E_x[\mu^{y,x}_{\varepsilon,\varepsilon'}(\Gamma_\varphi)] = E_x\Big[\int_{\Lambda_{p-1}\times\Lambda_p} \tilde\mu^{y,x}_{\varepsilon,\varepsilon'}(ds_1\dots ds_{p-1}dt_1\dots dt_p)\,\Big(\ell^y_\varepsilon((s_{p-1},\zeta)) + h_\varepsilon\Big)\Big]$$

where $\tilde\mu^{y,x}_{\varepsilon,\varepsilon'}(ds_1\dots dt_p)$ is a signed measure on $\Lambda_{p-1}\times\Lambda_p$, supported on $\{(s_1,\dots t_p)\ ;\ t_p<s_{p-1}$ and $|B_{s_{p-1}} - y| = \varepsilon\}$ and such that the bounded variation process $t \longrightarrow \tilde\mu^{y,x}_{\varepsilon,\varepsilon'}(\{s_{p-1}\le t\})$ is predictable. Replacing $\ell^y_\varepsilon((s_{p-1},\zeta))$ by its predictable projection gives :

$$E_x[\mu^{y,x}_{\varepsilon,\varepsilon'}(\Gamma_\varphi)] = E_x\Big[\int_{\Lambda_{p-1}\times\Lambda_p} \tilde\mu^{y,x}_{\varepsilon,\varepsilon'}(ds_1\dots ds_{p-1}dt_1\dots dt_p)(E_{B_{s_{p-1}}}[\ell^y_\varepsilon(\zeta)] + h_\varepsilon)\Big].$$

By the very definition of h_ε,

$$E_{y_\varepsilon}[\ell^y_\varepsilon(\zeta)] = -\,h_\varepsilon \quad \text{if} \quad |y_\varepsilon - y| = \varepsilon.$$

Therefore $E_x[\mu^{y,z}_{\varepsilon,\varepsilon'}(\Gamma_\varphi)] = 0$ in this case.

We now turn to the general case where we only assume $\varphi \ne \varphi_1$, $\varphi \ne \varphi_2$. We may restrict our attention to the case when , for some $k \in \{1,\dots,p-1\}$,

$$\Gamma_\varphi \subset \{(s_1,\dots,s_p,t_1,\dots,t_p)\ ;\ t_k< s_k< s_{k+1}< t_{k+1}< s_{k+2}< t_{k+2}<\dots< s_p< t_p\}$$

(one should also consider the case

$$\Gamma_\varphi \subset \{t_{k+1} < s_k < s_{k+1} < t_{k+2} < s_{k+2} < \ldots < t_p < s_p\}$$

and the symmetric cases where the roles of s_i and t_i are interchanged ; all these cases however are treated in the same way). Then,

$$E_x[\mu^{y,z}_{\varepsilon,\varepsilon'}(\Gamma_\varphi)] = E_x\Big[\iint_{\Lambda_k\times\Lambda_k} \tilde\mu^{y,z}_{\varepsilon,\varepsilon'}(ds_1..ds_k dt_1..dt_k)\int_{s_k}^{\zeta} (\ell^y_\varepsilon(ds_{k+1})+h_\varepsilon \delta_{(s_k)}(ds_{k+1}))$$

$$\times \int_{s_{k+1}}^{\zeta} \ell^z_{\varepsilon'}(dt_{k+1})\int_{t_{k+1}}^{\zeta} \ell^y_\varepsilon(ds_{k+2}) \ldots \int_{t_{p-1}}^{\zeta} \ell^y_\varepsilon(ds_p)\int_{s_p}^{\zeta} \ell^z_{\varepsilon'}(dt_p)\Big]$$

where $\tilde\mu^{y,z}_{\varepsilon,\varepsilon'}$ is a measure on $\Lambda_k \times \Lambda_k$, supported on $\{(s_1,\ldots,t_k) ; t_k < s_k ,$ $|B_{s_k} - y| = \varepsilon\}$ and such that the bounded variation process $t \longrightarrow$ $\tilde\mu^{y,x}_{\varepsilon,\varepsilon'}(\{s_k \le t\})$ is predictable. Crude bounds show that the total variation $|\tilde\mu^{y,z}_{\varepsilon,\varepsilon'}|$ of $\tilde\mu^{y,z}_{\varepsilon,\varepsilon'}$ satisfies :

$$E_x\Big[\iint_{\Lambda_k\times\Lambda_k} |\tilde\mu^{y,x}_{\varepsilon,\varepsilon'}|(ds_1\ldots ds_k dt_1\ldots dt_k)\ (\ell^y_\varepsilon((s_k,\zeta)) + |h_\varepsilon|)\Big]$$

$$\le C\ |\log \varepsilon|^{2k}\ (G(\tfrac{y-x}{2}) + G(\tfrac{z-x}{2}))$$

(use the bounds $E_x[\ell^y_\varepsilon(\zeta)] \le G(\frac{y-x}{2})$ and $\sup_{w\in C} E_w[\ell^y_\varepsilon(\zeta)] \le C'\ |\log \varepsilon|$).

Next , in the previous formula for $|E_x[\mu^{y,z}_{\varepsilon,\varepsilon'}(\Gamma_\varphi)]$ we replace $\Big(\int_{s_{k+1}}^{\zeta} \ldots\Big)$ by its predictable projection, which coincides with

$$H^{2(p-k)-1}_{\varepsilon',\varepsilon}(B_{s_{k+1}}, z, y),$$

in the notation of Lemma 3. By Lemma 3, for $w \in C(y,\varepsilon)$,

$$|H^{2(p-k)-1}_{\varepsilon',\varepsilon}(w,z,y) - G(z-y)^{2(p-k)-1}| \le C\ \varepsilon\ |\log \varepsilon|^{2(p-k)-2}\ g(|z-y|)$$

(use again the bound $G(\frac{y-x}{4}) \le C\ |\log \varepsilon|$) and it follows that

$$\Big|E_x[\mu^{y,z}_{\varepsilon,\varepsilon'}(\Gamma_\varphi)] - G(z-y)^{2(p-k)-1}\ E_x\Big[\iint_{\Lambda_k\times\Lambda_k} \tilde\mu^{y,z}_{\varepsilon,\varepsilon'}(ds_1\ldots dt_k)\ \ (\ell^y_\varepsilon((s_k,\zeta)) + h_\varepsilon)\Big]\Big|$$

$$\le C'\ \varepsilon\ |\log \varepsilon|^{2p-2}\ (G(\tfrac{y-x}{2}) + G(\tfrac{z-x}{2}))\ g(|z-y|)$$

by our previous bound on $|\mu^{y,z}_{\varepsilon,\varepsilon'}|$. This completes the proof since

$$E_x[\int \tilde{\mu}^{y,z}_{\varepsilon,\varepsilon'}(ds_1\ldots dt_k)\,(\ell^y_\varepsilon((s_k,\zeta)) + h_\varepsilon)] = 0,$$

by the same arguments as above (that is, by replacing $\ell^y_\varepsilon((s_k,\zeta))$ by its predictable projection), using the fact that $\mu^{y,z}_{\varepsilon,\varepsilon'}$ is supported on $\{t_k < s_k\}$.□

We now need to bound the contribution of pairs (y,z) that do not satisfy the assumption of Lemma 4.

Lemma 5. *There exists a constant* C'_p *such that for every* $\varepsilon, \varepsilon' \in (0,1/2)$ *such that* $\varepsilon/2 \le \varepsilon' \le 2\varepsilon$ *and every* $x, y, z \in \mathbf{C}$,

$$E_x[|Q^p_\varepsilon(\ell^y_\varepsilon(\zeta))\ Q^p_{\varepsilon'}(\ell^z_{\varepsilon'}(\zeta))|] \le C'_p\ |\log\ \varepsilon|^{2p-2}\ \left(G(\tfrac{y-x}{2}) + G(\tfrac{z-x}{2})\right)^{1/2} G(\tfrac{z-y}{2})^{1/2}.$$

Proof : We use the easy bound

$$|Q^p_\varepsilon(\ell^y_\varepsilon(\zeta))| \le \ell^y_\varepsilon(\zeta)\ (\ \ell^y_\varepsilon(\zeta) + |h_\varepsilon|\ )^{p-1}$$

and we observe that $\ell^y_\varepsilon(\zeta) = 0$ unless $T_\varepsilon(y) < \zeta$, where

$$T_\varepsilon(y) = \inf\{\ s\ ;\ |B_s - y| \le \varepsilon\ \}.$$

Then the Cauchy-Schwarz inequality yields:

$$E_x[|Q^p_\varepsilon(\ell^y_\varepsilon(\zeta))\ Q^p_{\varepsilon'}(\ell^z_{\varepsilon'}(\zeta))|] \le P_x[\ T_\varepsilon(y) < \zeta\ ;\ T_{\varepsilon'}(z) < \zeta\ ]^{1/2}$$

$$\times\ E_x[\ell^y_\varepsilon(\zeta)^2(\ell^y_\varepsilon(\zeta) + |h_\varepsilon|)^{2(p-1)}\ \ell^z_{\varepsilon'}(\zeta)(\ell^z_{\varepsilon'}(\zeta) + |h_{\varepsilon'}|)^{2(p-1)}]^{1/2}.$$

Next we make use of the bound

$$P_x[\ T_\varepsilon(y) < \zeta\ ;\ T_{\varepsilon'}(z) < \zeta\ ] \le C\ |\log\ \varepsilon|^{-2}\ \left(G(\tfrac{y-x}{2}) + G(\tfrac{z-x}{2})\right) G(\tfrac{z-y}{2})$$

which follows from the techniques of Chapter VI (to bound $P[T_\varepsilon(y) \le T_{\varepsilon'}(z) < \zeta]$, apply the strong Markov property at $T_\varepsilon(y)$ and use Lemma VI-1 (iii)) . Also notice that for every integer $m \ge 1$,

$$\sup_{y\in C}\ E_x[\ell^y_\varepsilon(\zeta)^m] = |h_\varepsilon|^m$$

(the supremum is attained for $y \in C(x,\varepsilon)$ and in this case the distribution of $\ell^y_\varepsilon(\zeta)$ is exponential with mean $|h_\varepsilon|$).

The previous bounds and another application of the Cauchy-Schwarz inequality lead to:

$$E_x[|Q^p_\varepsilon(\ell^y_\varepsilon(\zeta))Q^p_{\varepsilon'}(\ell^z_{\varepsilon'}(\zeta))|] \le C\ |\log\varepsilon|^{-2}\ (G(\tfrac{y-x}{2}) + G(\tfrac{z-x}{2}))^{1/2}\ G(\tfrac{z-y}{2})^{1/2}$$

$$\times\ C''_p|h_\varepsilon h_{\varepsilon'}|^p.$$

Lemma 5 follows. □

We now turn to the proof of Theorem 1. We note that:

$$E_x[T^p_\varepsilon f\ T^p_{\varepsilon'}f] = \int dy\ dz\ f(y)\ f(z)\ E_x[Q^p_\varepsilon(\ell^y_\varepsilon(\zeta))Q^p_{\varepsilon'}(\ell^z_{\varepsilon'}(\zeta))].$$

We then apply Lemma 4 and we use Lemma 5 to bound the contribution of the pairs (y,z) that do not satisfy the assumptions of Lemma 4. We get:

$$|E_x[T^p_\varepsilon f\ T^p_\varepsilon f] - 2\int dy\ dz\ f(y)\ f(z)\ G(y-x)\ G(z-y)^{2p-1}| \le C\ \varepsilon\ |\log\varepsilon|^{2p-2}$$

whenever $\varepsilon' \in [\varepsilon/2,\varepsilon]$, $\varepsilon \in (0,1/2)$ (notice that both functions $G(y)$, $g(|y|)$ are integrable avec $\mathbb{C}$). The previous bound implies that for $\varepsilon' \in [\varepsilon/2,\varepsilon]$,

$$E_x[(T^p_\varepsilon f - T^p_{\varepsilon'}f)^2] \le 4C\ \varepsilon\ |\log\varepsilon|^{2p-2}.$$

It follows that the sequence $T^p_{2^{-n}}f$ converges in the L^2-norm. If T^pf denotes its limit, it is then immediate that :

$$T^pf = L^2 - \lim_{\varepsilon\to 0} T^p_\varepsilon f.$$

This completes the proof of Theorem 1. □

4. Remarks.

The previous proof gives more information than is stated in Theorem 1. We get an estimate of the rate of convergence of $T^p_\varepsilon f$ towards T^pf :

$$\text{(8)}\qquad E[(T^p_\varepsilon f - T^pf)^2] \le C\ \|f\|^2_\infty\ \varepsilon\ |\log\varepsilon|^{2p-2}$$

for some constant C independent of f. We have also obtained the second moment of T^pf :

$$E_x[(T^pf)^2] = 2\int dy\ dz\ f(y)\ f(z)\ G(y-x)\ G(z-y)^{2p-1}$$

and, more generally,

$$E_x[T^pf\ T^pf'] = \int dy\ dz\ f(y)\ f'(z)\ (G(y-x) + G(z-x))\ G(z-y)^{2p-1}.$$

One can also check that

$$E_x[T^{p+1}f\ T^pf'] = \int dy\ dz\ f(y)\ f'(z)\ G(y-x)\ G(z-y)^{2p},$$

and that

$$E_x[T^pf\ T^qf'] = 0$$

whenever $|q-p| \geq 2$. These results are consequences of the following bounds, which hold under the assumptions of Lemma 4,

$$(9)\quad |E_x[Q_\varepsilon^{p+1}(\ell_\varepsilon^y(\zeta))\ Q_{\varepsilon'}^p(\ell_{\varepsilon'}^z(\zeta))] - G(y-x)G(z-y)^{2p}| \leq \varepsilon\ |\log \varepsilon|^{2p-1}\ F(y-x,z-x)$$

and

$$(10)\qquad |E[Q_\varepsilon^p(\ell_\varepsilon^y(\zeta))\ Q_{\varepsilon'}^q(\ell_{\varepsilon'}^z(\zeta))]| \leq \varepsilon\ |\log \varepsilon|^{p+q-2}\ F'(y-x,z-x)$$

where the functions $F(y,z)$, $F'(y,z)$ are integrable over $\mathbb{C}^2$. To prove these bounds, proceed as in the proof of Lemma 4. In the first case one needs to order $s_1,\dots,s_{p+1}$, $t_1,\dots,t_p$. The order $s_1<t_1<s_2<t_2<\dots<t_p<s_{p+1}$ is the only one that gives a nonnegligible contribution. In the second case all orders give negligible contributions.

Finally, it is easy to check that:

$$E_x[T^pf] = 0,$$

for $p \geq 2$ (indeed, $E_x[T_\varepsilon^pf] = 0$ for every $\varepsilon > 0$).

The proof of Theorem 1 can be adapted to yield L^n-convergence for any $n \geq 1$. The previous formulas have analogues for higher-order moments. For instance the nth-moment of T^pf is

$$E_x[(T^pf)^n] = \int dy_1\dots dy_n\ f(y_1)\dots f(y_n) \sum_\sigma \prod_{i=1}^{n} G(y_{\sigma(i)} - y_{\sigma(i-1)})$$

where the summation is over all mappings $\sigma : \{1,2,\dots,np\} \longrightarrow \{1,\dots,n\}$ such that $\sigma(i) \neq \sigma(i-1)$ for any $i \geq 2$, and $\operatorname{card} \sigma^{-1}(j) = p$ for $j \in \{1,\dots,n\}$ (by convention $y_{\sigma(0)} = x$).

As a final remark, one may wonder what is the role of the exponential time ζ. The estimates of the proof of Theorem 1 depend heavily on the fact that we are working with Brownian motion stopped at an exponential time. Note that changing λ would only change h_ε by an additive constant. Suppose that we replace h_ε by

$$\tilde{h}_\varepsilon = h_\varepsilon + c$$

for some constant $c \in \mathbb{R}$. Let $\tilde{T}_\varepsilon^pf$ be defined accordingly. Then it is immediately seen that :

$$\tilde{T}^p_\varepsilon f = \sum_{k=1}^{p} \binom{p-1}{k-1} c^{p-k}\, T^k_\varepsilon f$$

and therefore we may define

$$\tilde{T}^p f := \lim_{\varepsilon\to 0} \tilde{T}^p_\varepsilon f = \sum_{k=1}^{p} \binom{p-1}{k-1} c^{p-k}\, T^k f.$$

Note that $\tilde{T}^p f$ can also be considered as a renormalized version of $\beta_p(\Delta_p \cap [0,\zeta)^p)$. This corresponds to the non-uniqueness of the renormalization.

Bibliographical notes. *The renormalization for self-intersections of planar Brownian motion has been inspired by renormalization in field theory: see Dynkin* [Dy2] *and the references in this paper. The existence of the renormalized powers of the occupation field was derived by Dynkin* [Dy3] *(in a slightly more general setting) using his isomorphism theorem between the occupation field of a symmetric Markov process and a certain Gaussian field. Later (in* [Dy5], [Dy6], [Dy7]), *Dynkin proposed a different approach, based on a detailed combinatorial analysis. The material of this Chapter is taken from* [L12]. *It has been inspired by Dynkin's second method, but it avoids the combinatorial analysis of Dynkin's work. Our construction is however not as general as Dynkin's one in* [Dy6]. *See also Rosen* [R4] *for a different method of renormalization (whose relationship with Dynkin's work is not clear) and Rosen and Yor* [RY] *for an approach based on stochastic calculus in the case of triple self-intersections. The renormalized fields* $T^p\varphi$ *appear in certain limit theorems for planar random walks: see Dynkin* [Dy6].

CHAPTER XI

Asymptotic expansions for the planar Wiener sausage

1. A random field associated with the Wiener sausage.

Let $S_K(a,b)$ denote the Wiener sausage associated with a planar Brownian motion B and a nonpolar compact subset K of $\mathbb{R}^2$, on the time interval $[a,b]$. By definition,

$$S_K(a,b) = \bigcup_{a\le s\le b} (B_s + K).$$

Our goal in this chapter is to get a full asymptotic expansion for $m(S_{\varepsilon K}(0,t))$ as ε goes to 0. The different terms of this expansion will be the renormalized self-intersection local times introduced in Chapter X, for all multiplicity orders $p \ge 1$. Note that the expansion at the order 2 has already been derived in Chapter VIII,

$$(1)\qquad m(S_{\varepsilon K}(0,t)) = \frac{\pi}{\log 1/\varepsilon} + \frac{\pi}{(\log 1/\varepsilon)^2}\left(\frac{1+\kappa-\log 2}{2} - R(K) - \pi\,\gamma(\mathcal{T})\right) + o\left(\frac{1}{(\log 1/\varepsilon)^2}\right),$$

where κ denotes Euler's constant, and $\gamma(\mathcal{T})$ is the renormalized self-intersection local time that was defined in Section VIII-3 (note however that the proof of (1) required Spitzer's expansion of $E[m(S_{\varepsilon K}(0,1))]$).

The approach of this chapter depends heavily on the estimates of Chapter X, but is independent of the results of Chapter VI and VIII (except for the potential-theoretic rersults of Section VI-2). We will recover the expansion (1), as well as Spitzer's theorem, as a special case of Theorem 5 below.

From now on, we fix a compact subset K of $\mathbb{R}^2$. We assume that K has positive logarithmic capacity, that is

$$\mathrm{cap}(K) = \exp -\left(\inf_{\mu\in\mathcal{P}(K)} \iint_{K\times K} \mu(dx)\,\mu(dy)\,\log\frac{1}{|y-x|}\right) > 0$$

where $\mathcal{P}(K)$ is the set of all probability measures supported on K. By definition, $R(K) = \log \mathrm{cap}(K)$.

As in Chapter X, it will be convenient to deal with Brownian motion killed at an independent exponential time ζ with parameter λ. As previously, we let $G(y-x) = G_\lambda(x,y)$ denote the Green function of the killed process. Set

$$T_K = \inf\{t \geq 0,\ B_t \in K\}.$$

As was recalled in Chapter VI, we have for every $x \in \mathbb{R}^2\backslash K$

$$(2) \qquad P_x[T_K < \zeta] = \int \mu_K^\lambda(dw)\ G(w-x),$$

where μ_K^λ, the λ-equilibrium measure of K, is a finite measure supported on K. The λ-capacity of K is $C_\lambda(K) = \mu_K^\lambda(K)$, and we have :

$$(3) \qquad C_\lambda(K)^{-1} = \inf_{\mu \in \mathcal{P}(K)} \iint \mu(dx)\ \mu(dy)\ G(y-x).$$

An important role will be played by the constants a_ε defined for $\varepsilon > 0$ by

$$a_\varepsilon = - C_\lambda(\varepsilon K).$$

It easily follows from (3) and formula (4') of Chapter X that, as ε goes to 0,

$$\frac{1}{a_\varepsilon} = - \frac{1}{\pi} \log \frac{1}{\varepsilon} - \frac{1}{\pi} \left(\frac{\log 2/\lambda}{2} - \kappa - R(K) \right) + O(\varepsilon^2 \log \tfrac{1}{\varepsilon}).$$

For any bounded Borel function f on $\mathbb{R}^2$, we set :

$$S_\varepsilon^K f = \int dy\ f(y)\ 1_{S_{\varepsilon K}(0,\zeta)}(y).$$

<u>**Theorem 1**</u> : *Let* $n \geq 1$. *Then, for any bounded Borel function* f *on* $\mathbb{R}^2$,

$$S_\varepsilon^K f = - \sum_{p=1}^{n} (a_\varepsilon)^p\ T^p f + R_n(\varepsilon, f)$$

where the remainder $R_n(\varepsilon,f)$ *satisfies :*

$$\lim_{\varepsilon \to 0} |\log \varepsilon|^{2n}\ E[R_n(\varepsilon,f)^2] = 0.$$

In the special case $f = 1$, Theorem 1 provides an asymptotic expansion of $m(S_{\varepsilon K}(0,\zeta))$ in the L^2-norm. Using scaling arguments it is then possible to check that a similar expansion holds for $m(S_{\varepsilon K}(0,t))$, for any constant time $t > 0$. In fact, one can even get an almost sure expansion of $m(S_{\varepsilon K}(0,t))$ (see the end of this chapter).

Let us briefly outline the proof of Theorem 1. Thanks to the estimate (8) of Chapter X, it is enough to check that the given statement holds with $T^p f$ replaced by $T_\varepsilon^p f$. Then,

$$E\Big[(S^K_\varepsilon f + \sum_{p=1}^{n} (a_\varepsilon)^p\, T^p_\varepsilon f)^2\Big] = E\Big[\Big(\int dy\, f(y)\, (1_{S_{\varepsilon K}(0,\zeta)}(y) + \sum_{p=1}^{n} (a_\varepsilon)^p\, Q^p_\varepsilon(\ell^y_\varepsilon(\zeta))\Big)^2\Big]$$

$$= \iint dy\, dz\, f(y)\, f(z)$$

$$\times E\Big[\Big(1_{S_{\varepsilon K}(0,\zeta)}(y) + \sum_{p=1}^{n} (a_\varepsilon)^p\, Q^p_\varepsilon(\ell^y_\varepsilon(\zeta))\Big)\Big(1_{S_{\varepsilon K}(0,\zeta)}(z) + \sum_{p=1}^{n} (a_\varepsilon)^p\, Q^p_\varepsilon(\ell^z_\varepsilon(\zeta))\Big)\Big].$$

Expanding the product inside the expectation sign, we are led to study the following three quantities :

(a) $$E[Q^p_\varepsilon(\ell^y_\varepsilon(\zeta))\, Q^q_\varepsilon(\ell^z_\varepsilon(\zeta))]$$

This quantity was studied in detail in Chapter X, in the special case $p = q$. The general case offers no additional difficulty.

(b) $$P[y \in S_{\varepsilon K}(0,\zeta),\ z \in S_{\varepsilon K}(0,\zeta)].$$

Sharp estimates for this probability will be derived in Section 2.

(c) $$E[Q^p_\varepsilon(\ell^y_\varepsilon(\zeta))\, 1_{S_{\varepsilon K}(0,\zeta)}(z)].$$

This quantity will be studied in Section 4, after some preliminary estimates have been established in Section 3.

2. The probability of hitting two small compact sets.

From now on, we shall assume that the compact set K is contained in the closed unit disk $\bar{D}$ (this restriction can be removed by a scaling argument). To simplify notation, we set

$$T_\varepsilon(y) = T_{y-\varepsilon K} = \inf\{t \ge 0\ ;\ B_t \in y - \varepsilon K\}$$

so that

$$P[y \in S_{\varepsilon K}(0,\zeta), z \in S_{\varepsilon K}(0,\zeta)] = P[T_\varepsilon(y) < \zeta,\ T_\varepsilon(z) < \zeta].$$

Lemma 2 : *Let* $n \ge 2$. *There exists a function* $F_n \in L^1((\mathbb{R}^2)^2, dy\, dz)$, *such that, for any* $\varepsilon \in (0,1/2)$, $y,z \in \mathbb{R}^2$ *with* $|y| > 4\varepsilon$, $|z| > 4\varepsilon$, $|z-y| > 4\varepsilon$,

$$|P[T_\varepsilon(y) < \zeta,\ T_\varepsilon(z) < \zeta] - \sum_{p=2}^{n} (a_\varepsilon)^p\, (G(y)+G(z))G(z-y)^{p-1}| \le |\log\varepsilon|^{-n-1}\, F_n(y,z).$$

Proof : We will give details for $n = 2,3$. It will then be clear that the proof can be continued by induction on n. We first observe that

$$P[T_\varepsilon(y) < \zeta,\ T_\varepsilon(z) < \zeta] = P[T_\varepsilon(y) \le T_\varepsilon(z) < \zeta] + P[T_\varepsilon(z) \le T_\varepsilon(y) < \zeta].$$

Then,

$$P[T_\varepsilon(y) \le T_\varepsilon(z) < \zeta] = P[T_\varepsilon(y) \le T'_\varepsilon(z) < \zeta] - P[T_\varepsilon(z) \le T_\varepsilon(y) \le T'_\varepsilon(z) < \zeta]$$

where :

$$T'_\varepsilon(z) = \inf\{t \ge T_\varepsilon(y) \ ; \ B_t \in z-\varepsilon K\}.$$

By the Markov property at $T_\varepsilon(y)$,

$$P\Big[T_\varepsilon(y) \le T'_\varepsilon(z) < \zeta\Big] = E\Big[(T_\varepsilon(y) < \zeta)\ P_{B_{T_\varepsilon(y)}}\ [T_\varepsilon(z) < \zeta]\Big].$$

Notice that $B_{T_\varepsilon(y)} \in y - \varepsilon K \subset \bar{D}(y,\varepsilon)$. By (2) and formula (6) of Chapter X, we have for any $y_\varepsilon \in \bar{D}(y,\varepsilon)$,

$$\text{(4)} \qquad |P_{y_\varepsilon}[T_\varepsilon(z) < \zeta] + a_\varepsilon\ G(z-y)| = \Big|\Big(\int \mu^\lambda_{\varepsilon K}(dw)\ G(z-w-y_\varepsilon)\Big) + a_\varepsilon\ G(z-y)\Big|$$

$$\le |a_\varepsilon| \sup_{z' \in D(z,2\varepsilon)} |G(z'-y) - G(z-y)|$$

$$\le C\ \varepsilon\ |\log \varepsilon|^{-1}\ g(|z-y|)$$

where g is as in chapter X. Similarly,

$$|P[T_\varepsilon(y) < \zeta] + a_\varepsilon\ G(z-y)| \le C\ \varepsilon\ |\log \varepsilon|^{-1}\ g(|y|)$$

and (2) also gives

$$P[T_\varepsilon(y) < \zeta] \le |a_\varepsilon|\ G(\tfrac{y}{2}).$$

It follows from these estimates that

$$\text{(5)} \qquad |P[T_\varepsilon(y) \le T'_\varepsilon(z) < \zeta] - a^2_\varepsilon\ G(y)\ G(z-y)|$$

$$\le C\ \varepsilon\ |\log \varepsilon|^{-2}\ \Big(G(\tfrac{y}{2})\ g(|z-y|) + g(|y|)\ G(z-y)\Big)$$

On the other hand, by applying the Markov property at $T_\varepsilon(y)$ and then at $T_\varepsilon(z)$, one easily gets

$$\text{(6)} \qquad P[T_\varepsilon(z) \le T_\varepsilon(y) \le T'_\varepsilon(z) < \zeta] \le C\ |\log \varepsilon|^{-3}\ \Big(G(\tfrac{z}{2})\ G(\tfrac{z-y}{2})^2\Big).$$

This gives the case $n = 2$ of the Lemma.

In the case $n = 3$, we again use (5) but we replace (6) by :

$$P[T_\varepsilon(z) \le T_\varepsilon(y) \le T'_\varepsilon(z) < \zeta]$$

$$= P[T_\varepsilon(z) \le T'_\varepsilon(y) \le T''_\varepsilon(z) < \zeta] - P[T_\varepsilon(y) \le T_\varepsilon(z) \le T'_\varepsilon(y) \le T''_\varepsilon(z) < \zeta]$$

where

$$T'_\varepsilon(y) = \inf\{t \geq T_\varepsilon(z),\ B_t \in y - \varepsilon K\},$$

$$T''_\varepsilon(z) = \inf\{t \geq T'_\varepsilon(y),\ B_t \in z - \varepsilon K\}.$$

The bound (4) and the Markov property give :

$$|P[T_\varepsilon(z) \leq T'_\varepsilon(y) \leq T''_\varepsilon(z) < \zeta] - (a_\varepsilon)^3\, G(z)\, G(z-y)^2|$$

$$\leq C\, \varepsilon\, |\log \varepsilon|^{-3} \left[G(\tfrac{z}{2})\, G(\tfrac{z-y}{2})\, g(|z-y|) + g(|z|)\, G(\tfrac{z-y}{2})^2\right],$$

whereas it is easily checked that

$$P[T_\varepsilon(y) \leq T_\varepsilon(z) \leq T'_\varepsilon(y) \leq T''_\varepsilon(z) < \zeta] \leq C\, |\log \varepsilon|^{-4}\, G(\tfrac{y}{2})\, G(\tfrac{z-y}{2})^3. \ \square$$

Remark : It immediately follows from Lemma 2 that

$$E[m(S^K_\varepsilon f)^2] = \iint dy\, dz\, f(y)\, f(z)\, P[T_\varepsilon(y) < \zeta,\ T_\varepsilon(z) < \zeta]$$

$$= 2 \sum_{p=2}^{n} (a_\varepsilon)^p \int dy\, dz\ \ f(y)\, f(z)\, G(y)\, G(z-y)^{p-1} + O(|\log \varepsilon|^{-n-1}).$$

3. A preliminary lemma.

The study of the limiting behavior of the term $E[Q^p_\varepsilon(\ell^y_\varepsilon(\zeta))1_{S_{\varepsilon K}(0,\zeta)}(z)]$ requires the following lemma, which is analogous to Lemma X-3.

Lemma 3 : *Let* $n \geq 1$ *and* $n' = n$ *or* $n-1$. *Set :*

$$U^{n,n'}_\varepsilon(x,y,z) = E_x\left[\int\!\!\int_{\Lambda_n} \ell^z_\varepsilon(ds_1)\, \ell^z_\varepsilon(ds_2)\ldots\ell^z_\varepsilon(ds_n) \prod_{i=0}^{n'} 1_{S_{\varepsilon K}(s_i,s_{i+1})}(y)\right],$$

$$V^{n,n'}_\varepsilon(x,y,z) = E_x\left[\int\!\!\int_{\Lambda_n} \ell^y_\varepsilon(ds_1)\, \ell^y_\varepsilon(ds_2)\ldots\ell^y_\varepsilon(ds_n) \prod_{i=1}^{n'} 1_{S_{\varepsilon K}(s_i,s_{i+1})}(z)\right],$$

where by convention $s_0 = 0$, $s_{n+1} = \zeta$.

There exists a positive constant $C_{n,n'}$ *such that, for any* $x, y, z \in \mathbb{C}$, $\varepsilon \in (0,1/2)$, *with* $|y-x| \geq 4\varepsilon$, $|z-y| \geq 8\varepsilon$,

$$|U^{n,n'}_\varepsilon(x,y,z) - |a_\varepsilon|^{n'+1}\, G(y-x)\, G(z-y)^{n+n'}|$$

$$\leq C_{n,n'}\varepsilon\, |a_\varepsilon|^{n'+1}\, (g(2|y-x|)\, G(z-y)^{n+n'} + G(\tfrac{y-x}{2})\, g(|z-y|) G(\tfrac{z-y}{4})^{n+n'-1})$$

and, if $n' \geq 1$,

$$|V_\varepsilon^{n,n'}(x,y,z) - |a_\varepsilon|^{n'} G(y-x)\, G(z-y)^{n+n'-1}|$$

$$\leq C_{n,n'}\varepsilon\, |a_\varepsilon|^{n'} \left(g(2|y-x|)\, G(z-y)^{n+n'-1} + G(\tfrac{y-x}{2})\, g(|z-y|)G(\tfrac{z-y}{4})^{n+n'-2}\right).$$

Proof : We consider only the case of $U_\varepsilon^{n,n'}$ and we further assume that $n' = n-1$. The other cases are treated in a similar manner.

We argue by induction on n. For $n = 1$,

$$U_\varepsilon^{1,0}(x,y,z) = E_x\left[1_{(T_\varepsilon(y)<\zeta)}\ \ell_\varepsilon^z([T_\varepsilon(y),\zeta))\right]$$

$$= E_x\left[1_{(T_\varepsilon(y)<\zeta)}\ E_{B_{T_\varepsilon(y)}}[\ell_\varepsilon^z(\zeta)]\right]$$

by the Markov property. However, by Lemma X-3, for any $y_\varepsilon \in y - \varepsilon K \subset D(y,\varepsilon)$,

$$|E_{y_\varepsilon}[\ell_\varepsilon^z(\zeta)] - G(z-y)| \leq \varepsilon\, g(2|z-y|)$$

and

$$E_{y_\varepsilon}[\ell_\varepsilon^z(\zeta)] \leq G(\tfrac{z-y}{2}).$$

Moreover, by (2),

$$|P_x[T_\varepsilon(y) < \zeta] - |a_\varepsilon|\, G(y-x)| \leq \varepsilon\, |a_\varepsilon|\, g(2|y-x|)$$

and

$$|P_x[T_\varepsilon(y) < \zeta] \leq |a_\varepsilon|\, G(\tfrac{y-x}{2}).$$

The case $n = 1$ follows readily from these bounds. We also get the bound:

$$U_\varepsilon^{1,0}(x,y,z) \leq |a_\varepsilon|\, G(\tfrac{y-x}{2})\, G(\tfrac{z-y}{2}) \tag{7}$$

Next suppose that $n \geq 2$ and that the desired result holds at the order n-1. We have :

$$U_\varepsilon^{n,n-1}(x,y,z) = E_x\left[1_{(T_\varepsilon(y)<\zeta)} \int_{T_\varepsilon(y)}^{\zeta} \ell_\varepsilon^z(ds_1)\, U_\varepsilon^{n-1,n-2}(B_{s_1},y,z)\right]$$

where we have simply replaced

$$\int_{s_1}^{\zeta} \ell_\varepsilon^z(ds_2)\ldots\int_{s_{n-1}}^{\zeta} \ell_\varepsilon^z(ds_n)\ \prod_{i=1}^{n-1} 1_{S_{\varepsilon K}(s_i,s_{i+1})}(y)$$

by its predictable projection $U_\varepsilon^{n-1,n-2}(B_{s_1},y,z)$.

To get the desired result at the order n , it now suffices to use the induction hypothesis, the bound (7) and the bound

$$|G(y-B_{s_1}) - G(y-z)| \le \varepsilon\, g(|y-z|)$$

which holds when $|B_{s_1}-z| \le \varepsilon$. □

4. Proof of Theorem 1.

As in Chapter X, the proof of our main result depends on a basic lemma which we now state.

Lemma 4 : *Let* $p \ge 1$. *There exists a constant* C_p *such that, for any* $y,z \in \mathbb{R}^2$, $\varepsilon \in (0,1/2)$, *with* $|y| > 4\varepsilon$, $|z| > 4\varepsilon$, $|z-y| > 8\varepsilon$,

- *if* $p \ge 2$,

$$\Big|E[Q^p_\varepsilon(\ell^y_\varepsilon(\zeta))\, 1_{S_{\varepsilon K}(0,\zeta)}(z)] + a_\varepsilon^{p-1}\, G(y)\, G(z-y)^{2p-2}$$

$$+ a_\varepsilon^p\, (G(y) + G(z))\, G(z-y)^{2p-1} + a_\varepsilon^{p+1}\, G(z)\, G(z-y)^{2p}\Big|$$

$$\le C_p\, \varepsilon\, |\log \varepsilon|^{2p}\, \Big((g(|y|) + g(|z|))\, G(\tfrac{z-y}{4}) + (G(\tfrac{y}{2}) + G(\tfrac{z}{2}))\, g(|z-y|)\Big)$$

- *if* $p = 1$,

$$\Big|E[\ell^y_\varepsilon(\zeta)\, 1_{S_{\varepsilon K}(0,\zeta)}(z)] + a_\varepsilon(G(y) + G(z))\, G(z-y) + a_\varepsilon^2\, G(z)\, G(z-y)^2\Big|$$

$$\le C_1\, \varepsilon\, |\log \varepsilon|^2\, \Big((g(|y|) + g(|z|))\, G(\tfrac{z-y}{4}) + (G(\tfrac{y}{2}) + G(\tfrac{z}{2}))\, g(|z-y|)\Big)\ .$$

Proof : We assume that $p \ge 2$ (the case $p = 1$ is easier). By Lemma X-2,

$$E[Q^p_\varepsilon(\ell^y_\varepsilon(\zeta))\, 1_{S_{\varepsilon K}(0,\zeta)}(z)]$$

$$= E\Big[\int_{\Lambda_p} \ell^y_\varepsilon(ds_1) \Big(\prod_{i=2}^{p} (\ell^y_\varepsilon(ds_i) + h_\varepsilon\, \delta_{(s_{i-1})}(ds_i))\Big)\, 1_{S_{\varepsilon K}(0,\zeta}(z)\Big].$$

Now the key idea is to write :

$$S_{\varepsilon K}(0,\zeta) = \bigcup_{i=0}^{p} S_{\varepsilon K}(s_i, s_{i+1})$$

with the usual convention $s_o = 0$, $s_{p+1} = \zeta$. It follows that

$$1_{S_{\varepsilon K}(0,\zeta)}(z) = \sum_{L\in\mathcal{P}_p} (-1)^{|L|+1}\, 1_{(\bigcap_{i\in L} S_{\varepsilon K}(s_i, s_{i+1}))}(z)$$

where $\mathcal{P}_p$ denotes the set of all nonempty subsets of $\{0,1,\ldots,p\}$, and $|L| = \mathrm{Card}(L)$. Therefore,

$$E[Q^p_\varepsilon(\ell^y_\varepsilon(\zeta))\ 1_{S_{\varepsilon K}(0,\zeta)}(z)] = \sum_{L\in\mathcal{P}_p} (-1)^{|L|+1}\ \Phi_L(\varepsilon,y,z),$$

where

$$\Phi_L(\varepsilon,y,z) = E\Big[\int_{\Lambda_p} \ell^y_\varepsilon(ds_1) \prod_{i=2}^{p} (\ell^y_\varepsilon(ds_i) + h_\varepsilon \delta_{(s_{i-1})}(ds_i)) \prod_{i\in L} 1_{S_{\varepsilon K}(s_i,s_{i+1})}(z)\Big].$$

Suppose first that $\{1,\dots,p-1\} \subset L$, which happens only in the four cases:

$$L_1 = \{0,1,\dots,p\},\quad L_2 = \{1,\dots,p\},\quad L_3 = \{0,1,\dots,p-1\},\quad L_4 = \{1,\dots,p-1\}.$$

In each of these cases, we can use Lemma 3 to analyse the behavior of $\Phi_L(\varepsilon,y,z)$. Simply notice that

$$\Phi_{L_1}(\varepsilon,y,z) = U^{p,p}_\varepsilon(0,z,y),\qquad \Phi_{L_2}(\varepsilon,y,z) = V^{p,p}_\varepsilon(0,y,z)$$

$$\Phi_{L_3}(\varepsilon,y,z) = U^{p,p-1}_\varepsilon(0,z,y),\quad \Phi_{L_4}(\varepsilon,y,z) = V^{p,p-1}_\varepsilon(0,y,z).$$

Taking account of Lemma 3, we see that the proof of Lemma 4 will be complete once we have checked that the other choices of L give a negligible contribution. This is very similar to what we did in the proof of Lemma X-4. Set

$$k = \sup\{i \in \{1,\dots,p-1\},\ i \notin L\}$$

and assume for definiteness that $p \in L$. Then we may write

$$\Phi_L(\varepsilon,y,z) = E\Big[\int_{\Lambda_k} \mu(ds_1\dots ds_k) \int_{s_k}^{\zeta} (\ell^y_\varepsilon(ds_{k+1}) + h_\varepsilon\ \delta_{(s_k)}(ds_{k+1}))$$

$$\times \int_{s_{k+1}}^{\zeta} \ell^y_\varepsilon(ds_{k+2}) \int\dots\int_{s_{p-1}}^{\zeta} \ell^y_\varepsilon(ds_p) \prod_{i=k+1}^{p} 1_{S_{\varepsilon K}(s_i,s_{i+1})}(z)\Big]$$

(notice that the Dirac measures $\delta_{(t_i)}(dt_{i+1})$, for $i > k$, have been dropped). Here the random measure $\mu(ds_1\dots ds_k)$ is such that the process $t \longrightarrow \mu(\{s_k \le t\})$ is predictable ; furthermore it is easy to get the bound

$$E\Big[\int_{\Lambda_k} |\mu|(ds_1\dots ds_k)\ (\ell^y_\varepsilon(\zeta) + |h_\varepsilon|)\Big] \le C\ |\log\varepsilon|^k\ G(\tfrac{y}{2})$$

We may replace $\Big(\int_{s_{k+1}}^{\zeta} \dots\Big)$ by its predictable projection and get :

$\Phi_L(\varepsilon,y,z) =$

$$= E\Big[\int_{\Lambda_k} \mu(ds_1\dots ds_k) \int_{s_k}^{\zeta} (\ell^y_\varepsilon(ds_{k+1}) + h_\varepsilon \delta_{(s_k)}(ds_{k+1}))\ U^{p-k-1,p-k-1}_\varepsilon(B_{s_{k+1}},z,y)\Big],$$

with the convention $U_\varepsilon^{0,0}(x,y,z) = P_x[y \in S_{\varepsilon K}(0,\zeta)]$. The remaining part of the proof is entirely similar to the end of the proof of Lemma X-4. Simply use Lemma 3 instead of Lemma X-3. □

We may now complete the proof of Theorem 1. Write

$$u_\varepsilon(y,z) \simeq v_\varepsilon(y,z)$$

if there exists a function $F \in L^1(\mathbb{C}^2)$ such that for $\varepsilon \in (0,1/2)$,

$$|u_\varepsilon(y,z) - v_\varepsilon(y,z)| \le |\log \varepsilon|^{-2n-2} F(y,z).$$

It is enough to prove that

$$E\Big[\Big(1_{S_{\varepsilon K}(0,\zeta)}(y) + \sum_{p=1}^{n} a_\varepsilon^p\, Q_\varepsilon^p(\ell_\varepsilon^y(\zeta))\Big)\Big(1_{S_{\varepsilon K}(0,\zeta)}(z) + \sum_{p=1}^{n} a_\varepsilon^p\, Q_\varepsilon^p(\ell_\varepsilon^z(\zeta))\Big)\Big] \simeq 0.$$

By Lemma 2,

$$E\Big[1_{S_{\varepsilon K}(0,\zeta)}(y)\, 1_{S_{\varepsilon K}(0,\zeta)}(z)\Big] \simeq \sum_{p=2}^{2n+1} a_\varepsilon^p\, (G(y) + G(z))\, G(z-y)^{p-1}.$$

By Lemma 4 (and easy bounds when $x = 0$, y, z do not satisfy the assumptions of this lemma), if $p \ge 2$,

$$E[Q_\varepsilon^p(\ell_\varepsilon^y(\zeta))1_{S_{\varepsilon K}(0,\zeta)}(z)] \simeq - a_\varepsilon^{p-1}\, G(y)\, G(z-y)^{2p-2} - a_\varepsilon^p\, (G(y)+G(z))\, G(z-y)^{2p-1}$$
$$- a_\varepsilon^{p+1}\, G(z)\, G(z-y)^{2p},$$

and, if $p = 1$,

$$E[Q_\varepsilon^1(\ell_\varepsilon^y(\zeta))1_{S_{\varepsilon K}(0,\zeta)}(z)] \simeq - a_\varepsilon\, (G(y) + G(z))\, G(z-y) - a_\varepsilon^2\, G(z)\, G(z-y)^2.$$

Finally, Lemmas X-4, X-5 give

$$E[Q_\varepsilon^p(\ell_\varepsilon^y(\zeta))\, Q_\varepsilon^p(\ell_\varepsilon^z(\zeta))] \simeq (G(y) + G(z))G(z-y)^{2p-1}.$$

Furthermore, it was pointed out in Section X-4 (see formulas X-(9), X-(10)) that the proof of Lemma X-4 can be adapted to give:

$$E[Q_\varepsilon^p(\ell_\varepsilon^y(\zeta))\, Q_\varepsilon^{p+1}(\ell_\varepsilon^z(\zeta))] \simeq G(z)\, G(z-y)^{2p}$$

$$E[Q_\varepsilon^p(\ell_\varepsilon^y(\zeta))\, Q_\varepsilon^q(\ell_\varepsilon^z(\zeta))] \simeq 0 \quad \text{if} \quad |q-p| \ge 2.$$

The desired result follows. □

5. Further results.

Theorem 1 yields an asymptotic expansion of $m(S_{\varepsilon K}(0,\zeta))$ as ε goes to 0. A natural question is: can we replace ζ by a constant time t ? We first have to define random variables $T^p(t)$ in such a way that $T^p(t)$ coincides with T^p1 "conditionally on $\{\zeta = t\}$". The next theorem can be deduced from Theorem 1 by using the scaling properties of Brownian motion.

Theorem 5 : *There exists a sequence of processes* $T^p = (T^p(t), t \geq 0)$, *adapted to the natural filtration of* B, *such that*

$$T^p(\zeta) = T^p1 \quad a.s.$$

and the following holds. For every $n \geq 1$, $t \geq 0$,

$$m(S_{\varepsilon K}(0,t)) = - \sum_{p=1}^{n} a_\varepsilon^p \, T^p(t) + \mathcal{R}_n(\varepsilon),$$

where

$$\lim_{\varepsilon \to 0} |\log \varepsilon|^n \, \mathcal{R}_n(\varepsilon) = 0$$

in the L^2*-norm, and a.s. when* K *is star-shaped.*

Remark : Both the constants a_ε and the random variables $T^p(t)$ depend on the choice of λ (but not on the choice of ζ). Changing the value of λ leads to different equivalent expansions of $m(S_{\varepsilon K}(0,t))$. This corresponds to the non-uniqueness of the renormalization, which was already pointed out in Chapter X.

The case $n = 1$ of Theorem 5 is exactly Theorem VI-6. The case $n = 2$ is equivalent to Theorem VIII-7. If we compare these two results we get :

$$T^2(1) = \gamma(\mathcal{T}) + C_\lambda$$

for some constant C_λ depending on λ. Of course we could have proved this more directly, by comparing the approximations of $T^2 1$ and $\gamma(\mathcal{T})$.

By taking expectations in Theorem 5, one gets a full asymptotic expansion of $E[m(S_{\varepsilon K}(0,t)]$ and by scaling an asymptotic expansion of $E[m(S_K(0,t)]$ as t goes to infinity. The latter expansion refines a theorem of Spitzer in [Sp2]. These expansions involve the quantities $E[T^p(1)]$, which can be computed by induction, using the fact that $E[T^p(\zeta)] = 0$ (see Chapter X) and the scaling properties of $T^k(t)$. It is worth noting that the coefficients of the expansion for $E[m(S_{\varepsilon K}(0,t)]$ depend on K only through the constant R(K) . This should be compared with the similar results in higher dimensions [L11].

Bibliographical notes. The material of this Chapter is taken from [L12]. In particular we refer to this paper for a detailed proof of Theorem 5. A recent paper of Rosen [R7] gives analogues of Theorems 1 and 5 for the sausage associated with certain stable processes. See also Feldman and Rosen [FR] for an extension of Theorem 1 to Brownian motion on Riemannian surfaces. The work of Le Jan [LJ] may provide an alternative approach to the results of this Chapter.

REFERENCES

[A1] **O. Adelman.** Le brownien pique-t-il ? Exposé au Séminaire de Probabilités de l'Université Paris VI (1982). *Unpublished.*

[A2] **O. Adelman.** *Private communication.*

[AD] **O. Adelman, A. Dvoretzky.** Plane Brownian motion has strictly n-multiple points. *Israel J. Math.* 52, 361-364 (1985).

[Bi] **C.J. Bishop.** Brownian motion in Denjoy domains. Preprint (1990).

[B1] **K. Burdzy.** Brownian paths and cones. *Ann. Probab.* **13**, 1006-1010 (1985).

[B2] **K. Burdzy.** *Multidimensional Brownian Excursions and Potential Theory.* Longman, New York, 1987.

[B3] **K. Burdzy.** Geometric properties of two-dimensional Brownian paths. *Probab. Th. Rel. Fields* **81**, 485-505 (1989).

[B4] **K. Burdzy.** Cut points on Brownian paths. *Ann. Probab.* **17**, 1012-1036.

[BL1] **K. Burdzy, G.F. Lawler.** Non-intersection exponents for Brownian paths. Part I. Existence and an invariance principle. *Probab. Th. Rel. Fields* **84**, 393-410 (1990).

[BL2] **K. Burdzy, G.F. Lawler.** Non-intersection exponents for Brownian paths. Part II. Estimates and application to a random fractal. *Ann. Probab.* **18**, 981-1009 (1990).

[BSM] **K. Burdzy, J. San Martin.** Curvature of the convex hull of planar Brownian motion near its minimum point. *Stoch. Process. Appl.* 33, 89-103 (1989).

[CML] **M. Chaleyat-Maurel, J.F. Le Gall.** Green function, capacity and sample path properties for a class of hypoelliptic diffusion processes. *Probab. Th. Rel. Fields* **83**, 219-264 (1989).

[CF1] **I. Chavel, E.A. Feldman.** The Wiener sausage, and a theorem of Spitzer in Riemannian manifolds. *Probability and Harmonic Analysis, J. Chao, W.A. Woyczinski eds*, pp. 45-60. Dekker, New York, 1986.

[CF2] **I. Chavel, E.A. Feldman.** The Lenz shift and Wiener sausage in Riemannian manifolds. *Compositio Math.* **60**, 65-84 (1986).

[CFR] **I. Chavel, E.A. Feldman, J. Rosen.** Fluctuations of the Wiener sausage for manifolds. *To appear.*

[CT] **Z. Ciesielski, S.J. Taylor.** First passage times and sojourn times for Brownian motion in space and the exact Hausdorff measure of the sample path. *Trans. Amer. Math. Soc.* **103**, 434-450 (1963).

[CHM] **M. Cranston, P. Hsu, P. March.** Smoothness of the convex hull of planar Brownian motion. *Ann. Probab.* **17**, 144-150 (1989).

[Da] **B. Davis.** Brownian motion and analytic functions. *Ann. Probab.* 7, 913-932 (1979).

[DV] **M.D. Donsker, S.R.S. Varadhan.** Asymptotics for the Wiener sausage. *Comm. Pure Appl. Math.* **28**, 525-565 (1975).

[Du1] **R. Durrett.** A new proof of Spitzer's result on the winding of two-dimensional Brownian motion. *Ann. Probab.* **10**, 244-246 (1982).

[Du2] **R. Durrett.** *Brownian Motion and Martingales in Analysis.* Wadsworth, Belmont Ca. , 1984.

[DE] **A. Dvoretzky, P. Erdös.** Some problems on random walk in space. *Proc. Second Berkeley Symposium on Math. Statistics and Probability,* pp 353-367. University of California Press, Berkeley, 1951.

[DK1] **A. Dvoretzky, P. Erdös, S. Kakutani.** Double points of paths of Brownian motion in n-space. *Acta Sci. Math. (Szeged)* 12, 74-81 (1950).

[DK2] **A. Dvoretzky, P. Erdös, S. Kakutani.** Multiple points of Brownian motion in the plane. *Bull. Res. Council Israel Sect. F* 3, 364-371 (1954).

[DK3] **A. Dvoretzky, P. Erdös, S. Kakutani.** Points of multiplicity c of plane Brownian paths. *Bull. Res. Council Israel Sect. F* 7, 175-180 (1958).

[DKT] **A. Dvoretzky, P. Erdös, S. Kakutani, S.J. Taylor.** Triple points of Brownian motion in 3-space. *Proc. Cambridge Philos. Soc.* 53, 856-862 (1957).

[Dy1] **E.B. Dynkin.** Additive functionals of several time-reversible Markov processes. *J. Funct. Anal.* 42, 64-101 (1981).

[Dy2] **E.B. Dynkin.** Local times and quantum fields. *Seminar on Stochastic Processes 1983*, pp 69-84. Birkhäuser, Boston, 1984.

[Dy3] **E.B. Dynkin.** Polynomials of the occupation field and related random fields. *J. Funct. Anal.* 42, 64-101 (1984).

[Dy4] **E.B. Dynkin.** Random fields associated with multiple points of the Brownian motion. *J. Funct. Anal.* 62, 397-434 (1985).

[Dy5] **E.B. Dynkin.** Functionals associated with self-intersections of the Brownian motion. *Séminaire de Probabilités XX, Lecture Notes in Math.* 1204, 553-571. Springer, Berlin, 1986.

[Dy6] **E.B. Dynkin.** Self-intersection gauge for random walks and for Brownian motion. *Ann. Probab.* 16, 1-57 (1988).

[Dy7] **E.B. Dynkin.** Regularized self-intersection local times of the planar Brownian motion. *Ann. Probab.* 16, 58-74 (1988).

[E] **S.F. Edwards.** The statistical mechanics of polymers with excluded volume. *Proc. Phys. Sci.* 85, 613-624 (1965).

[EB] **M. El Bachir.** L'enveloppe convexe du mouvement brownien. Thèse de Troisième Cycle, Université Paul Sabatier, Toulouse, 1983.

[Ev1] **S.N. Evans.** On the Hausdorff dimension of Brownian cone points. *Math. Proc. Camb. Phil. Soc.* 98, 343-353 (1985).

[FR] **E.A. Feldman, J. Rosen.** An asymptotic development of the area of the Wiener sausage for surfaces. *To appear.*

[F] **W.E. Feller.** *An Introduction to Probability Theory and Its Applications, vol II, 2nd edition.* Wiley, New York, 1971.

[GHR] **D. Geman, J. Horowitz, J. Rosen.** A local time analysis of intersections of Brownian paths in the plane. *Ann. Probab.* 12, 86-107 (1984).

[GS] **R.K. Getoor, M.J. Sharpe.** Conformal martingales. *Invent. Math.* 16, 271-308 (1972).

[H] **J. Hawkes.** Some geometric aspects of potential theory. *Stochastic Analysis and Applications. Lecture Notes in Math.* 1095, 130-154. Springer, New York, 1984.

[IMK] **K. Itô, H.P. McKean.** *Diffusion Processes and their Sample Paths.* Springer, New York, 1965.

[JP] **N.C. Jain, W.E. Pruitt.** The range of random walk. *Proc. Sixth Berkeley Symposium on Math. Statistics and Probability, vol. 3,* pp 31-50. University of California Press, Berkeley, 1973.

[K] **M. Kac.** Probabilistic methods in some problems of scattering theory. *Rocky Mountain J. Math.* **4**, 511-537 (1974).

[Kh] **J.P. Kahane.** Points multiples des processus de Lévy symétriques stables restreints à un ensemble de valeurs du temps. Séminaire d'Analyse Harmonique. Public. Math. de l'Université d'Orsay **83-02**, p. 74-105 (1983).

[KR] **G. Kallianpur, H. Robbins.** Ergodic property of the Brownian motion process. *Proc. Nat. Acad. Sci. U.S.A.* **39**, 525-533 (1953).

[Ka] **R. Kaufman.** Une propriété métrique du mouvement brownien. *C.R. Acad. Sci. Paris, Série I* , **268**, 727-728 (1969).

[Kn] **F.B. Knight.** *Essentials of Brownian Motion and Diffusion.* A.M.S., Providence, 1981.

[L1] **J.F. Le Gall.** Sur la mesure de Hausdorff de la courbe brownienne. *Séminaire de Probabilités XIX. Lecture Notes in Math.* **1123**, 297-313. Springer, New York, 1985.

[L2] **J.F. Le Gall.** Sur le temps local d'intersection du mouvement brownien plan et la méthode de renormalisation de Varadhan. *Séminaire de Probabilités XIX. Lecture Notes in Math.* **1123**, 314-331. Springer, New York 1985.

[L3] **J.F. Le Gall.** Sur la saucisse de Wiener et les points multiples du mouvement brownien. *Ann. Probab.* **14**, 1219-1244 (1986).

[L4] **J.F. Le Gall.** Propriétés d'intersection des marches aléatoires, I. *Comm. Math. Physics* **104**, 471-507 (1986).

[L5] **J.F. Le Gall.** Propriétés d'intersection des marches aléatoires, II. *Comm. Math. Physics* **104**, 509-528 (1986).

[L6] **J.F. Le Gall.** Le comportement du mouvement brownien entre les deux instants où il passe par un point double. *J. Funct. Anal.* **71**, 246-262 (1987).

[L7] **J.F. Le Gall.** Mouvement brownien, cônes et processus stables. *Probab. Th. Rel. Fields* **76**, 587-627 (1987).

[L8] **J.F. Le Gall.** Temps locaux d'intersection et points multiples des processus de Lévy. *Séminaire de Probabilités XXI. Lecture Notes in Math.* **1247**, 341-375. Springer, New York, 1987.

[L9] **J.F. Le Gall.** The exact Hausdorff measure of Brownian multiple points. *Seminar on Stochastic Processes 1986*, pp. 107-137. Birkhäuser, Boston, 1987.

[L10] **J.F. Le Gall.** Fluctuation results for the Wiener sausage. *Ann. Probab.* **16**, 991-1018 (1988).

[L11] **J.F. Le Gall.** Sur une conjecture de M. Kac. *Probab. Th. Rel. Fields* **78**, 389-402 (1988).

[L12] **J.F. Le Gall.** Wiener sausage and self-intersection local times. *J. Funct. Anal.* **88**, 299-341 (1990).

[L13] **J.F. Le Gall.** On the connected components of the complement of a two-dimensional Brownian path. *Festschrift in Honor of Frank Spitzer.* Birkhäuser, Boston, 1991.

[LR] **J.F. Le Gall, J. Rosen.** The range of stable random walks. *Ann. Probab.*, *to appear*.

[LY] **J.F. Le Gall, M. Yor.** Etude asymptotique de certains mouvements browniens complexes avec drift. *Probab. Th. Rel. Fields* **71**, 183-229 (1986).

[LJ] **Y. Le Jan.** On the Fock space representation of functionals of the occupation field and their renormalization. *J. Funct. Anal.* **80**, 88-108 (1988).

[Lé1] **P. Lévy.** Le mouvement brownien plan. *Amer. J. Math.* **62**, 487-550 (1940).

[Lé2] **P. Lévy.** La mesure de Hausdorff de la courbe du mouvement brownien. *Giornale dell'Istituto Italiano degli Attuari* **16**, 1-37 (1953).

[Lé3] **P. Lévy.** Le caractère universel de la courbe du mouvement brownien et la loi du logarithme itéré. *Rendiconti del circolo matematico di Palermo, II*, 4, 337-366 (1955).

[Lé4] **P. Lévy.** *Processus Stochastiques et Mouvement Brownien.* Gauthier-Villars, Paris, 1965.

[Ma] **N.G. Makarov.** On the distorsion of boundary sets under conformal mappings. *Proc. London Math. Soc. (3)* 51, 369-384 (1985).

[M] **B.B. Mandelbrot.** *The Fractal Geometry of Nature.* W.H. Freeman and Co., New York, 1982.

[MK] **H.P. McKean.** *Stochastic Integrals.* Academic Press, New York, 1969.

[MM] **J.E. McMillan.** Boundary behaviour of a conformal mapping. *Acta Math.* 123, 43-68 (1969).

[Mo] **T.S. Mountford.** On the asymptotic number of small components created by planar Brownian motion. *Stochastics* 28, 177-188 (1989).

[PY1] **J.W. Pitman, M. Yor.** Asymptotic laws of planar Brownian motion. *Ann. Probab.* **14**, 733-779 (1986).

[PY2] **J.W. Pitman, M. Yor.** Level crossings of a Cauchy process. *Ann. Probab.* **14**, 780-792 (1986).

[PY3] **J.W. Pitman, M. Yor.** Further asymptotic laws of planar Brownian motion. *Ann. Probab.* **17**, 965-1011 (1989).

[Po] **C. Pommerenke.** *Univalent Functions.* Vandenhoeck & Ruprecht, Göttingen, 1975.

[PS] **S.C. Port, C.J. Stone.** *Brownian Motion and Classical Potential Theory.* Academic Press, New York, 1978.

[ReY] **D. Revuz, M. Yor.** *Continuous Martingales and Brownian Motion.* Springer, 1991.

[RoW] **L.C.G. Rogers, D. Williams.** *Diffusions, Markov Processes, and Martingales, Vol. II.* Wiley, Chichester, 1987

[R1] **J. Rosen.** A local time approach to the self-intersections of Brownian paths in space. *Comm. Math. Physics* **88**, 327-338 (1983).

[R2] **J. Rosen.** Self-intersections of random fields. *Ann. Probab.* 12, 108-119 (1984).

[R3] **J. Rosen.** Tanaka formula and renormalization for intersections of planar Brownian motion. *Ann. Probab.* **14**, 1245-1251 (1986).

[R4] **J. Rosen.** A renormalized local time for multiple intersections of planar Brownian motion. Séminaire de Probabilités XX. *Lecture Notes in Math.* 1204, 515-531. Springer, Berlin, 1986.

[R5] **J. Rosen.** Joint continuity of the intersection local time of Markov processes. *Ann. Probab.* **15**, 659-675 (1987).

[R6] **J. Rosen.** Random walks and intersection local time. *Ann. Probab.* 18, 959-977 (1990).

[RY] **J. Rosen, M. Yor.** Tanaka formulae and renormalization for triple intersections of Brownian motion in the plane. *Ann. Probab., to appear.*

[Sh1] **M. Shimura.** A limit theorem for two-dimensional conditioned random walk. *Nagoya Math. J.* 95, 105-116 (1984).

[Sh2] **M. Shimura.** Excursions in a cone for two-dimensional Brownian motion. *J. Math. Kyoto Univ.* 25, 433-443 (1985).

[Sh3] **M. Shimura.** Meandering points of two-dimensional Brownian motion. *Kodai Math. J.* **11**, 169-176 (1988).

[S1] **F. Spitzer**. Some theorems concerning two-dimensional Brownian motion. *Trans. Amer. Math. Soc.* **87**, 187-197 (1958).

[S2] **F. Spitzer**. Electrostatic capacity, heat flow and Brownian motion. *Z. Wahrsch. verw. Gebiete* **3**, 110-121 (1964).

[S3] **F. Spitzer**. *Principles of Random Walk*. Princeton, Van Nostrand 1964.

[S4] **F. Spitzer**. Appendix to: Subadditive ergodic theory, by J.F.C. Kingman. *Ann. Probab.* **1**, 883-909 (1973).

[Sy] **K. Symanzik**. Euclidean quantum field theory. *Local Quantum Theory, R. Jost ed.* Academic, New York, 1969.

[Sz] **A.S. Sznitman**. Some bounds and limiting results for the Wiener sausage of small radius associated with elliptic diffusions. *Stochastic Process. Appl.* **25**, 1-25 (1987).

[T1] **S.J. Taylor**. The exact Hausdorff measure of the sample path for planar Brownian motion. *Proc. Cambridge Philos. Soc.* **60**, 253-258 (1964).

[T2] **S.J. Taylor**. Sample path properties of processes with stationary independent increments. *Stochastic Analysis, D. Kendall, E. Harding eds.* Wiley, London, 1973.

[T3] **S.J. Taylor**. The measure theory of random fractals. *Math. Proc. Camb. Phil. Soc.* **100**, 383-406 (1986).

[V] **S.R.S. Varadhan**. Appendix to "Euclidean quantum field theory", by K. Symanzik. In *Local Quantum Theory* (R. Jost ed.). Academic, New York, 1969.

[VW] **S.R.S. Varadhan, R.J. Williams**. Brownian motion in a wedge with oblique reflection. *Commun. Pure Appl. Math.* **38**, 405-443 (1985).

[W1] **S. Weinryb**. Etude asymptotique par des mesures de $\mathbb{R}^3$ de saucisses de Wiener localisées. *Probab. Th. Rel. Fields* **73**, 135-148 (1986).

[W2] **S. Weinryb**. Asymptotic results for independent Wiener sausages. Application to generalized intersection local times. *Stochastics* **27**, 99-127 (1989).

[Wi1] **R.J. Williams**. Recurrence classification and invariant measure for reflected Brownian motion in a wedge. *Ann. Probab.* **13**, 758-778 (1985).

[Wi2] **R.J. Williams**. Reflected Brownian motion in a wedge: semi-martingale property. *Z. Wahrsch. verw. Gebiete* **69**, 161-176 (1985).

[Wi3] **R.J. Williams**. Local time and excursions of reflected Brownian motion in a wedge. *Publ. Res. Inst. Math. Sci. Kyoto University* **23**, 297-319 (1987).

[Wo] **R. Wolpert**. Wiener path intersections and local time. *J. Funct. Anal.* **30**, 329-340 (1978).

[Y1] **M. Yor**. Compléments aux formules de Tanaka-Rosen. *Séminaire de Probabilités XIX. Lecture Notes in Math.* **1123**, 332-349. Springer, Berlin, 1985.

[Y2] **M. Yor**. Renormalisation et convergence en loi pour les temps locaux d'intersection du mouvement brownien dans $\mathbb{R}^3$. *Séminaire de Probabilités XIX. Lecture Notes in Math.* **1123**, 350-365. Springer, Berlin, 1985.

[Y3] **M. Yor**. Précisions sur l'existence et la continuité des temps locaux d'intersection du mouvement brownien dans $\mathbb{R}^d$. Séminaire de Probabilités XX. *Lecture Notes in Math.* **1204**, 532-541. Springer, Berlin, 1986.

[Y4] **M. Yor**. Sur la représentation comme intégrales stochastiques des temps d'occupation du mouvement brownien dans $\mathbb{R}^d$. Séminaire de Probabilités XX. *Lecture Notes in Math.* **1204**, 543-552. Springer, Berlin, 1986.

SUBJECT INDEX

EXPOSES 1990

AZEMA Jacques

Some properties of the linear Brownian motion

BENASSI Albert

Martingales hiérarchiques à plusieurs paramètres

BERNARD Pierre

Régularité C∞ des noyaux de Wiener d'une diffusion

BIRGE Lucien

A propos du maximum de vraisemblance

BOUGEROL Philippe

Filtre de Kalman avec temps d'observations aléatoires

CHASSAING Philippe

Contrôle des chaînes de Markov : application aux structures auto-adaptatives

DOISY Michel

Conditions de saturation dans les réseaux de Petri

DONATI-MARTIN Catherine

Equations différentielles stochastiques avec conditions au bord

ESTRADE Anne

Une formule de Girsanov pour les groupes de Lie

FLORCHINGER Patrick

Filtrage non linéaire avec bruits de dimension infinie. Existence d'une densité régulière pour le filtre

GRUET Marie-Anne

Consistance de l'estimateur non paramétrique d'une fonction de régression dans l'espace de Sobolev $W_2^2\langle[0,1]\rangle$

HARISON Victor

Jonction maximale en distribution

HONGLER Max-Olivier

Phase errors distribution obtained for a class of clocks

HU Ying

Adapted solution of a backward linear equation and stochastic maximum principle for control systems

IOFFE Dimitry

On a class of bounded solutions to the exterior Dirichlet problem

ISTAS Jacques

Etude de la compression d'un signal ou d'une image par ondelettes

JACQUOT Sophie

Comportement asymptotique de la seconde valeur propre des processus de Kolmogorov

KERKYACHARIAN Gérard et PICARD Dominique

Estimation de densité par méthodes d'ondelettes

LAREDO Catherine

Estimation non paramétrique de la variance d'une diffusion par méthodes d'ondelettes

LEGLAND François

Consistance de l'estimateur du maximum de vraisemblance pour des diffusions partiellement observées

MACGIBBON Brenda

Estimation minimax d'un vecteur de la loi de Poisson avec contraintes

MANSMANN Ulrich

Localisation in a simple Polaron-model

NUALART David

Randomized stopping points and optimal stopping on the plane

PICARD Jean

Martingales à valeurs dans les variétés riemanniennes

RIO Emmanuel

Vitesses de convergence pour des estimateurs de type moindres carrés dans des régressions en non paramétrique

ROUAULT Alain

Deux résultats relatifs à l'équation KPP

SANZ SOLE Marta

Large deviations for a class of anticipating stochastic differential equations

SOWERS Richard

Large deviations results for a reaction-diffusion equation with non-Gaussian perturbations

SZEKLI Ryszard

Martingale approach to stochastic ordering of point processes on R_+

TRAN HUNG THAO Tran

A problem of filtering from point process observations

VINH Nham Thi

A limit theorem in Banach spaces

WU Liming

Un traitement unifié de la représentation des fonctionnelles sur l'espace de Wiener

LISTE DES AUDITEURS

Mr. ABRAHAM Romain	Ecole Normale Supérieure, Paris
Mr. ALABERT Aureli	Université de Catalunya (Espagne)
Mr. AZEMA Jacques	Université de Paris VI
Mr. BADRIKIAN Albert	Université Blaise Pascal (Clermont II)
Mr. BALDI Paolo	Université de Catane (Italie)
Mr. BEN AROUS Gérard	Université de Paris-Sud, Orsay
Mr. BENASSI Albert	Université Blaise Pascal (Clermont II)
Mr. BERNARD Pierre	Université Blaise Pascal (Clermont II)
Mr. BIRGE Lucien	Université de Paris X, Nanterre
Mr. BOUGEROL Philippe	Université de Nancy I
Mr. BRUNAUD Marc	Université de Paris-Sud, Orsay
Mme CHALEYAT-MAUREL Mireille	Université de Paris VI
Mr. CHASSAING Philippe	Université de Nancy I
Mle CHEVET Simone	Université Blaise Pascal (Clermont II)
Mr. CIPRIANI Fabio	International School for Advanced Studies Trieste, Italie
Mr. DACUNHA-CASTELLE Didier	Universite de Paris-Sud, Orsay
Mr. DARWICH Abdul	Université de Rennes I
Mle DELGADO DE LA TORRE Rosario	Université de Catalunya (Espagne)
Mr. DERMOUNE Azzouz	Université Blaise Pascal (Clermont II)
Mr. DOISY Michel	I.U.T. / STID, Pau
Mme DONATI-MARTIN Catherine	Université de Provence, Aix-Marseille
Mr. DOUKHAN Paul	Université de Paris-Sud, Orsay
Mr. EL-HOSSEINY Hany	Institut Fourier, Grenoble I
Mle ESTRADE Anne	Université d'Orléans
Mr. FELLAH Dominique	Université de Provence, Aix-Marseille
Mr. FLORCHINGER Patrick	Université de Reims
Mr. GALLARDO Léonard	Université de Bretagne Occidentale, Brest
Mme GASSIAT Elisabeth	Université de Paris-Sud, Orsay
Mme GRUET Marie-Anne	I.N.R.A. de Jouy-en-Josas
Mr. HARISON Victor	Université d'Antanarivo (Madagascar)
Mr. HENNEQUIN Paul-Louis	Université Blaise Pascal (Clermont II)
Mle HESSEL Pascale	Université de Paris-Sud (Orsay)
Mr. HONGLER Max-Olivier	Université de Genève (Suisse)
Mr. HU Ying	Université de Provence, Aix-Marseille
Mr. IOFFE Dimitry	Technion Haifa (Israël)
Mr. ISTAS Jacques	I.N.R.A. de Jouy-en-Josas
Mle JACQUOT Sophie	Université d'Orléans
Mr. KERKYACHARIAN Gérard	Université de Nancy I
Mme KOWALSKA Anna	Université de Nancy I
Mr. LAPEYRE Bernard	ENPC - CERMA, Noisy-le-Grand
Mme LAREDO Catherine	I.N.R.A. de Jouy-en-Josas
Mle LAURENT Béatrice	Université de Paris-Sud (Orsay)
Mr. LE GLAND François	I.N.R.I.A. Sophia Antipolis, Valbonne
Mme MacGIBBON Brenda	Université du Québec à Montréal (Canada)
Mr. MANSMANN Ulrich	Université de Berlin (R.F.A.)
Mr. MATHIEU Pierre	Ecole Normale Supérieure, Paris
Mr. NOBLE John	Université de Warwick (Grande-Bretagne)
Mr. NUALART David	Université de Barcelone (Espagne)
Mle PICARD Dominique	Université de Paris VII
Mr. PICARD Jean	I.N.R.I.A. Sophia Antipolis, Valbonne

Mr. POSILIGANO Andrea	International School for Advanced Studies Trieste, Italie
Mr. RIO Emmanuel	Université de Paris-Sud (Orsay)
Mr. ROUAULT Alain	Université de Paris-Sud (Orsay)
Mr. ROUX Daniel	Université Blaise Pascal, Clermont-Ferrand
Mr. ROYNETTE Bernard	Université de Nancy II
Mme SANZ SOLE Marta	Université de Barcelone (Espagne)
Mle SAVONA Catherine	Université Blaise Pascal, (Clermont II)
Mr. SOWERS Richard	University of Maryland (U.S.A.)
Mr. SZEKLI Ryszard	Université de Wroclaw (Pologne)
Mme THIEULLEN Michèle	Université de Paris-Sud, Orsay
Mr. TOUIJAR Driss	Université de Lille
Mr. TRAN HUNG THAO Tran	Université de Hanoi (Vietnam)
Mme VINH Nham Thi	Université de Hanoi (Vietnam)
Mr. WERNER Wendelin	Ecole Normale Supérieure, Paris
Mr. WONG Henri	Université d'Ottawa (Canada)
Mr. WU Liming	Université de Wuhan (Chine)

LIST OF PREVIOUS VOLUMES OF THE "Ecole d'Eté de Probabilités"

1971 - J.L. Bretagnolle (LNM 307)
"Processus à accroissements indépendants"
S.D. Chatterji
"Les martingales et leurs applications analytiques"
P.A. MEYER
"Présentation des processus de Markov"

1973 - P.A. MEYER (LNM 390)
"Transformation des processus de Markov"
P. PRIOURET
"Processus de diffusion et équations différentielles stochastiques"
F. SPITZER
"Introduction aux processus de Markov à paramètres dans Z_v"

1974 - X. FERNIQUE (LNM 480)
"Régularité des trajectoires des fonctions aléatoires gaussiennes"
J.P. CONZE
"Systèmes topologiques et métriques en théorie ergodique"
J. GANI
"Processus stochastiques de population"

1975 A. BADRIKIAN (LNM 539)
"Prolégomènes au calcul des probabilités dans les Banach"
J.F.C. KINGMAN
"Subadditive processes"
J. KUELBS
"The law of the iterated logarithm and related strong convergence theorems for Banach space valued random variables"

1976 J. HOFFMANN-JORGENSEN (LNM 598)
"Probability in Banach space"
T.M. LIGGETT
"The stochastic evolution of infinite systems of interacting particles"
J. NEVEU
"Processus ponctuels"

1977 D. DACUNHA-CASTELLE (LNM 678)
"Vitesse de convergence pour certains problèmes statistiques"
H. HEYER
"Semi-groupes de convolution sur un groupe localement compact et applications à la théorie des probabilités"
B. ROYNETTE
"Marches aléatoires sur les groupes de Lie"

1978 R. AZENCOTT (LNM 774)
"Grandes déviations et applications"
Y. GUIVARC'H
"Quelques propriétés asymptotiques des produits de matrices aléatoires"
R.F. GUNDY
"Inégalités pour martingales à un et deux indices : l'espace H^p"

1979 J.P. BICKEL (LNM 876)
"Quelques aspects de la statistique robuste"
N. EL KAROUI
"Les aspects probabilistes du contrôle stochastique"
M. YOR
"Sur la théorie du filtrage"

1980 J.M. BISMUT (LNM 929)
"Mécanique aléatoire"
L. GROSS
"Thermodynamics, statistical mechanics and random fields"
K. KRICKEBERG
"Processus ponctuels en statistique"

1981 X. FERNIQUE (LNM 976)
"Régularité de fonctions aléatoires non gaussiennes"
P.W. MILLAR
"The minimax principle in asymptotic statistical theory"
D.W. STROOCK
"Some application of stochastic calculus to partial differential equations"
M. WEBER
"Analyse infinitésimale de fonctions aléatoires"

1982 R.M. DUDLEY (LNM 1097)
"A course on empirical processes"
H. KUNITA
"Stochastic differential equations and stochastic flow of diffeomorphisms"
F. LEDRAPPIER
"Quelques propriétés des exposants caractéristiques"

1983	D.J. ALDOUS "Exchangeability and related topics" I.A. IBRAGIMOV "Théorèmes limites pour les marches aléatoires" J. JACOD "Théorèmes limite pour les processus"	(LNM 1117)
1984	R. CARMONA "Random Schrödinger operators" H. KESTEN "Aspects of first passage percolation" J.B. WALSH "An introduction to stochastic partial differential equations"	(LNM 1180)
1985-87	S.R.S. VARADHAN "Large deviations" P. DIACONIS "Applications of non-commutative Fourier analysis to probability theorems H. FOLLMER "Random fields and diffusion processes" G.C. PAPANICOLAOU "Waves in one-dimensional random media" D. ELWORTHY Geometric aspects of diffusions on manifolds" E. NELSON "Stochastic mechanics and random fields"	(LNM 1362)
1986	O.E. BARNDORFF-NIELSEN "Parametric statistical models and likelihood"	(LNS M50)
1988	A. ANCONA "Théorie du potentiel sur les graphes et les variétés' D. GEMAN "Random fields and inverse problems in imaging" N. IKEDA "Probabilistic methods in the study of asymptotics"	(LNM 1427)
1989	D.L. BURKHOLDER "Explorations in martingale theory and its applications" E. PARDOUX "Filtrage non linéaire et équations aux dérivées partielles stochastiques associées" A.S. SZNITMAN "Topics in propagation of chaos"	(LNM 1464)

1990 M.I. FREIDLIN (LNM 1527)
"Semi-linear PDE's and limit theorems for large deviations"
J.F. LE GALL
"Some properties of planar Brownian motion"

Printing: Druckhaus Beltz, Hemsbach
Binding: Buchbinderei Schäffer, Grünstadt